Functional Materials for the Oil and Gas Industry

Functional Materials for the Oil and Gas Industry: Characterization and Applications discusses the latest techniques in characterization and applications of functional materials in the oil and gas industry. It provides an expert review of recent developments in a variety of materials, such as ceramics, composites, and alloys, and covers all major aspects relevant to the industry, including asset management (corrosion), operation (pipeline engineering), energy management, and applications in extreme environments.

This book:

- Discusses modern characterization techniques, such as *in situ* TEM, SAXS, SANS, X-ray, and neutron tomography.
- Covers conventional and advanced nondestructive techniques (NDTs), such as ultrasonic testing and radiography for asset integrity checking in oil and gas sectors.
- Describes advanced properties of a variety of functional materials and their applications to the oil and gas field.
- Explains self-cleaning coating technologies and their applications and materials for renewable energy sources.
- Details advances in synthesis methods for functional materials.
- Features industrial aspects of afunctional materials application in each chapter.

Written for an interdisciplinary audience of industrial practitioners, academics, and researchers in petroleum, materials, chemical, and related disciplines of engineering, this work offers significant insight into the state-of-the-art in the development and characterization of advanced functional materials.

Emerging Trends and Technologies in Petroleum Engineering
Series Editor: Abhijit Y. Dandekar

Wax Deposition: Experimental Characterizations, Theoretical Modeling, and Field Practicess
Zhenyu Huang, Sheng Zheng, and H. Scott Fogler

Hydraulic Fracturing
Michael Berry Smith and Carl Montgomery

Unconventional Oil and Gas Resources: Exploitation and Development
Usman Ahmed and D. Nathan Meehan

Petroleum Fluid Phase Behavior: Characterization, Processes, and Applications
Raj Deo Tewari, Abhijit Y. Dandekar, and Jaime Moreno Ortiz

Practical Aspects of Flow Assurance in the Petroleum Industry
Jitendra S. Sangwai and Dandekar Abihijit

Functional Materials for the Oil and Gas Industry: Characterization and Applications
Deepak Dwivedi, Amit Ranjan, and Jitendra S. Sangwai

For more information about this series, please visit: https://www.routledge.com/ Emerging-Trends-and-Technologies-in-Petroleum-Engineering/book-series/ CRCEMETRETEC

Functional Materials for the Oil and Gas Industry

Characterization and Applications

Edited by
Deepak Dwivedi, Amit Ranjan,
and Jitendra S. Sangwai

CRC Press
Taylor & Francis Group
Boca Raton London New York

CRC Press is an imprint of the
Taylor & Francis Group, an **informa** business

First edition published 2024
by CRC Press
6000 Broken Sound Parkway NW, Suite 300, Boca Raton, FL 33487-2742

and by CRC Press
4 Park Square, Milton Park, Abingdon, Oxon, OX14 4RN

CRC Press is an imprint of Taylor & Francis Group, LLC

© 2024 selection and editorial matter, Deepak Dwivedi, Amit Ranjan, and Jitendra S. Sangwai; individual chapters, the contributors

ISBN: 978-1-032-15100-7 (hbk)
ISBN: 978-1-032-15101-4 (pbk)
ISBN: 978-1-003-24255-0 (ebk)

DOI: 10.1201/9781003242550

Typeset in Times
by codeMantra

Dedication

I would like to dedicate to my the late parents, Mr. Mithilesh Kumar Dwivedi and Mrs. Prabha Dwivedi, and grandparents. Late Mr. Dwarika Prasad Dwivedi, Mrs. Choti Bai Dwivedi, Mr. P.L. Dubey, and Mrs. Karuna Dubey. I would also like to dedicate this book to the new entrant of the family, my loving fiancée, Dr. Namrata Pandey.

Deepak Dwivedi
Amethi, Uttar Pradesh, India

Dedicated to my beloved family, whose unwavering love and support have fueled my journey towards success. This book is dedicated to you, my greatest source of inspiration and strength.

Amit Ranjan
Indian Institute of Technology, Madras, Tamilnadu, India

Dedicated to my sweet family, my wife Priti, and my children, Saket and Akshit, for making my life enjoyable.

Jitendra S. Sangwai
Chennai, Tamil Nadu, India

Contents

This petroleum engineering book series includes works on all aspects of petroleum science and engineering, but with a special focus on emerging trends and technologies that pertain to the paradigm shift in the petroleum engineering field. It deals with the increased exploitation of technically challenged and atypical hydrocarbon resources that are receiving a lot of attention from today's petroleum industry, as well as the potential use of advanced nontraditional or nonconventional technologies such as nanotechnology in diverse petroleum engineering applications. These areas have assumed a position of prominence in today's petroleum engineering field. However, although scientific literature exists on these emerging areas in the form of various publications, much of it is scattered and highly specific. The purpose of this book series is to provide a centralized and comprehensive collection of reference books and textbooks covering fundamentals but paying close attention to these emerging trends and technologies from the standpoint of the main disciplines of drilling engineering, production engineering, and reservoir engineering.

Given the dwindling supply of easy-to-produce conventional oil, rapidly climbing energy demands, the sustained ~$100/bbl oil price, and technological advances, the petroleum industry is increasingly in pursuit of E & P of atypical or unconventional and technically challenged oil and gas resources, which may eventually become the future of the petroleum industry. Unconventional resources typically include (1) coal bed methane or CBM gas; (2) tight sand gas in ultra-low-permeability formations; (3) shale gas and shale oil in very-low-permeability shales; (4) oil shales; (5) heavy and viscous oils; (6) tar sands; and (7) methane hydrates. Compared to the world's proven conventional natural gas reserves of 6600+ trillion cubic feet (TCF), the combined CBM, shale gas, tight sands gas, and methane hydrate resource estimates are in excess of 730,000 TCF.[1-3] Similarly, out of the world's total 9–13 trillion barrels of oil resources, the conventional (light and medium) oil is only 30%, whereas heavy oil, extra heavy oil, tar sands, and bitumen combined make up the remaining 70%.[4] In addition, shale-based oil resources worldwide are estimated to be between 6 and 8 trillion barrels.[5] As a case in point, shale-based oil production in North Dakota has increased from a mere 3000 barrels/d in 2005 to a whopping 400,000+ barrels/d in 2011.[6] Even the most conservative technical and economic recovery estimates of the unconventional resources represent a very substantial future energy portfolio that dwarfs the conventional gas and oil reserves. However, to a large extent, these particular resources, unlike the conventional ones, do not fit the typical profile and are, to some extent, in the stages of infancy, thus needing a different and unique approach from the drilling, production, and reservoir engineering perspective.

The petroleum engineering academic and industry community also is aggressively pursuing nanotechnology with the hope of identifying innovative solutions for problems faced in the overall process of oil and gas recovery. In particular, a big spurt in this area in the last decade or so is evident from the significant activities in terms of research publications, meetings, formation of different consortia, workshops, and dedicated sessions in petroleum engineering conferences. A simple literature search for a keyword *nanotechnology* on www.onepetro.org, managed by the Society of Petroleum Engineers (SPE), returns over 250 publications dating from 2001 onwards with the bulk of them in the last 5 or 6 years. Since 2008, SPE has also organized three different applied technology workshops specifically focused on nanotechnology in the E & P industry. An Advanced Energy Consortium (AEC) with sponsorships from some major operators and service companies also was formed in 2007 with the mission of facilitating research in "micro and nanotechnology materials and sensors having the potential to create a positive and disruptive change in the recovery of petroleum and gas from new and existing reservoirs." Companies such as Saudi Aramco have taken the lead in taking the first strides in evaluating the potential of employing nanotechnology in the E & P industry. Their trademarked Resbots™ are designed for deployment with the injection fluids for *in situ* reservoir sensing (temperature, pressure, and fluid type) and intervention, eventually leading to more accurate reservoir characterization once fully developed. Following successful laboratory core flood tests, they conducted the industry's first field trial of reservoir nanoagents.[7]

The foregoing is clearly a statement of the new wave in the petroleum engineering field that is being created by emerging trends in unconventional resources and new technologies. The publisher and its series editor are fully aware of the rapidly evolving nature of these key areas and their long-lasting influence on the current state and future of the petroleum industry. The series is envisioned to have a very broad scope that includes, but is not limited to, analytical, experimental, and numerical studies and methods and field cases, thus delivering readers in both academia and industry an authoritative information source of trends and technologies that have shaped and will continue to impact the petroleum industry.

Abhijit Dandekar
University of Alaska Fairbanks, October 2022

REFERENCES

1. http://www.eia.gov/analysis/studies/worldshalegas/.
2. Kawata, Y. and Fujita, K., Some Predictions of Possible Unconventional Hydrocarbons Availability Until 2100, Society of Petroleum Engineers (SPE) paper number 68755.
3. http://www.netl.doe.gov/kmd/cds/disk10/collett.pdf.
4. https://www.slb.com/~/media/Files/resources/oilfield_review/ors06/sum06/heavy_oil.ashx.
5. Biglarbigi, K., Crawford, P., Carolus, M. and Dean, C., Rethinking World Oil-Shale Resource Estimates, Society of Petroleum Engineers (SPE) paper number SPE 135453.
6. Mason, J., http://www.sbpipeline.com/images/pdf/Mason_Oil%20Production%20Potential%20of%20the%20North%20Dakota%20Bakken_OGJ%20Article_10%20February%202012.pdf.
7. Kanj, M.Y., Rashid, M.H. and Giannelis, E.P., Industry First Field Trial of Reservoir Nanoagents, Society of Petroleum Engineers (SPE) paper number SPE 142592.

Preface

The focus of this book is to cover the functional materials available for the application in oil and gas industries. A range of materials have been discussed in this book such as self-healing coating materials, functional ceramics, functional refractory materials, insulating materials, and nanomaterials, which can be utilized for different purposes in oil and gas industries. In addition to this, our ambition in writing and editing the chapters of this book was also to illustrate the methods/techniques available for characterizing these materials using various destructive and nondestructive techniques. This book also highlights some of the recent developments that happened in the front of functional materials' characterization using advanced analytical techniques with their limitations.

This book covers the fundamentals associated with the functional materials' usage for different applications concerning oil and gas sectors and also tries to link the industrial requirements with ongoing research and development activities. However, it is worth noting that with ongoing R&D activities on the front of functional materials' development and discovery, the task of covering all the developments in this book became challenging for us. Although we tried our best to include chapters of a wide spectrum relevant to oil and gas industries, we don't deny its continuous revisions to make this book relevant throughout.

This book may not only be useful for scientists and scholars, but it would also be useful for working professionals who are working in the oil and gas sectors, particularly in the field of operation and maintenance.

As an instructor of courses like Corrosion Engineering, Materials Science, and Energy Resources and Utilization at Rajiv Gandhi Institute of Petroleum Technology, one of the editors, Dr. Deepak Dwivedi, believes this book could be a good reference book for B.Tech, M.Tech, and Ph.D. students.

Dr. Deepak Dwivedi
Prof. Amit Ranjan
RGIPT Jais, Uttar Pradesh, India

Prof. Jitendra S. Sangwai
Indian Institute of Technology, Madras, Tamilnadu, India
January/February 2023

Acknowledgments

We would like to thank all the authors who contributed to this book. We also express our gratitude to our colleagues; well-wishers; and students of RGIPT, Jais, Amethi, and IIT Madras.

Authors also acknowledge the support and guidance received from Prof. A. S. K. Sinha (Director, RGIPT, who himself is a leading Chemical Engineer) and Prof., Kamakoti Veezhinathan (Director, IIT Madras).

Editors

Dr. Deepak Dwivedi completed his Ph.D. (Chemical Engineering) from WASM Curtin University, Perth Australia (QS world rank-2) in collaboration with the University of Cambridge, UK and Australian Nuclear Science and Technology Organization (ANSTO), Sydney (as Australian Institute of Nuclear Science and Engineering [AINSE] Postgraduate Research Awardee). He has completed his short postdoc at ESRF, France. Prior to this, he was selected as a Research Scientist at Stanford University, USA. He has been a recipient of various awards including the prestigious "Balshree Award" by H.E. The President of India, Child Scientist Award (2006 and 2007) by H.E. The President of India (Department of Science and Technology, Govt. of India) Dr. A.P.J. Abdul Kalam, National Talent Search Award by MHRD, National Merit scholarship by MHRD, Recommendation for MEXT fellowship at Japan by MHRD, New Delhi, and Award by Institution of Engineers (India) for outstanding contribution to research at an early career. He had secured state merit rank in the high school certificate examination (95.5%) (with a Governor medal), state merit rank in the higher secondary examination (94%) (with a Governor medal), and university merit rank in Engineering (with a Governor medal) and outstanding thesis declaration by AINSE, Australia apart from the district merit ranks in district-level primary (96%) and district-level middle school examinations (98.6%). He has been involved in funded projects worth $128,000 AUD.

Currently, he is serving as an Assistant Professor (Chemical Engineering and Biochemical Engineering) and Assistant Dean (Research and Development) at Rajiv Gandhi Institute of Petroleum Technology (RGIPT) (Ministry of Petroleum and Natural Gas, Government of India) (joined in 2021). He has also been appointed as an advisor by a company named Geon Engineers Private Limited, Noida. The company is establishing biogas, compressed biogas (CBG), and biomass-based power plants and working closely with PSUs, and central and state governments.

He is also serving as a board member in various international journals such as *Journal of Loss Prevention in the Process Industries* (Elsevier) (IF: 3.660), *Scientific Reports* (Nature) (IF: 5.134), *Corrosion Engineering, Science and Technology* (Taylor and Francis) (IF: 2.087), *Journal of Pipeline Science and Engineering* (Elsevier) (IF: Pending), and *Green Technology, Resilience, and Sustainability* (Springer) (IF: Pending).

His areas of interest are corrosion science and engineering, surface engineering, energy resources, and utilization and advanced materials. In addition to his Assistant Professorship, Dr. Deepak has also been offered the position of Senior Visiting Researcher at the University of Cambridge, UK; Adjunct Lecturer at WASM, Department of Chemical Engineering, Perth, WA Australia; and International Invited Member (Research) at Dublin City College, Ireland Honorary Regional Secretary, Indian Institute of Chemical Engineers (IICHE)-Amethi Regional Centre, UP, India.

Dr. Amit Ranjan is currently a *full* Professor at the Department of Chemical Engineering and Biochemical Engineering at RGIPT, Jais, Amethi. He holds an M.Tech. and a Ph.D. from IIT Kanpur and the University of Minnesota, respectively. His research interest lies mainly in the broad areas of Chemical and Materials Engineering. He was also heading the Department of Chemical Engineering and Biochemical Engineering at RGIPT. He has guided two Ph.D. students under his supervision and is currently supervising three more students at RGIPT.

Dr. Jitendra S. Sangwai is currently a *full* Professor at the Department of Chemical Engineering at the Indian Institute of Technology Madras. He holds an M.Tech. and a Ph.D. in Chemical Engineering from IIT Kharagpur and IIT Kanpur, respectively. Dr. Sangwai worked with Schlumberger for a brief period before joining academia. Dr. Sangwai's research interest lies mainly in broad areas of earth, energy, and environmental engineering. His primary focus is on carbon capture and sequestration, gas hydrates, and enhanced oil recovery. He has published approximately 150 peer-reviewed international journal papers, 100 conference publications, and filed/ granted 20 patents. He has supervised 20 Ph.D. and several master degree students. Dr. Sangwai is the recipient of the *prestigious* National Geoscience Award of the Government of India, Society of Petroleum Engineers' Distinguished Achievement Award for Petroleum Engineering Faculty of the South Asia and Pacific region, the National Award for Technology Innovation from the Government of India, the Young Faculty Recognition Award for excellence in teaching and research, Institute Research and Development Awards (both at Early- and Mid-Career levels) and Shri. J. C. Bose Patent Award from IIT Madras, and SPE Regional Service Award (2015). Dr. Sangwai has been highlighted as "One among 25 Emerging Investigators," "Top 1% Highly Cited Author" by ACS journals, and the "top 3% highly cited authors" from India by the American Chemical Society.

Contributors

Luqman Abidoye
Engineering Department
International Maritime College Oman
National University of Science and
    Technology
Muscat, Oman

Jimoh Adewole
Engineering Department
International Maritime College Oman
National University of Science and
    Technology
Muscat, Oman

Muna Al-Ajmi
Engineering Department
University of Technology and Applied
    Sciences
Sohar, Oman

Khadija Al-Balushi
Engineering Department
International Maritime College Oman
National University of Science and
    Technology
Muscat, Oman

Maryam Al Buraiki
Engineering Department
International Maritime College Oman
National University of Science and
    Technology
Muscat, Oman

Ahoud Al-Hamadani
Engineering Department
University of Technology and Applied
    Sciences
Sohar, Oman

Muna Al-Hinai
Engineering Department
International Maritime College Oman
National University of Science and
    Technology
Muscat, Oman

Asma Al kharousi
Engineering Department
International Maritime College Oman
National University of Science and
    Technology
Muscat, Oman

Suraj Aryan
Department of Chemical Engineering
    and Biochemical Engineering
Rajiv Gandhi Institute of Petroleum
    Technology
Amethi, India

Siddharth Atal
Department of Chemical Engineering
    and Biochemical Engineering
Rajiv Gandhi Institute of Petroleum
    Technology
Amethi, India

Enna Chakervarty
Department of Chemical Engineering
Manipal University Jaipur
Jaipur, India

Swati Chaudhary
Department of Chemical Engineering
    and Biochemical Engineering
Rajiv Gandhi Institute of Petroleum
    Technology
Amethi, India

Love Dashairya
Department of Ceramic Engineering
National Institute of Technology,
   Rourkela
Rourkela, India

Deepak Dwivedi
Department of Chemical Engineering
   and Biochemical Engineering
Rajiv Gandhi Institute of Petroleum
   Technology
Amethi, India

Tahereh Jafary
Engineering Department
International Maritime College Oman
National University of Science and
   Technology
Muscat, Oman

Abdulhameed Khalifullah
Engineering Department
International Maritime College Oman
National University of Science and
   Technology
Muscat, Oman

Ganesh Kumar
Enhanced Oil Recovery Laboratory
Petroleum Engineering Program
Department of Ocean Engineering
Indian Institute of Technology Madras
Chennai, India

Pranav Kumar
Department of Chemical Engineering
   and Biochemical Engineering
Rajiv Gandhi Institute of Petroleum
   Technology
Amethi, India

Saurabh Kumar
Department of Chemical Engineering
   and Biochemical Engineering
Rajiv Gandhi Institute of Petroleum
   Technology
Amethi, India

Thirumalai Kumar
Engineering Department
International Maritime College Oman
National University of Science and
   Technology
Muscat, Oman

Vivek Kumar
Department of Chemical Engineering
   and Biochemical Engineering
Rajiv Gandhi Institute of Petroleum
   Technology
Amethi, India

Yogendra Kumar
Department of Chemical Engineering
Indian Institute of Technology Madras
Chennai, India

Lalita Ledwani
Department of Chemistry
Manipal University Jaipur
Jaipur, India

Kamal Nayan
Department of Chemical Engineering
   and Biochemical Engineering
Rajiv Gandhi Institute of Petroleum
   Technology
Amethi, India

Hakeem Niyas
Renewable Energy Department
Energy Institute Bengaluru
Centre of Rajiv Gandhi Institute of
   Petroleum Technology
Bengaluru, India

Habeebllah Oladipo
Engineering Department
International Maritime College Oman
National University of Science and
   Technology
Muscat, Oman

Govardhan Pandurangappa
Department of Chemical Engineering
Indian Institute of Technology Madras
Chennai, India

Karan Kumar Paswan
Department of Chemical Engineering
   and Biochemical Engineering
Rajiv Gandhi Institute of Petroleum
   Technology
Amethi, India

Ajeet Kumar Prajapati
Department of Chemical Engineering
   and Biochemical Engineering
Rajiv Gandhi Institute of Petroleum
   Technology
Amethi, India

Bijoy K. Purohit
Department of Chemical Technology
Loyola Academy Degree and PG
   College
Secunderabad, India

Mainak Ray
R&D and Innovation
HPCL-Mittal Energy Ltd
Bhatinda, India

Partha Saha
Department of Ceramic Engineering
Centre for Nanomaterials
National Institute of Technology,
   Rourkela
Rourkela, India

Jitendra S. Sangwai
Department of Chemical Engineering
Indian Institute of Technology Madras
Chennai, India

Nellya Serikova
Safety and Risk Engineer
DNV France SARL
Paris, France

Abhishek Sharma
Department of Chemical Engineering
Manipal University Jaipur
Jaipur, India

Sumit Sharma
Department of Chemical and
   Biomolecular Engineering
Ohio University
Athens, Ohio

Himanshu Singh
Department of Chemical and
   Biomolecular Engineering
Ohio University
Athens, Ohio

Sukriti Singh
Department of Chemical Engineering
   and Biochemical Engineering
Rajiv Gandhi Institute of Petroleum
   Technology
Amethi, India

A.S.K. Sinha
Department of Chemical Engineering
   and Biochemical Engineering
Rajiv Gandhi Institute of Petroleum
   Technology
Amethi, India

Naveen Mani Tripathi
Mechanical Engineering
Rajiv Gandhi Institute of Petroleum
   Technology, Amethi (Assam Center)
Sivasagar, India

Prerna Tripathi
Department of Basic Science and
   Humanities
Rajiv Gandhi Institute of Petroleum
   Technology
Amethi, India

Anshika Verma
Department of Chemical Engineering
    and Biochemical Engineering
Rajiv Gandhi Institute of Petroleum
    Technology
Amethi, India

Yogendra Yadawa
Department of Chemical Engineering
    and Biochemical Engineering
Rajiv Gandhi Institute of Petroleum
    Technology
Amethi, India

Anteneh Mesfin Yeneneh
Engineering Department
International Maritime College Oman
National University of Science and
    Technology
Muscat, Oman

Prerna Yogeshwar
Department of Chemical Engineering
Indian Institute of Technology Madras
Chennai, India

1 Introduction to Functional Materials

Synthesis, Properties, Environmental Sustainability, and General Applications

Yogendra Kumar
Indian Institute of Technology Madras

Karan Kumar Paswan and Kamal Nayan
Rajiv Gandhi Institute of Petroleum Technology

Govardhan Pandurangappa
Indian Institute of Technology Madras

Deepak Dwivedi
Rajiv Gandhi Institute of Petroleum Technology

Jitendra S. Sangwai
Indian Institute of Technology Madras

CONTENTS

DOI: 10.1201/9781003242550-1

1

ABBREVIATIONS

CVD	Chemical vapor deposition
MOFs	Metal-organic frameworks
MWCNT	Multiwalled carbon nanotubes
NPs	Nanoparticles
PE	Polyethylene
PET	Polyethylene terephthalate
PP	Polypropylene
PS	Polystyrene
PTFE	Polytetrafluoroethylene

1.1 INTRODUCTION

The nanomaterials gained much attention among researchers owing to their tunability, multifunctional characteristics, and nanostructural surface. Nanoparticles-related research opens up new avenues for researchers who are currently working to make nanomaterial more effective. However, there is still scope for making them more effective through functionalization via some post-processing routes. Researchers are making microstructural and surface changes in raw nanoparticles to make them more effective. These small changes can be carried out during synthesis by tuning synthesis parameters or via post-synthesis coating or immobilization. NPs can be controlled by reaction parameters such as pressure, time, temperature, and pH, so it is imperative to optimize these parameters in order to achieve specific characteristic

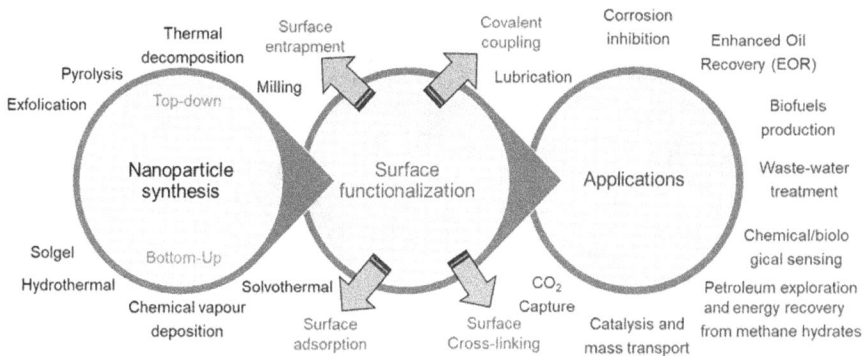

FIGURE 1.1 Pictorial depiction of proposed nanoparticle synthesis to application process route with different synthesis, functionalization strategies for multifarious applications.

products with unique features [1]. Functional nanomaterials are tuned or post-processed chemically or physically to attain specific chemical and physical properties that pristine nanomaterials did not exhibit. The general process route of nanoparticles preparation to application of functionalized nanoparticles is depicted in Figure 1.1. However, the immobilization and surface growth chemistry of nanoparticles during synthesis/functionalization play a key role.

The nanomaterials' functionalization is all about microstructural changes in surface behavior, porosity, surface passivation, or functional-group immobilization to improve nanoparticle specificity and functional features. Nanoparticles are functionalized via covalent or physical attachment over nanoparticle surfaces. Surface adsorption, covalent attachment, crosslinking, and entrapment are some of the functionalization strategies used in the last decade to create functionalized surfaces [2]. A nanoparticle is not a simple molecule; therefore, it consists of three layers: one on the surface, another inside, and a third layer underneath. A shell layer provides space for functional groups, biomolecules, and passivation liquids to improve dispersion stability, which is chemically different from the nanoparticle core. The core is essentially the central portion of the NP and usually refers to the NP itself consisting of core substances [3].

This chapter discusses different functional nanomaterials, their features, types, and synthesis, and functionalization strategies aiming to exploit the fullest potential of current technology. This chapter also offers key insights into environment-friendly and sustainable nanoparticle synthesis strategies such as green synthesis and nanoparticle synthesis from waste. The proposed scheme is depicted in Figure 1.2.

1.2 FEATURES AND PROPERTIES

Nanomaterials have a complicated definition, and their meaning varies based on their size and composition. Depending on the uses and materials, researchers have varied perspectives on how to describe nanoparticles. According to most studies, nanoparticles are thought to range in size from a few nanometers to hundreds of nanometers. Nanoparticles should have a size range of 1–100nm to be useful in various applications, according to the ISO-2008 standard [4]. Although there are

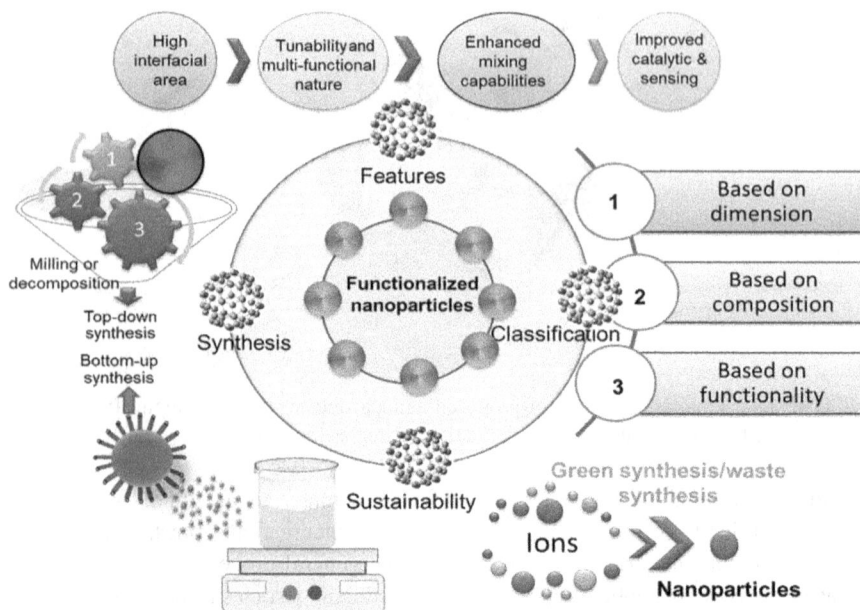

FIGURE 1.2 The pictorial representation of general scheme of this chapter, presenting summarized depiction in terms of functionalized nanoparticles classification, features, synthesis, and sustainability.

exceptions to the present criteria, bigger-size nanomaterials are also successfully utilized in the drug delivery, nano sensing, and semiconductor domains. Thus, some believe that nanomaterials are anything smaller than a microscopic scale. Nanomaterials' physicochemical, mechanical, biochemical, and chemical properties depend on the composition, size, geometry, and synthetic approach [5–7]. The nanoparticles have higher interfacial area, chemical/biological sensing ability, enhanced physicochemical properties, improved stability, magneto crystalline anisotropy, high coercive forces, and high thermal conductivity. Owing to these properties, nanomaterials are effectively utilized in targeted drug delivery, wastewater treatment, chemical and biological sensing, semiconductors, battery technology, nano-catalysts, and heat and mass transfer stimulants. The nanoparticles also play an influential role in the sustainable development and commercialization of processes and help to protect the environment. The use of nanomaterials in biofuel production to increase product yield/mass transfer coefficient and large-scale carbon capture technologies have also been extensively researched in the last decennium [8,9]. Several nanomaterials have been investigated to facilitate applications in various domains in the previous decade. Newly researched nanomaterials such as core-shell nanoparticles, nanocages, and metal-organic frameworks (MOFs) play an essential role in sensing and catalysis applications. On the other hand, enzymes, nano capsules, hydrogels, and nanobots also play a role in targeted drug delivery and biochemical applications.

1.3 TYPES OF NANOMATERIALS

Nanoparticles are classified depending on their shape, structure, and composition of nanoclusters. The nanoparticles are classified in 0D, 1D, 2D, and 3D depending on aspect ratio, geometry, and shape. Particles as tiny as 1–10 nm are classified in 0D; nanowires are classified in 1D; nanofilms are classified in 2D; and nanorods, nanosphere, nanocages, and nanocapsules are classified as 3D nanoparticles (Figure 1.3) [10]. Distinct classification approaches are used to classify nanoparticles depending on their shape, dimension, composition, and behavior:

- **According to geometry:** (1) Nanosphere, (2) Nanorods, (3) Nanoprism, (4) Nanowires, (5) Nanodots, (6) Nanocapsules, (7) Nanoplates, (8) Nanocages.
- **According to dimensions:** (1) Zero dimensional (dots), (2) One dimensional (Nanowires), (3) Two dimensional (nanofilms), (4) Three-dimensional (other).
- **According to behavior:** (1) Nanoparticles, (2) Nano Enzymes, (3) Micelles, (4) Xenocells.
- **According to composition:** (1) Metallic, (2) Polymeric, (3) Carbon-based, (4) Metal oxide, (5) Ceramic.
- **Hybrid materials-conjugated:** (1) Nanobots, (2) Bimetallic, (3) Core-shell, (4) Metal oxide frameworks, (5) Biological, (6) Janus, (7) Functionalized NPs, (8) Aerogel/hydrogels.

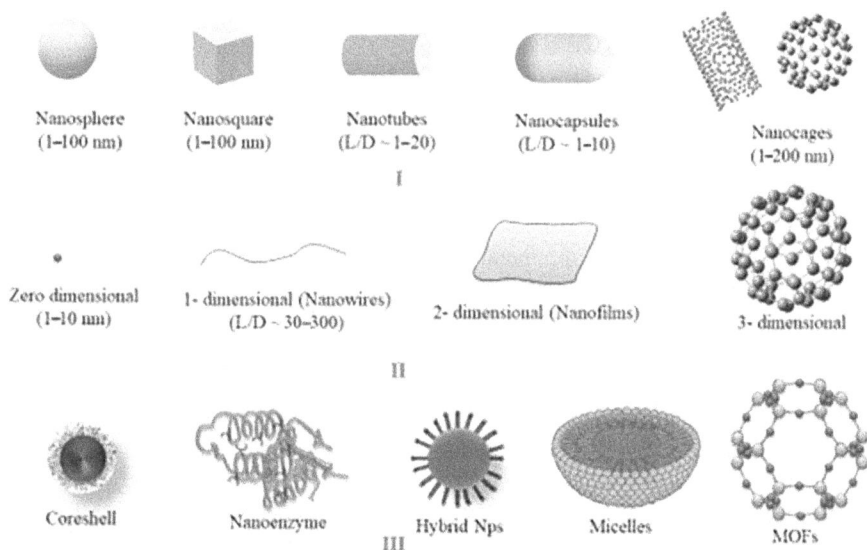

Nanosphere (1–100 nm)   Nanosquare (1–100 nm)   Nanotubes (L/D ~ 1–20)   Nanocapsules (L/D ~ 1–10)   Nanocages (1–200 nm)

I

Zero dimensional (1–10 nm)   1- dimensional (Nanowires) (L/D ~ 30–300)   2- dimensional (Nanofilms)   3- dimensional

II

Coreshell   Nanoenzyme   Hybrid Nps   Micelles   MOFs

III

FIGURE 1.3 Types of nanoparticles: (I) classification based on geometry, (II) classification based on dimensionality, and (III) other nanomaterials.

TABLE 1.1

Different Nanomaterial/Conjugated Materials, Their Examples, and Industrial Applications

Nanomaterial/ Conjugated Particles	Examples	Applications
Metal nanoparticles	Au, Ag, Pt, Mo, Cu	Nanocatalyst, CO_2 sequestration, EOR, biofuels, heat transfer application
Carbon NPs	CNT, MWCNT, Fullerene, Graphene	Gas absorption, CO_2 sequestration, EOR, heat transfer applications, energy recovery from gas hydrates, wastewater treatment
Biological nanoparticles	Enzymes, microbe	Biochemical reactions, biofuels, etc.
Metal oxide NPs	Al_2O_3, SiO_2, Fe_3O_4, ZnO, NiO	Heat transfer application, EOR, cancer treatment, catalysis, biofuels, wastewater treatment
Hydrogels	Drug-loaded PAA, PHEMA, PVA, PEG and poly (ethylene oxide) (PEO) hydrogels	Drug delivery, contact lenses, wound dressing, hygiene products, and nanoparticles-immobilized hydrogel are used as catalysts
Core-shell nanoparticles	Au/SiO_2, Ag/SiO_2, Ag/PEG, $SiO_2/PMMA$, Fe_3O_4/PEG etc.	Multifunctional properties, lower deactivation rate, improved inertness, higher residence time for catalysis, high immobilization efficiency
Bimetallic	Ni-Au, Pt-Rh, Au-Pt, Ag-Ni, Pt-Ag, and Pd-Au	Synergistic functional properties, tunable characteristics, improved electronic, catalytic nature
Hybrid NPs	NPs loaded aerogels, conjugated NPs, and functionalized NPs	Catalysis, heat transfer applications
Metal oxide frameworks	Mg-MOF-74, MOF-5, HKUST-1, ZTF-1, and MOF-200	Chemical and biological sensing, adsorbent, micro-nano robotics, supercapacitor, water purification, CO_2 capture

Nanoparticles are classified into metallic, metal oxide, ceramic, biomolecules, polymeric, and carbon-based depending on composition and materials specifications. Several hybrid nanomaterials such as Janus, bimetallic, core-shell, micelles, and metal oxide frameworks are designed for specific functions and applications [11]. Nanomaterials such as lipids, hydrogels, enzymes, and microbes are also classified in biological NPs, having high human biocompatibility and specificity to perform a specific task (Table 1.1). 2D nanomaterials such as nanofilms find their broad application in membrane-based separation, corrosion inhibition, and semiconductor industry. Emerging nanomaterials such as xenobots and nanobots demonstrate their effectiveness in several biomedical and sensing applications. These nanomaterials come in the category of smart materials because they have memory, self-healing, and reorganizing capabilities and can be tuned to perform specific tasks such as targeted drug delivery, scraping clogged arteries, and cancer cell detection. The brief classification of functional materials is shown in Figure 1.4.

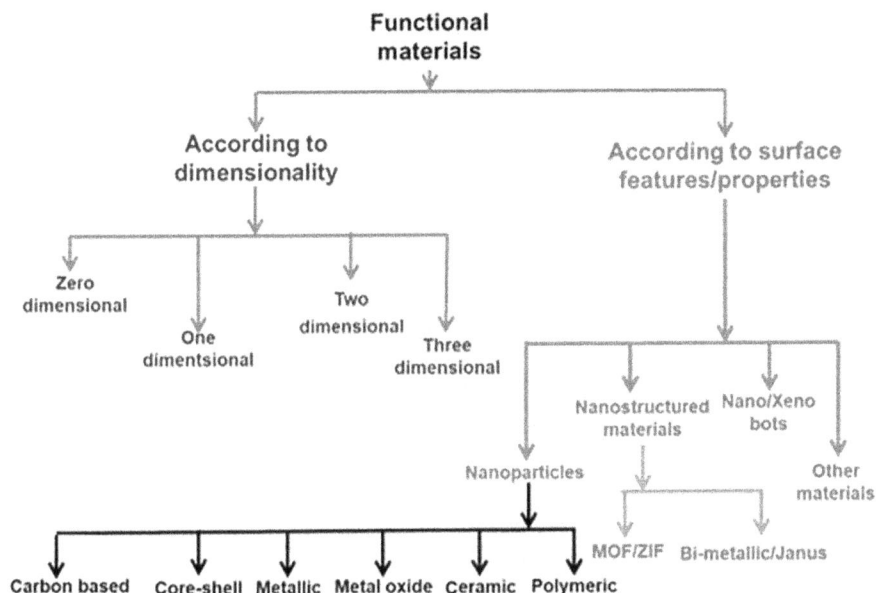

FIGURE 1.4 Brief classification of functional materials.

1.4 NANOMATERIALS AND THEIR TYPES

1.4.1 CONVENTIONAL NANOPARTICLES

Conventional nanoparticles such as carbon-based, metallic, polymeric, and metal oxides are widely used in multifarious applications. However, these nanoparticles require surface charge passivation to remain stable in dispersion. The functionalization of these nanoparticles will enhance dispersion stability and reduce aggregation tendencies. Surfactants and dispersants are common functionalizing agents used with these nanoparticles.

1.4.1.1 Carbon-Based

Carbon-based nanoparticles (such as multiwalled carbon nanotubes (MWCNTs), carbon nanotubes (CNTs), graphene and fullerene) demonstrate inert behavior, high thermal conductivity, and corrosion-resistance properties. Carbon-based nanoparticles are frequently used as particle reinforcement, nanofluids for heat and mass transfer, gas absorption, and corrosion-resistance materials.

1.4.1.2 Metallic

Metallic nanoparticles have tunable catalytic characteristics, and antibacterial and microbial properties. Most metallic nanoparticles are unstable and oxidize quickly in the atmosphere. The metallic nanoparticles are further classified as (1) magnetic metallic particles (Fe, Ni, Mn) and (2) nonmagnetic metallic particles (Ag, Au, Pt, Al, Cu). Transition metal nanoparticles are frequently used as a catalyst in chemical processes and sensing applications.

1.4.1.3 Metal Oxide

Metal oxide nanoparticles are relatively stable compared to metallic nanoparticles and demonstrate both catalytic and optical properties. Metal oxide nanoparticles are used in solar cells, coating, optoelectronic packaging, transistors, light-emitting diodes, and biofuels. The metal oxide nanoparticle is also classified as (1) magnetic (Fe_3O_4, Fe_2O_3, NiO) and (2) nonmagnetic (SiO_2, Al_2O_3, ZnO, CaO, TiO_2, etc.).

1.4.1.4 Functionalized Nanoparticles

Enzymes, biomolecules, and functional groups are commonly immobilized over porous or charged nanoparticles to facilitate specific functions. Doping of DNA, dye, aptamers, and functional groups is more common in biomedical and targeted delivery applications. Amino-functionalized nanoparticles recently gained much attention in CO_2 capture and water treatment applications.

1.4.1.5 Core-Shell

The core-shell nanoparticle has multifunctional nature, improved thermal stability, and dispersive nature. Core-shell nanoparticles can be used as catalysts, electrocatalysts, or sensing materials [12]. Surface passivation through coating reduces the poisoning of core material, and improves wear resistance and multifunctional properties. Due to the features of core-shell micro/nanoparticles, such as their diverse compositions, morphologies, particle sizes, and synergistic effects among components, they have been widely used for adjusting or enhancing material properties, including optical and electronic properties, interface bonds, catalytic activity, and magnetic thermal induction [13].

1.4.2 Nanostructured Materials

Nanostructured materials consist of nano-size features, bi- or trimetallic regions, and a nanoporous framework that offers higher interfacial surface area and enhanced porosity. Nanostructure materials also facilitate super hydrophobicity, and enhanced catalytic and optical properties owing to their multifunctional nature.

1.4.2.1 Bimetallic and Janus Nanoparticles

To make bimetallic nanoparticles, two metal precursors are typically reduced simultaneously, forming bimetallic alloy cores. One example of a stable dry bimetallic nanoparticle is monolayer-protected alloy clusters [14]. The composition of metal salts used in nanoparticle synthesis may differ significantly from the composition of metal salts in and on the core surface. Alternatively, a second metal can be deposited onto the surface of preformed nanoparticles, resulting in core-shell nanoparticles whose thickness can be varied by varying experimental parameters [15]. The metal elements of the original nanoparticles can also be replaced using galvanic exchange reactions, forming new metallic nanostructures. For example, noble metals (e.g., Au, Pt, and Pd) have been used to replace Cu or Ag particles encapsulated in dendrimers, and bi- and trimetallic nanoparticles have also been prepared using such reactions [16].

1.4.2.2 Nano-Metal-Organic Frameworks

MOFs are unique in that they are highly tunable and structurally diverse and unique in their chemical and physical properties. In MOFs, metal cations or clusters of cations ("nodes") are linked by multitopic organic "struts" or "linkers" ions or molecules [17]. MOFs are conjugated materials that offer very high microporosity, catalytic, and sensing properties [18,19]. MOFs possess intriguing properties due to the synergistic action between the inorganic and organic components. MOFs are crystalline that offer extremely high porosity (up to 90% free volume) and large internal surface areas exceeding 6000 m²/g [20]. The MOFs structures are classified into first generation (rigid structure), second generation (breathing structure) and other structural classifications. The crystalline order of the underlying coordination network is combined with the cooperative structural transformability of MOFs. The flexible MOFs are important in the so-called swelling and breathing phenomenon during operation due to mechanical stimuli, thermal, photochemical, and Guest adsorption/desorption trigger phase transitions [21]. Upon being subjected to external stimuli, a highly ordered structure may undergo a crystal-to-amorphous or crystal-to-crystal phase transformation with drastic volume changes. In these materials, porous forms are converted between open-pore and closed-pore (cp) or narrow-pored (np) and large-pored (lp). Yet, fewer than 100 compounds exhibit substantial breathing behavior or similar stimuli-responsive properties among the about 20,000 coordination network structures listed in the Cambridge Structural Database (CSD) [21], and some of them are represented in Figure 1.5. It is necessary that the magnetic, dimensional, optical, density, and electrical properties of the material can be controlled if the MOF is to be used as an adsorbent, sensor, or switching device [22]. The MOFs have gained popularity in vivid domains such as gas storage, separation, chemical sensors, separation, heterogeneous catalysis, biomedicine, and proton conductivity [23].

MOFs are prepared via the homogenization of organic ligands with metal clusters/ions followed by self-assembly.

1.4.3 OTHER FUNCTIONAL MATERIALS

1.4.3.1 Hydrogels and Aerogels

Hydrogels have very high biocompatibility, high drug-loading capacity, and low toxicities. Hydrogels are effectively used as drug delivery materials, cancer treatments, contact lenses, and optical applications. The term hydrogel refers to a network of hydrophilic polymer chains that swells in water and can hold a lot of water while maintaining its structural integrity due to chemical or physical crosslinking between the polymer chains. Physical and chemical stimuli such as electric field, pressure, temperature, sound, and light significantly alter the volume of hydrogels, while ionic strength, solvent composition, molecular species, and pH substantially alter the volume of hydrogels [24]. Hydrogels facilitate the slow release of desired drugs owing to these physical and chemical stimuli and can be used in combinational therapy [25,26]. Hydrogels also have prospective solar water purification technology due to high water transport and improved solar thermal conversion efficiency [27]. Three-dimensional hydrogels with high mechanical strength and shape recovery properties

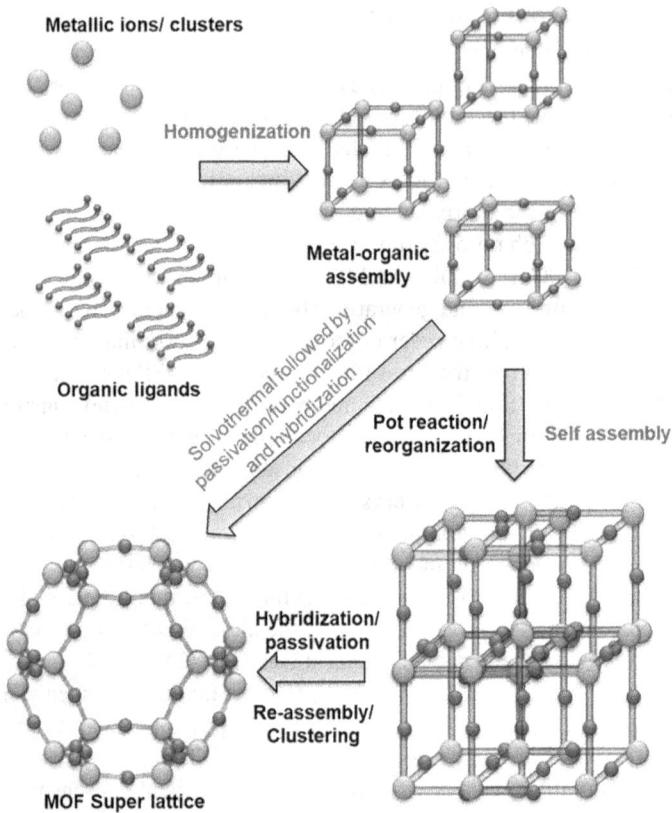

FIGURE 1.5 Pictorial depiction of synthesis route of MOFs (Reprinted with permission from Chowdhury et al. [49]).

are used in many fields, such as tissue engineering, implant devices, soft robots, and other sustainable materials [28].

1.5 SYNTHESIS OF ADVANCED FUNCTIONAL MATERIALS

The field of advanced functional materials has attracted a lot of interest due to its targeted applications. These applications cover various fields such as medicine, energy storage devices, magnetics, optoelectronics, electronics, environmental remediation, and catalysis. It is possible to tailor the particle size of materials, especially nanomaterials, to satisfy the targeted applications with specific properties. Chemical properties such as surface energy, chemical potential, and surface reactivity and physical properties such as size effect, crystalline structure, lattice parameters, and morphology vary from nanomaterial to nanomaterial [29]. Moreover, optical properties such as absorption, emission, reflection, and transmission and mechanical properties such as porosity, adhesion, and friction vary for nanomaterials compared to their bulk materials. Any material can be classified as nanostructured material if at least one

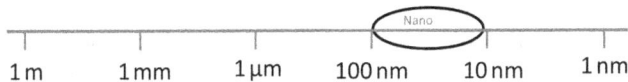

FIGURE 1.6 Scaling of nanomaterials and their prospective sizes.

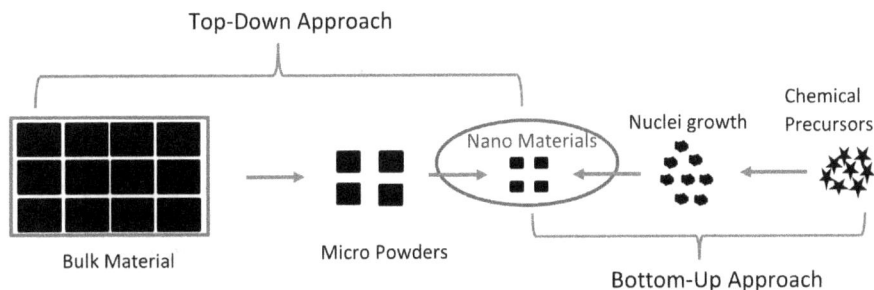

FIGURE 1.7 Pictorial depiction of top-down and bottom-up methods.

dimension is less than 100 nm [30]. The properties of the nanomaterials are different from those of the bulk materials due to their high surface-to-volume ratio. The surface-to-volume ratio of a nanosphere of 100 nm is 107:1 as shown in Figure 1.6. There are many different shapes and dimensions of nanomaterials, including nanoclusters, quantum dots, nanospheres, nanowires, nanotubes, nanocomposites, nanosheets, nanofibers, nanofilms, core shells, nanofilms, and nanocomposites. Nanomaterials are dimensionally dependent, so their dimensionality often determines their activity.

Two approaches can be used to synthesize nanomaterials: the top-down method, where successive size reductions are carried out from the bulk material using the physical methods. The bottom-up method involves growing nanomaterials from their chemical precursors in either liquid or gas phases, starting with atoms and molecules. These two approaches have their advantages and disadvantages based on the technique used to produce the nanomaterials. These two approaches are pictorially depicted in Figure 1.7.

1.5.1 TOP-DOWN APPROACH

This technique of nanoparticle synthesis for functional materials involves successive size reduction from its bulk material through various physical methods such as lithography, milling, electrospinning, sputtering, laser ablation, and chemical vapor deposition. These physical methods are extensively energy-dependent, and the final obtained product does not have a uniform size distribution compared to that of the bottom-up approach [7]. Many factors influence the end product's size, including the speed of rotation during ball milling, lithography, etching, sputtering, laser ablation, and thermal evaporation. It is crucial to adjust several operating parameters during ball milling in order to achieve uniform product size distribution, such as the speed of rotation, temperature control, noise, and contamination. Furnace temperature and substrate distance from the source are key parameters in thermal evaporation.

However, beam energy, transmission, and light absorption efficiency of the targeted material are important in laser ablation process. Further, sputtering process is more dependent on annealing time, temperature, angle of substrate and source and nanoparticle thickness. The main advantages of this synthesis approach are that it can be easily scalable, can deposit on large area substrates, and does not require chemical purification. The major disadvantages are the broad distribution of product size, higher cost, and control over the operating parameters.

1.5.2 BOTTOM-UP APPROACH

This technique of nanoparticle synthesis for functional materials involves the synthesis of direct atoms and molecules either in the gas phase or in the liquid phase. The processing methods involve both chemical and physical methods. The starting chemicals are allowed to dissolve in the liquid or vaporize and condense to form particles. The synthesis methods include supercritical synthesis, co-precipitation, sol-gel process, laser pyrolysis, chemical reduction, solvo thermal and hydrothermal method, chemical vapor deposition, molecular condensation, self-assembly, and green synthesis [30]. The size of the end product depends on the various operating parameters involved during the process, such as the distance between the target and substrate vacuum condition during the chemical vapor deposition. Hydrothermal and solvothermal methods depend on temperature, solvent, and pressure, whereas co-precipitation method depends on pH, reaction time, precursor concentration, stirring speed, and temperature [7,29].

The major advantages of this approach are that fine nanoparticles with uniform size distribution can be obtained, and deposition parameters can be tuned easily to obtain different shapes and sizes. There is a lot of scope for tuning parameters as the synthesis can be carried out in the gas or liquid phase. The major disadvantages include the fact that chemical purification is necessary, and large area deposition cannot be done.

1.5.3 EMERGING AND SUSTAINABLE ROUTES FOR THE FUNCTIONAL NANOPARTICLES SYNTHESIS

Nanoparticle synthesis is a complex, costly, and less environment-friendly process. So the focus of researchers is now shifted toward adopting greener, cost-effective, and environment-friendly approaches. The regeneration and recycling of nanoparticles have also been researched in the last decade. The cost of renewal and recycling is confronting this approach. However, emerging nanoparticle synthesis approaches such as green synthesis and waste-synthesized nanomaterials can be the best alternatives to the current physical and chemical nanoparticle synthesis routes.

1.5.3.1 Green Synthesis

Green synthesis is a low-cost, simple, newer, and environment-friendly nanoparticle synthesis approach. The green synthesis approach uses plant extract, microbes, enzymes, and other biological precursors to synthesize nanoparticles as shown in Figure 1.8. Plant extracts are rich in phytochemicals such as amides, carboxylic acid,

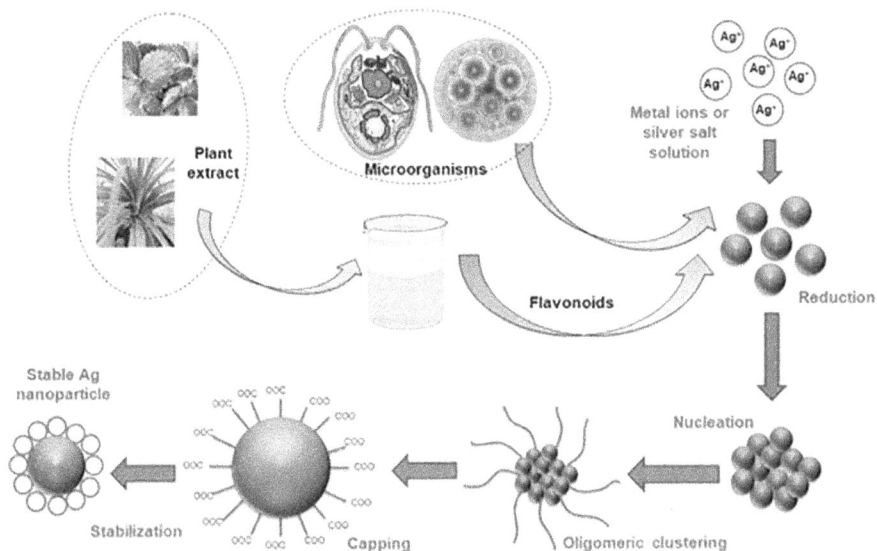

FIGURE 1.8   Green synthesis of silver nanoparticles using plant extract or microorganism.

phenols, and flavonoids. Amides and ascorbic acids play an indispensable role in reducing metals to metal nanoparticles [31]. Different nanomaterials such as core-shell, Janus, aerogels, hydrogels, and metal-organic frameworks can be prepared using green synthesis. Several green synthesis approaches are used to synthesize metal and metal oxide nanoparticles [32].

  i. Bacteria-mediated
  ii. Fungi-mediated
 iii. Yeast-mediated
  iv. Actinomycetes-mediated
   v. Algae-mediated
  vi. Plant-mediated
 vii. Biomolecular templating

In most green synthesis approaches, biological mediated species can be used as reductants and act as capping and stabilizing agents. Researchers studied different biological species such as unicellular Protista (binucleous), plantae (autotrophs, eukaryotic), fungi (saprophyte or parasite, eukaryotic), plantae (autotrophs, eukaryotic), and prokaryotic monera (binucleous) to synthesize silver nanoparticles [33]. The silver nanoparticle is derived through sequential steps: (1) biological reduction of silver ions, (2) formation of metallic silver, (3) aggregation in oligomeric clusters, and (4) formation of metallic colloidal silver nanoparticles [34,35].

Silver nanoparticles are usually synthesized via green synthesis routes using biological reduction process. The unstable oligomeric clusters are formed owing to reduction of metallic ions via bio-extracts (plant, leaves, flower, etc.). The benefits of

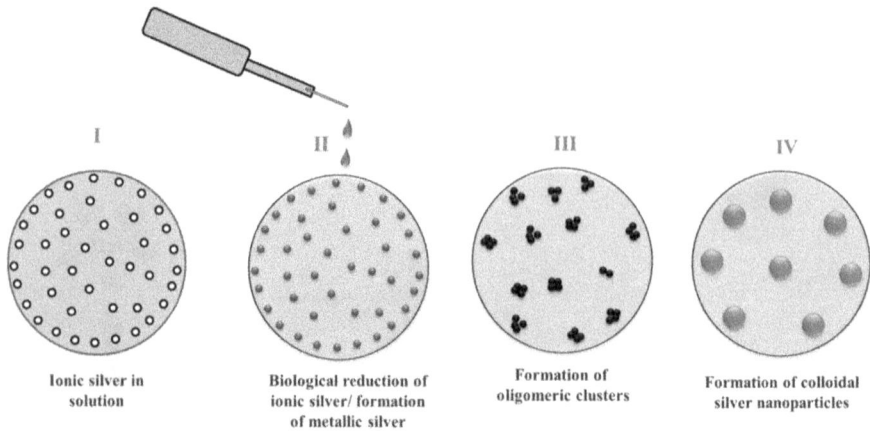

I	II	III	IV
Ionic silver in solution	Biological reduction of ionic silver/ formation of metallic silver	Formation of oligomeric clusters	Formation of colloidal silver nanoparticles

FIGURE 1.9   Process of green synthesis of silver nanoparticle using biological extract.

green synthesis are that the oligomeric clusters are capped by green extract itself and a stable colloidal solution of nanoparticles has been produced as shown in Figure 1.9. Owing to efficient capping performance and slow kinetics of biological extracts during growth phase, the control over size and shape during the process is possible.

1.5.3.2   Nanoparticle Synthesis from Waste

The cost competitiveness of nanoparticle-mediated processes is another roadblock confronting the large-scale utilization of nanomaterials on industrial scales. Nanoparticle synthesis from polymeric, rubber, metal scraps, and biological wastes via different top-down and bottom-up approaches can solve this issue. The extraction of noble metals from electronic scrap wastes and solar panels through pyro-/hydrometallurgical routes can also help minimize waste and process beneficiation. The nanoparticles derived from coconut shells via mechanical milling can be used as particle reinforcement and fillers in different cosmetic products. Similarly, eggshells rich in calcium can be used to derive CaO, $CaCO_3$, and hydroxyapatite nanoparticles. The eggshell-derived nanoparticles can be used as alkaline heterogeneous catalysts for different base-catalyzed reactions (transesterification). Carbon-based nanoparticles such as CNTs, MWCNTs, and graphene are derived via thermal decomposition, pyrolysis, and catalytic carbonization techniques [36]. Moreover, magnetite (Fe_3O_4), zero-valent irons (ZVIs), and iron oxide nanoparticles from steel scrap, iron pickling waste, rust, and slag gained momentum in hybridizing low-cost energy alternatives such as biogas, biodiesel, and biohydrogen production. The extraction of precious metals from electronics and solar panel waste is always challenging and hazardous. The recycling approaches require sophisticated control over process parameters and human interference. The fumes of recycled materials are highly toxic and poisonous, and thus require adequate measures in dumping or recycling these wastes. Emerging advanced pyro-/hydrometallurgical routes can extract precious and semi-precious metals from these electronic and solar panel waste. The synthesis methodologies of different nanoparticles by using industrial and biological waste are summarized in Table 1.2.

TABLE 1.2

Different Nanomaterials and Their Preparation Approaches

Nanoparticles	Waste Precursor Material	Synthesis Methodologies	Remarks
ZnO	Zn-C battery spent material	Thermal technique in horizontal quartz tube at 900°C	Size < 50 nm Nanocatalyst, nanoadsorbant, water treatment
Graphene	Polymeric waste PP PTFE PE & PS PET	Catalytic carbonization Epitaxial growth over SiC Chemical vapor deposition Thermal decomposition or pyrolysis is followed by graphitization and exfoliation	<100 nm Improved thermal performance, wear and tear resistance and anti-corrosion properties, nano adsorbent, water treatment, CO_2 sequestration
Zn, carbon, S and minerals	Natural rubber or tire waste	Ball milling 5–8 consecutive hours with silica waste	<100 nm It can be used as fertilizer and particle reinforcement
MWCNT	Hydrocarbon tires have low boiling point	Pyrolysis in the presence of a ferrocene catalyst over a quartz substrate	<200 nm Improve thermal performance, nanoadsorbant
Cu, Ni, Al, Pb, and Sn	Electronic waste	1. The mechanical operation is followed by separation and chemical treatment 2. Pyro- and hydrometallurgical routes are innovative approaches to extracting precious metals	<100 nm Nanocatalyst for sensing and reactor systems
Fe	Steel pickling waste	Sodium borohydride reduction method	Nanocatalyst
Fe_2O_3, Fe_3O_4	Iron rust, steel slag	Hydrothermal synthesis in an autoclave reactor at 150°C	Heterogeneous acid catalyst
Reinforcing nanoparticles	Coconut shell	Planetary ball milling	<100 nm Improving wear & tear resistance of materials
Hydroxyapatite, $Ca_{10}(PO_4)_6(OH)_2$	Eggshell	Calcination and sol-gel method	<100 nm Nanocatalyst for biodiesel, base-catalyzed reactions
Ag, AgO	$AgNO_3$	Combined synthesis of eggshell collagen reduced Ag and AgO nanoparticles	<100 nm Nanocatalyst, water treatment
Fe_3O_4	Papaya leaves	Thermal decomposition	
SiO_2	Rice husk	Acid pretreatment followed by pyrolysis of rice husk	<50 nm, Nanoadsorbant, water treatment, extraction of chemicals

The high-temperature synthesis route (top-down) is commonly used to obtain nano-materials from electronics, metal scrape, hard shells, rubbers, and polymers, whereas bottom-up approaches are used with soft, biological wastes. The biological and soft waste only processed through bottom-up approaches owing to the fact that process requires moderate T, P condition during synthesis using solvothermal, hydrothermal, and sol-gel strategies. However, CVD can operate at higher temperatures.

1.6 SURFACE TREATMENT AND FUNCTIONALIZATION STRATEGIES

Some fundamental functionalization strategies include surface adsorption, covalent attachment, surface entrapment, and crosslinking. Van der Waals' forces usually carry out the adsorption of species on absorbent surfaces. Hydrogen bonding and dipolar attraction can also induce physical adsorption through other forces. When a solid surface whose critical temperature Tc is higher than T meets a vapor whose critical temperature Tc is higher than T. The multilayer absorption occurs since the forces keeping molecules in the liquid phase are electrostatic or van der Waals' forces. A relatively thick film forms as a result of this phenomenon, which depends on the relative pressure of the vapor, but generally falls within the thickness range of 3%–15%. The relative pressure is calculated by dividing the partial pressure P by the saturation pressure P_0 at the given temperature [2,37].

Polymer–polymer covalent bonding (grafting) provides a new strategy for forming long-term coating composites (polymer/nanosilver). It was found that different chemical functional groups could be embedded in the polymer coatings by utilizing organic synthesis methods for incorporation of nanosilver particles. However, this approach has a drawback that it can only cover a certain number of substrates, and new surface attachment techniques will be required to cover other substrate types [38]. The electrostatic force is used to adsorb different building blocks in order to fabricate functional ultra-thin film coatings using layer-by-layer self-assembly (LBL) [39,40]. This approach has the following advantages: (1) polyelectrolytes, nanoparticles, and biomacromolecules can be utilized as different building blocks; (2) practically any substrate with any geometry and chemical composition can be used; (3) chemical composition and nanostructure can be observed at the molecular level; and (4) it is an inexpensive, eco-friendly, and simple method [2,41].

1.7 APPLICATIONS

Nanoparticles and their functionalized forms are now being used in the domain of biofuel, CO_2 capture, gas hydrates, oil and natural gas exploration, clean water, drug delivery, and heat and mass transport process. Moreover, the manufacture of scratch-proof eyeglasses, stain-repellent fabrics, crack-resistant paints, transparent sunscreens, anti-graffiti coatings for walls, ceramic coatings for solar cells, and self-cleaning windows also uses nanotechnology to improve luster and optical properties. The potential application of waste-synthesized nanomaterials in biofuel and clean water sectors with potential synthesis methodologies is depicted in Figure 1.10.

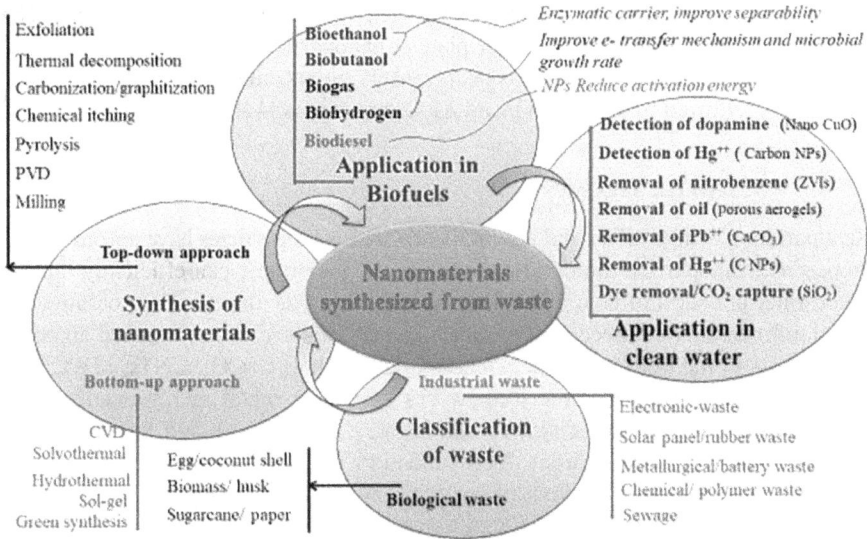

FIGURE 1.10 Nanomaterials synthesis and their applications in clean water and biofuels.

1.7.1 KINETIC PROMOTER IN CO_2 SEQUESTRATION

Amine-functionalized nanoparticles help in reducing transport barriers and facilitate the chemisorption of CO_2 on immobilized amine functional groups along with physisorption during gas absorption. Moreover, nanoparticles also improve gas hydrate formation kinetics which eventually helps the research community to develop large-scale CO_2 capture and sequestration strategies using seawater.

1.7.2 STIMULATION OF BIOFUELS PRODUCTION

Enzyme-functionalized nanoparticles have tremendous scope in biodiesel, biobutanol, biogas, and bioethanol production. The nanoparticle core in the structure enhances the recoverability, separability, and catalytic tunability of nanoenzymes [42]. Nanoparticle incorporation also reduces kinetics and mass transport barriers during the chemical reaction. Nanoparticles have a distinct role in stimulating the yield of different biofuels. Kumar et al. [29,43] described nanomaterials acting as nano-catalysts in biodiesel production and reducing activation energy for effective molecular collisions for reaction to happen. Biohydrogen and biogas nanoparticles help accelerate the electron transfer mechanism, improve microbial growth rate, and decrease the reduction pathway. In bioethanol and biobutanol, nanoparticles act as an enzymatic carrier and co-factor for enzymatic activities, improve the separability and structural stability of enzymes, and modulate the oxidation and reduction potential of the system [29].

1.7.3 HEAT TRANSFER AGENTS IN THERMAL APPLICATION

Nanoparticles in dispersion offer higher heat transfer rates owing to Brownian motion and thermophoresis of nanoparticles. However, the heat transfer efficacy of nanoparticles is geometry, material, and their diffusion kinetics dependent. The highest heat transfer was recorded for platelet nanoparticles, followed by cylinders, bricks, blades, and spheres, whereas heat transfer enhancement was highest for SiO_2 nanofluids, followed by ZnO, CuO, and Al_2O_3 nanofluids [44].

1.7.4 WASTEWATER TREATMENT

Nanoparticles such as silver and Ag/SiO_2 core-shell nanoparticles have antimicrobial properties. Nanoparticle immobilization over the membrane can efficiently tap the clean water potential of nanotechnology by enhancing its antimicrobial nature. The use of amino-functionalized nanoparticles in wastewater treatment gained apparent acceleration in the last few years. Nanomaterials such as $CaCO_3$, CNTs, MWCNTs, SiO_2, Fe_3O_4, and metal-organic frameworks have high metal and ions adsorption capacity due to high interfacial area and high catalytic nature and improved pore volume. Nanomaterials are used in suspended or membrane-embedded conditions to improve the selective sorption of specific ions, oils, dyes, and other impurities. Nanomaterials are effectively used to remove the dye, Pb^{++}, Hg^{++}, nitrobenzene, and oils from wastewater. Using waste-generated nanomaterials in chemical and biological sensing applications is also feasible.

1.7.5 KINETIC PROMOTER IN EXTRACTION AND PRODUCTION OF CHEMICALS

Nanostructured particles, MOFs, and immobilized nanoparticle reactors have the greater prospect and will open up new avenues in this field. Higher porosity and enhanced perturbation in these systems are major advantages. Silane immobilization in microchannels and the capillary reactor will serve the continuous production scheme better without the requirement of additional separation and recovery costs [43,45]. The perturbation of fluid streams around immobilized nanoparticles facilitates mixing, which offers higher separation and catalytic conversion efficiency.

1.7.6 CORROSION INHIBITOR

Nanoparticles can be used to make hydrophobic, anti-cracking, anti-reflective, self-cleaning, and crack-resistance surfaces. These properties help substrate materials to gain corrosion-resistance properties. Nanoparticle coating also blocks ionic migration across the coating, thus offering a barrier against corrosion. Nano-encapsulated corrosion inhibitors, carbon dots, and other nanoparticles help reduce photodegradation and increase barrier resistance and hydrophobicity of the surface [46]. Some nanoparticle coatings demonstrated self-healing and self-cleaning properties. Detection of hot spots in high-temperature reactors using nanoparticles-assisted thermal sensitized paint is possible, and these coatings avoid any thermal corrosion or accidents during operation.

1.7.7 CHEMICAL AND BIOLOGICAL SENSING

In addition to their optoelectronic properties, functional nanoparticles exhibit excellent biocompatibility when they are ligated with appropriate ligands. Changing the size, shape, and chemical environment of NPs can readily tune these properties. Functional nanoparticles have many properties such as conductivity, plasmon resonance absorption, and redox behavior, which are conducive to their ability to generate detectable responses [47,48]. Janus, core-shell, and MOFs are other potential candidates for sensing applications owing to their unparallel tunability, multifunctional characteristics, structural diversity, and higher interfacial area [49].

1.8 CONCLUSION

Nanoparticle tuning and functionalization are essential for improving the characteristics of pristine nanomaterials. Coating, conjugation, and immobilization are three important processes to make pristine nanoparticles more effective. Surface functionalization was demonstrated to have improved interfacial area, higher catalytic activities, high sensing, and optical characteristics. Small size allows NPs to have large surface areas, with functional groups making them suitable for a variety of applications. As these materials have a large interfacial area, their optical, magnetic, and catalytic properties are prominent, further increasing their utility as catalysts, sensors, and carriers. However, nanoparticle's cost-effectiveness also plays a vital role in commercializing the current hybrid systems. Synthesis of nanomaterials consumes a major cost fraction in nanoparticle-functionalized systems. Additionally, environmental issues need to be considered before using these materials for any application, especially in the case of heavy metals, which are environmentally hazardous and can also affect human health. Environmental-friendly synthesis approaches may play an essential role in reducing the overall cost of functional nanomaterials. Environmentally cordial approaches such as waste nanomaterial synthesis and green synthesis are explored to have good potential to exploit the current potential. This chapter discusses different aspects such as features, types, functionalization, and environmental cordial synthesis approaches to functionalized nanomaterials with their potential applications.

REFERENCES

[1] Khan I, Saeed K, Khan I. Nanoparticles: Properties, applications and toxicities. *Arab J Chem* 2019;12:908–31. https://doi.org/10.1016/j.arabjc.2017.05.011.
[2] Fahmy HM, Mosleh AM, Elghany AA, Shams-Eldin E, Abu Serea ES, Ali SA, et al. Coated silver nanoparticles: Synthesis, cytotoxicity, and optical properties. *RSC Adv* 2019;9:20118–36. https://doi.org/10.1039/c9ra02907a.
[3] Shin S, Um E, Sabass B, Ault JT, Rahimi M, Warren PB, et al. Size-dependent control of colloid transport via solute gradients in dead-end channels. *Proc Natl Acad Sci U S A* 2016;113:257–61. https://doi.org/10.1073/pnas.1511484112.
[4] ISO/TS 27687:2008. Nanotechnologies—Terminology and definitions for nano-objects nanofibre and nanoplate (ISO/TS 27687:2008). *Management* 2008;1:7.
[5] Jiang J, Oberdörster G, Elder A, Gelein R, Mercer P, Biswas P. Does nanoparticle activity depend upon size and crystal phase? *Nanotoxicology* 2008;2:33–42. https://doi.org/10.1080/17435390701882478.
[6] Gatoo MA, Naseem S, Arfat MY, Mahmood Dar A, Qasim K, Zubair S. Physicochemical properties of nanomaterials: Implication in associated toxic manifestations. *Biomed Res Int* 2014:498420. https://doi.org/10.1155/2014/498420.

[7] Baig N, Kammakakam I, Falath W, Kammakakam I. Nanomaterials: A review of synthesis methods, properties, recent progress, and challenges. *Mater Adv* 2021;2:1821–71. https://doi.org/10.1039/d0ma00807a.

[8] He Y, Wang F. Hydrate-based CO_2 capture: Kinetic improvement: Via graphene-carried-SO3- and Ag nanoparticles. *J Mater Chem A* 2018;6:22619–25. https://doi.org/10.1039/c8ta08785g.

[9] Chaturvedi S, Dave PN, Shah NK. Applications of nano-catalyst in new era. *J Saudi Chem Soc* 2012;16:307–25. https://doi.org/10.1016/j.jscs.2011.01.015.

[10] Tiwari JN, Tiwari RN, Kim KS. Zero-dimensional, one-dimensional, two-dimensional and three-dimensional nanostructured materials for advanced electrochemical energy devices. *Prog Mater Sci* 2012;57:724–803. https://doi.org/10.1016/j.pmatsci.2011.08.003.

[11] Cao A, Zhang T, Liu D, Xing C, Zhong S, Li X, et al. Electrostatic self-assembly of 2D Janus PS@Au nanoraspberry photonic-crystal array with enhanced near-infrared SERS activity. *Mater Adv* 2022;3:1512–7. https://doi.org/10.1039/d1ma01033f.

[12] Gawande MB, Goswami A, Asefa T, Guo H, Biradar AV, Peng DL, et al. Core-shell nanoparticles: Synthesis and applications in catalysis and electrocatalysis. *Chem Soc Rev* 2015;44:7540–90. https://doi.org/10.1039/c5cs00343a.

[13] Chen H, Zhang L, Li M, Xie G. Synthesis of core–shell micro/nanoparticles and their tribological application: A review. *Materials (Basel)* 2020;13:1–29. https://doi.org/10.3390/ma13204590.

[14] Hostetler MJ, Zhong C-J, Yen BKH, Anderegg J, Gross SM, Evans ND, et al. Stable, monolayer-protected metal alloy clusters. *J Am Chem Soc* 1998;120:9396–7. https://doi.org/10.1021/ja981454n.

[15] Steinbrück A, Csáki A, Festag G, Schüler T, Fritzsche W. Preparation and optical characterization of core-shell bi-metallic nanoparticles. *Opt InfoBase Conf Pap* 2007:66332I. https://doi.org/10.1117/12.728625.

[16] Hoover NN, Auten BJ, Chandler BD. Tuning supported catalyst reactivity with dendrimer-templated Pt-Cu nanoparticles. *J Phys Chem B* 2006;110:8606–12. https://doi.org/10.1021/jp060833x.

[17] Kreno LE, Leong K, Farha OK, Allendorf M, Van Duyne RP, Hupp JT. Metal-organic framework materials as chemical sensors. *Chem Rev* 2012;112:1105–25. https://doi.org/10.1021/cr200324t.

[18] Lee J, Farha OK, Roberts J, Scheidt KA, Nguyen ST, Hupp JT. Metal-organic framework materials as catalysts. *Chem Soc Rev* 2009;38:1450–9. https://doi.org/10.1039/b807080f.

[19] Fu Y, Yan X. Metal-organic framework composites. *Prog Chem* 2013;25:221–32.

[20] Zhou HC, Long JR, Yaghi OM. Introduction to metal-organic frameworks. *Chem Rev* 2012;112:673–4. https://doi.org/10.1021/cr300014x.

[21] Schneemann A, Bon V, Schwedler I, Senkovska I, Kaskel S, Fischer RA. Flexible metal-organic frameworks. *Chem Soc Rev* 2014;43:6062–96. https://doi.org/10.1039/c4cs00101j.

[22] Pathak DP, Kumar Y, Yadav S. Effectiveness of metal-organic framework as sensors: Comprehensive review. In *Sustainable Materials for Sensing and Remediation of Noxious Pollutants* 2022:47–64. https://doi.org/10.1016/b978-0-323-99425-5.00002-5.

[23] Jiao L, Seow JYR, Skinner WS, Wang ZU, Jiang HL. Metal–organic frameworks: Structures and functional applications. *Mater Today* 2019;27:43–68. https://doi.org/10.1016/j.mattod.2018.10.038.

[24] Sardar M, Ahmad R. Enzyme immobilization: An overview on nanoparticles as immobilization matrix. *Biochem Anal Biochem* 2015;04(2):1000178. https://doi.org/10.4172/2161-1009.1000178.

[25] Bastiancich C, Bozzato E, Luyten U, Danhier F, Bastiat G, Préat V. Drug combination using an injectable nanomedicine hydrogel for glioblastoma treatment. *Int J Pharm* 2019;559:220–7. https://doi.org/10.1016/j.ijpharm.2019.01.042.

[26] Wang Q, Dong X, Zhang H, Li P, Lu X, Wu M, et al. A novel hydrogel-based combination therapy for effective neuroregeneration after spinal cord injury. *Chem Eng J* 2021;415:128964. https://doi.org/10.1016/j.cej.2021.128964.

[27] Mao S, Johir MAH, Onggowarsito C, Feng A, Nghiem LD, Fu Q. Recent developments of hydrogel based solar water purification technology. *Mater Adv* 2022;3:1322–40. https://doi.org/10.1039/d1ma00894c.

[28] Zhao T, Chen L, Wang P, Li B, Lin R, Abdulkareem Al-Khalaf A, et al. Surface-kinetics mediated mesoporous multipods for enhanced bacterial adhesion and inhibition. *Nat Commun* 2019;10:4387. https://doi.org/10.1038/s41467-019-12378-0.

[29] Kumar Y, Yogeshwar P, Bajpai S, Jaiswal P, Yadav S, Pathak DP, et al. Nanomaterials: Stimulants for biofuels and renewables, yield and energy optimization. *Mater Adv* 2021;2:5318–43. https://doi.org/10.1039/d1ma00538c.

[30] Abid N, Khan AM, Shujait S, Chaudhary K, Ikram M, Imran M, et al. Synthesis of nanomaterials using various top-down and bottom-up approaches, influencing factors, advantages, and disadvantages: A review. *Adv Colloid Interface Sci* 2022;300:102597. https://doi.org/10.1016/j.cis.2021.102597.

[31] Soni V, Raizada P, Singh P, Cuong HN, Selvasembian R, Saini A, et al. Sustainable and green trends in using plant extracts for the synthesis of biogenic metal nanoparticles toward environmental and pharmaceutical advances: A review. *Environ Res* 2021;202:111622. https://doi.org/10.1016/j.envres.2021.111622.

[32] Pal G, Rai P, Pandey A. Green synthesis of nanoparticles: A greener approach for a cleaner future. In *Green Synthesis, Characterization and Applications of Nanoparticles* 2019:1–26. https://doi.org/10.1016/b978-0-08-102579-6.00001-0.

[33] Mohanpuria P, Rana NK, Yadav SK. Biosynthesis of nanoparticles: Technological concepts and future applications. *J Nanoparticle Res* 2008;10:507–17. https://doi.org/10.1007/s11051-007-9275-x.

[34] Iravani S, Korbekandi H, Mirmohammadi SV, Zolfaghari B. Synthesis of silver nanoparticles: Chemical, physical and biological methods. *Res Pharm Sci* 2014;9:385–406.

[35] Iravani S. Green synthesis of metal nanoparticles using plants. *Green Chem* 2011;13:2638–50. https://doi.org/10.1039/c1gc15386b.

[36] Karakoti M, Pandey S, Tatrari G, Dhapola PS, Jangra R, Dhali S, et al. A waste to energy approach for the effective conversion of solid waste plastics into graphene nanosheets using different catalysts for high performance supercapacitors: A comparative study. *Mater Adv* 2022;3:2146–57. https://doi.org/10.1039/d1ma01136g.

[37] Yoo HS, Kim TG, Park TG. Surface-functionalized electrospun nanofibers for tissue engineering and drug delivery. *Adv Drug Deliv Rev* 2009;61:1033–42. https://doi.org/10.1016/j.addr.2009.07.007.

[38] Wan D, Yuan S, Neoh KG, Kang ET. Surface functionalization of copper via oxidative graft polymerization of 2,2′-bithiophene and immobilization of silver nanoparticles for combating biocorrosion. *ACS Appl Mater Interfaces* 2010;2:1653–62. https://doi.org/10.1021/am100186n.

[39] Li Z, Lee D, Sheng X, Cohen RE, Rubner MF. Two-level antibacterial coating with both release-killing and contact-killing capabilities. *Langmuir* 2006;22:9820–3. https://doi.org/10.1021/la0622166.

[40] Salah AbdelHamied Mohamed S, Fahmy HM, Mohamed Shams-Eldin EM, Mahmoud MoslehSelim A. Metal oxide coating of silver nanoparticles to improve their physicochemical and optical properties. *Biophys J* 2019;116:447a. https://doi.org/10.1016/j.bpj.2018.11.2407.

[41] Malcher M, Volodkin D, Heurtault B, André P, Schaaf P, Möhwald H, et al. Embedded silver ions-containing liposomes in polyelectrolyte multilayers: Cargos films for antibacterial agents. *Langmuir* 2008;24:10209–15. https://doi.org/10.1021/la8014755.

[42] Kumar Y, Das L, Biswas KG. *Biodiesel: Features, Potential Hurdles, and Future Direction*, In *Status and Future Challenges for Non-conventional Energy Sources*, Volume 2, 2022: 99–122. https://doi.org/10.1007/978-981-16-4509-9_5.

[43] Kumar Y, Jaiswal P, Panda D, Nigam KDP, Biswas KG. A critical review on nanoparticle-assisted mass transfer and kinetic study of biphasic systems in millimeter-sized conduits. *Chem Eng Process - Process Intensif* 2022;170:108675. https://doi.org/10.1016/j.cep.2021.108675.

[44] Alawi OA, Abdelrazek AH, Aldlemy MS, Ahmed W, Hussein OA, Ghafel ST, et al. Heat transfer and hydrodynamic properties using different metal-oxide nanostructures in horizontal concentric annular tube: An optimization study. *Nanomaterials* 2021;11:1979. https://doi.org/10.3390/nano11081979.

[45] Jaiswal P, Kumar Y, Shukla R, Nigam KDP, Panda D, Guha Biswas K. Covalently immobilized nickel nanoparticles reinforce augmentation of mass transfer in millichannels for two-phase flow systems. *Ind Eng Chem Res* 2022;61:3672–84. https://doi.org/10.1021/acs.iecr.1c04419.

[46] Bhadouria AS, Kumar A, Raj D, Verma A, Singh S, Tripathi P, et al. Corrosion mitigation in oil reservoirs during CO_2 injection using nanomaterials. In *Nanotechnology for CO_2 Utilization in Oilfield Applications* 2022:127–146. https://doi.org/10.1016/B978-0-323-90540-4.00014-4.

[47] Saha K, Agasti SS, Kim C, Li X, Rotello VM. Gold nanoparticles in chemical and biological sensing. *Chem Rev* 2012;112:2739–79. https://doi.org/10.1021/cr2001178.

[48] Kumar Y, Sinha ASK, Nigam KDP, Dwivedi D, Sangwai J. Functionalized nanoparticles: Tailoring properties through surface energetics and coordination chemistry for advanced biomedical applications. *Nanoscale* 2023;15:6075–104. https://doi.org/10.1039/D2NR07163K.

[49] Chowdhury S, Kumar Y, Shrivastava S, Patel S, Sangwai J. A review on the recent scientific and commercial progresses on the direct air capture technology to manage atmospheric CO_2 concentrations and future perspectives. *Energy & Fuels*, 2023. https://doi.org/10.1021/acs.energyfuels.2c03971.

2 Application of Functional Materials and Corrosion Inhibitors for Downstream Offshore Oil and Gas Industry

Anteneh Mesfin Yeneneh,
Tahereh Jafary, and Asma Al kharousi
National University of Science and Technology

Muna Al Ajmi
University of Technology and Applied Sciences

Habeebllah Oladipo, Thirumalai Kumar,
Jimoh Adewole, Khadija Al-Balushi, and Muna Al Hinai
National University of Science and Technology

Ahoud Al Hamadani
University of Technology and Applied Sciences

CONTENTS

DOI: 10.1201/9781003242550-2

23

ABBREVIATIONS

DFT   Fluid dynamics simulator
EDX   Energy dispersive X-ray
H$_2$S   Hydrogen sulfide
MOF   Metal organic framework
OMS   Ordered mesoporous silica
PIL   Polymeric ionic liquids
PSA   Pressure swing adsorption

2.1   FUNCTIONAL MATERIAL FOR H$_2$S TREATMENT

Hydrogen sulfide (H$_2$S) is arguably the main toxic gas present in offshore oil and gas industries. H$_2$S in hydrocarbon reservoir originates from three main sources: the thermal decomposition of sulfur-containing compounds, bacterial sulfate reduction reactions, and thermochemical sulfate reduction. Each of these sources has its peculiarities; for instance, reservoirs where both thermal decompositions of sulfur-containing compounds and bacterial sulfate reduction are the main H$_2$S sources typically have a low level of H$_2$S (less than 5%). On the contrary, H$_2$S composition can reach up to 50% in reservoirs where H$_2$S is produced by thermochemical sulfate reduction [1]. Obviously, the presence of H$_2$S in petroleum crude oil and gas is highly undesirable due to the challenges it presents. Some of the most common challenges posed by H$_2$S gas are catalyst poisoning, pipeline corrosion and fouling. Another challenge, although rare but catastrophic, is the possibility of leakage in the offshore facility. Considering the limited space availability in offshore, H$_2$S leakage on crude oil production platform could lead to death of several workers in such industry. An example of such incidence is the leakage of H$_2$S-laden natural gas in Kab-101 platform in October 2007, which resulted in the death of 22 workers [2]. H$_2$S is very harmful at concentration levels as low as 1 ppm. Besides the deleterious effect that H$_2$S has on offshore oil and gas industry equipment and workers, it could also result in acid rain if released into the atmosphere. Thus, if not managed properly, H$_2$S could drastically affect the economic viability of oil and gas exploration. Several H$_2$S gas removal technologies have been extensively researched into especially since the present technology for treating H$_2$S gas (i.e., Claus process) is of low economic benefit. Researchers have investigated the use of various functional materials as potential technology for H$_2$S gas treatment. H$_2$S absorption in amine is the most popular technique for removing H$_2$S in the industry. However, this technology is highly energy intensive and also unable to remove traces of H$_2$S, which is potentially dangerous [3]. In this respect, H$_2$S removal using functional materials represents a potentially economical and effective technique for the removal of

H_2S gas. In general, functional materials used for H_2S gas treatment can be broadly divided into two categories: those that physically adsorb H_2S and those that adsorb and transform H_2S into another compound.

2.2   FUNCTIONAL MATERIALS FOR H_2S ADSORPTION

There are several functional materials that have been widely reported in the literature as adsorbent, including zeolites, metal organic framework (MOF), ordered mesoporous silica (OMS), and activated carbon. zeolites are multi-cavity crystalline structures with framework built from AlO_4 and SiO_4 tetrahedrons linked together by a common oxygen atom, giving them their unique ordered pore channel suitable for selectively adsorbing gases. Zeolite exists naturally and has also been synthesized in the laboratory. The use of both synthetic and naturally occurring zeolites as H_2S adsorbent has been widely demonstrated to be highly efficient and cost-effective. This is owing to their unique porous structure, which is not only favorable for H_2S gas adsorption but also desorbs the gas easily. It is noteworthy that the pore structure of zeolite is very sensitive to the treatment method. In a recent study, the effectiveness of natural zeolite modified by thermal treatment and chemical treatment was compared [4]. Although both techniques were found to improve the adsorption capacity of zeolite, those modified by thermal treatment outperformed the acid/base treated ones. There is a plethora of naturally occurring zeolites, which are very cheap and widely available such as ferrierite, clinoptilolite, phillipsite, mordenite, and erionite. However, natural zeolites have limited industrial application due to the restrictions in their crystal structures. Therefore, there is a need to synthesize a tailor-made zeolite for H_2S removal application.

2.3   SYNTHESIS AND CHARACTERIZATION

The pressure swing adsorption (PSA) method, which requires a high-pressure gas, is the most preferred method due to its low cost and low energy requirement [5]. Briefly, PSA has four steps: pressurization, adsorption, depressurization, and regeneration. A number of efforts have been invested in utilizing zeolite (natural and synthetic) for H_2S adsorption. As a usual practice in studying the effectiveness of zeolite material for H_2S adsorption, the procedure is to pretreat zeolite by washing off soluble impurities followed by calcination to remove volatile compounds. However, calcination of natural zeolite has been demonstrated to affect the crystallinity of clinoptilolite, which is the most abundant natural zeolite, by introducing some amorphous structure in the material [6]. Higher temperature was found to favor the rate of H_2S adsorption on zeolite, which comes with a trade-off between adsorption capacity and adsorption rate.

2.4   FUNCTIONAL MATERIALS FOR H_2S OXIDATION

The Claus process is the most popular H_2S removal technique in the oil and gas industry. However, the Claus process has a very slow kinetics and thus requires the use of catalysts. In this regard, some functional materials have also been investigated for this application. Hellmut and Janos investigated the mechanism of H_2S

adsorption on a series of faujasite-type zeolite having different Si/Al ratios under low and high H_2S pressure. It was found that there is a correlation between the nature of H_2S adsorption and Si/Al ratio, wherein dissociative adsorption prevails at Si/Al ratio below 2.5 [6].

2.5 FUNCTIONAL MATERIALS FOR H_2S VALORIZATION

H_2S adsorption on functional material has been demonstrated as a way to remove H_2S from oil and gas streams. After H_2S adsorption, the next step is the regeneration of the adsorbent by releasing the trapped gas. H_2S valorization offers a more promising and sustainable approach for H_2S gas treatment in that it has a dual benefit in that it generates fuel (H_2). Pipelines are the most efficient means of transporting materials or chemicals from a source to a destination, such as a storage tank or another processing and operation point. That's inshore and offshore, with inshore being within 2 miles of land and offshore being further out to sea. Offshore pipeline systems are safer, more efficient, and environmentally friendly, and the impact of a pipeline failure is greatly reduced compared with onshore systems because they are located within the plant. The pipeline's material should be carefully chosen, with the precise design of the wall thickness and others, because a failure would have a significant impact on the marine ecosystem and its inhabitants, posing a high level of danger and difficult sequences in the future. In order to ensure a theoretically low failure rate throughout the pipeline's lifespan, extensive analysis of the selection of pipes and fittings is performed during the design phase.

A lot of efforts have been made to avoid and prevent corrosion, and a pipeline is electrically isolated by providing a high-resistance external coating material and protective measure. These highly resistant coating materials and inhibitors have applied an important role to save the pipelines for the long-life term. The researchers have improved the inhibitors and functionalized them to improve it is properties for the use of offshore oil and gas industries. Furthermore, the selection of the corrosion inhibitor is very essential, which should consider many factors such as stability of the inhibitor on change of the temperature and under corrosive medium, compatibility of inhibitor in corroding system, cost of the inhibitor, friendly to the environment, and the efficiency [7]. On the other hand, another protection method has been used such as cathode protection [8] and paint-based corrosion inhibitor [9]. Topics of primary interest in this section include recent developments in corrosion inhibitors and advanced functional materials. They are crucial to the advancement of technology in petroleum engineering because they create favorable conditions for the intelligentsia and facilitate the automation of related processes. New findings on advanced functional materials and corrosion inhibitors, as well as their strengths and weaknesses, and their potential in offshore oil and gas industries, will be presented and discussed. To begin, it is important to understand that all materials can be classified into two broad categories: structural and functional. Since it ushers in new trajectories for use in petroleum engineering, functional materials are the materials that power cutting-edge technology across sectors [10].

The functional materials and corrosion inhibitors are categorized in three sides to make the discussion in useful classification: inorganic inhibitors, organic inhibitors, sealers and barrier inhibitors, and green corrosion inhibitors as it is classified in the corrosion review by Materials like piezoelectrics, thermoelectrics, low-temperature superconductors, shape-memory metals, functional polymers, nanocomposites, and functional fibers are all available commercially [7]. The basic form of the functional materials and corrosion inhibitors is from a variety of sources such as chemical-based inhibitors, which are and green-based inhibitors which may be consist of composite organic or inorganic materials. The organic materials are working through adsorption techniques to reduce the corrosion. They are mainly composed of nitrogen, oxygen, and sometimes sulfur because they are more capable of decreasing mental disintegration. The strength of the organic materials is that they can work with all acid concentrations without causing any poisoning to the refinery's products. Low treatment concentrations of inhibitors, often made from common organic compounds like azole and pyrimidine byproducts, are injected into the system to prevent or slow the corrosion of metals and alloys [3,11]. The inhibitor used to prevent corrosion in heat exchangers was shown to be toxic to marine life at concentrations as low as 3 ppm [12]. Because of the negative impacts on the environment caused by the disposal of used organic inhibitors for industrial corrosion, the trend toward utilizing green inhibitors is desirable. Moreover, the use of organic materials as an inhibitor in industries is extremely costly, since their productivity decreases with time in the presence of acids, and they are not resistant to temperatures beyond 95°C. Olajire [13] conducted an in-depth review of the topic of organic compound inhibitors for preventing corrosion in offshore oil and gas production facilities. The author described the synthesis of organic corrosion inhibitors in detail, including the relevant chemical reaction and chemical structure. However, inorganic compounds can inhibit corrosion by reacting with either the anodic or cathodic steps of the process. Salts containing arsenic, zinc, copper, nickel, and other compounds with arsenic are commonly utilized. The cathode cell of the unprotected metal surfaces is scraped with a mixture of the arsenic compound and the corrosive solution. Iron sulfide forms among the steel and acids, creating a barrier that decreases the fraction of hydrogen ions being exchanged. What happens when acid meets iron sulfide is a dynamic process. Some of the benefits of utilizing inorganic materials include their resistance to high acidic concentrations, their ability to withstand high temperatures for extended periods of time, and their lower cost compared with organic inhibitors. They may seem harmless at first, but their long-term effects on the ecosystem are undeniable. An attractive material used in the oil field is diamond-like carbon (DLC) nanocomposite film, which was reported as an excellent material that it has a low efficiency of friction of 0.05, good chemical stability, high hardness, corrosion resistance with wear resistance, electrical insulation, and is environmentally acceptable. The DLC was developed by multi-functional composite vapor deposition technology. The standard thickness of the DLC film is around 1–5 μm in the oil field the thickness should be around 10–50 μm, and the tests shows that the thicker the material films, the more excellent properties for a long period [6,14,15]. Still, this material required some economic analysis in the oil field, selection, and optimization design. On the other side, other materials were introduced such as polymers [1] and plant-based inhibitors [2]. The type of polymer needs secondary refinement, which can increase the process cost, and the plant-based inhibitor requires more studies for the easy types for extraction with sufficient cost [16,17].

2.6 SUSTAINABLE GREEN INHIBITORS FOR OFFSHORE OIL AND GAS FACILITIES

The development of Deep Ocean inquiry and underwater activity equipment has been spurred by the recent surge in scientific exploration and extraction of crude oil at the ocean's depth. Offshore operators have constructed a variety of offshore facilities, including platforms, pipelines, ships, undersea storage, and shore facilities, to facilitate their exploration and drilling operations in recent years. It has been claimed, however, that the structural integrity of these offshore constructions is not entirely understood [18]. Many marine and offshore structures, from drilling rigs to production facilities, are made of steel, making them vulnerable to internal and external corrosion in extreme conditions. These conditions include ultra-high temperatures and high-pressure environments [19–21]. Mechanical qualities, including strength, ductility, and impact strength, are lost, and in extreme cases, these structures break entirely as a result of degradation and wear. Offshore oil and gas installations are susceptible to structural damage from factors like chemical attacks, wave erosion, and microbial attacks. Costs for new construction, maintenance on aging/corroding equipment, inspections, structural integrity evaluations, and the correction of corrosion-related failures and wear have been reported to run into billions of dollars per year [22], and the annual cost of corrosion could be as much as $US2.4 trill [22].

There is a wide variety of compounds that may be used as inhibitors, but the introduction of stringent new environmental restrictions has limited the usage of many of these substances. For this reason, eco-friendly options for preventing corrosion of industrial facilities should prioritize the use of non-toxic, natural, and biodegradable chemicals. Over the past two decades, scientists have worked to perfect corrosion inhibitors that are both effective and economical, while having little effects on the surrounding ecosystem. Plant extracts [23–27] and biomaterials [28] are only a few examples of the various types of green corrosion inhibitors that have been developed.

It is possible to find organic and inorganic varieties of green corrosion inhibitor. Organic compounds include things like plants, medicines, amino acids, surfactants, ionic liquids, and biopolymers, whereas inorganic compounds like rare earth elements function as green corrosion inhibitors. There has been a lot of research into polymers because of their low cost, durability, and security. Due to the high charge density on the polymer structure, only a small fraction of such polymers show strong contact with metals. Multi-component synthetic organic compounds and phytochemicals of green inhibitors with S, N, and O functionalities in their structure are discussed in detail in this chapter for use as corrosion inhibitors against the corrosion of offshore oil and gas production facilities like pipelines and storage containers in corrosive media, with an emphasis on their synthesis (where applicable), properties, and inhibitory efficacy. Focusing on offshore oil and gas production, this chapter also discusses future issues in flow assurance due to gas hydrate development in the flowline [13]. Chitosan has great promise as an active shield against corrosion for a wide range of metals in environmental conditions. Numerous publications on the topic of chitosan and related components have been written because of their potential for use in eliminating corrosion. One of chitosan's distinguishing features is how simple it is to chemically functionalize, which opens up a way to boost its adhesiveness to metal

[29]. Even if certain works have been reported on the effectiveness of pure chitosan to minimize the steel corrosion, chitosan derivatives including functional groups and chitosan composites are used for protection reasons instead [30,31]. Attaching polar functional groups to the chitosan backbone chain increases its adhesiveness to the steel surface. New modifications and uses for chitosan are made possible by its amine and hydroxyl groups. Chitosan modification methods typically focus on either the 2-site amine group or the 6-site hydroxyl group. To this end, several chitosan analogues were synthesized and tested for their ability to protect steel alloys from corrosive environments.

2.6.1   CHITOSAN SCHIFF BASES

The simple synthesis of chitosan Schiff bases is a major factor in their selection as corrosion inhibitors. Imine linkage (HC=N) insertion into the chitosan structure is predicted to enhance both its anti-corrosion performance and its film-forming capability. For the purpose of protecting Q235 steel [32] and mild steel [33,34] from corrosion in 1.0 M HCl solution, the chitosan salicylaldehyde Schiff base has been produced and described as an inhibitor. The condensation process between chitosan and salicylaldehyde yielded this compound. For Q235 steel, the inhibition efficiency ranged from 84.9% to 92.7% when the chitosan Schiff base was added at concentrations between 100 and 2,000 ppm [32]. It was determined that the adsorption mechanism of the synthesized chitosan Schiff base on Q235 and mild steels followed the Langmuir and Temkin isotherms, respectively. Chitosan Schiff bases (CSB-1, CSB-2, and CSB-3) were reported to be synthesized by microwave irradiation from benzaldehyde, 4-(dimethylamino) benzaldehyde, and 4-hydroxy-3-methoxybenzaldehyde, respectively. The mild steel in a 1.0 M HCl solution showed superior inhibitory potentials for the produced inhibitors. CSB-1, CSB-2, and CSB-3 all had their maximal inhibition efficiency of 84.6%, 87.3%, and 90.6% at a concentration of just 50 ppm. Results for chitosan benzaldehyde (CSB-1) were consistent with those previously published [35].

2.6.2   CHITOSAN POLYMERIC SALTS

The research and development of polymeric ionic liquids (PILs) have received significant attention because of the chemical and physical features they exhibit. The anti-corrosion abilities of polymeric ionic liquids based on chitosan were tested. Chitosan polymeric ionic liquids (CSPTAlauric, CSPTA-myristic, CSPTA-palmitic, and CSPTA-stearic) were prepared by reacting the chitosan amine group with P-(PTSA), followed by amidation via stearic, palmitic, myristic, and lauric acid with the addition of dicyclohexyl carbodi [36]. As a potential corrosion inhibitor for X65 carbon steel in 1.0 M HCl solution, the produced components were put through their paces. At 250 ppm, CSPTA-lauric efficiency was 98.2%, the highest of any tested compound. A recent paper reported the synthesis and application of N-benzyl chitosan quaternary ammonium chloride and N-propyl chitosan quaternary ammonium chloride for the inhibition of P110 steel in a 3.5% NaCl solution saturated with CO_2 at 80°C, continuing the investigation of chitosan quaternary ammonium salts [15]. It was clear that the inhibitors had been adsorbed onto the steel surface, as evidenced by the EDX and contact angle measurements. According to the data on the loss of mass, the dynamic corrosion rate was significantly higher than the static corrosion rate under the identical conditions. The corrosion resistance was shown to improve with the addition of more inhibitor. As the concentration of the inhibitor grew, the double-layer capacitance also dropped. Because the inhibitor molecules displace water molecules on the steel surface, the Cdl values shift. At a concentration of 100 ppm for both PHC and BHC, inhibition efficiencies of 91.6% and 93.3% were achieved. In summary, it is possible to declare that chitosan polymeric salts demonstrated inhibitory efficiency more than 90% at a practical concentration of 300 ppm or lower, thereby introducing them as an inexpensive inhibitor.

2.6.3 PEG CROSS-LINKED CHITOSAN

In an acidic environment, PEG has shown to be a powerful inhibitor of steel [4]. It has recently been revealed that PEG cross-linked chitosan (CS-PEG) can block mild steel [37,38]. Formaldehyde was employed in its synthesis, with formic acid serving as a solvent for dissolving chitosan. Schiff base intermediates were formed when formaldehyde reacted with free amine groups in chitosan. After that, crosslinking occurred because the two terminal hydroxyl groups of PEG interacted with two distinct intermediates. As a corrosion inhibitor for mild steel, the produced CS-PEG was evaluated using electrochemical and weight loss tests in 1.0 M HCl [37] and 1.0 M sulfamic acid [38] solutions. At the optimal concentration of 200 ppm, the corrosion-inhibiting efficiency was 93.9%. Researchers looked at what happens to corrosion when temperatures change, and they found that the metal dissolved faster when the temperature was raised. The results of the polarization measurements showed that the tested inhibitor had an impact on cathodic as well as anodic reactions. According to DFT simulations, CS-PEG exhibited stronger adsorption than PEG and chitosan alone due to its higher EHOMO and lower ELUMO values and, as a result, lower E. Based on these results, it was possible to draw the conclusion that PEG insertion into the chitosan backbone enhanced the inhibitory performance of chitosan in a surprising way.

2.6.4 CARBOXYMETHYL HYDROXYPROPYL CHITOSAN

As a corrosion inhibitor for mild steel in 1.0 M HCl solution, carboxymethyl hydroxypropyl chitosan (CHPCS) was produced and tested [39]. Polarization measurements showed that a 1,000 ppm concentration of the produced chitosan derivative significantly reduced corrosion. Cdl values were found to be 191.8, 116.2, and 69.4 F cm^2 in the absence, presence, and 1,000 ppm of the inhibitor, respectively. A greater double-layer thickness brought about by inhibitor adsorption on the metal surface was responsible for this declining tendency. Images taken using a scanning electron microscope (SEM) revealed a cleaner, smoother metal surface on the shielded substrates when CHPCS was present. When CHPCS and HPCS (hydroxypropyl chitosan) were put head-to-head, it was clear that CHPCS offered superior protection. It followed that the addition of hydroxypropyl groups to the chitosan structure significantly enhanced the polymer's inhibitory activity.

2.6.5   ACID GRAFTED CHITOSAN

Polyaspartic acid grafted chitosan (PASP/CS) was another component in this category. It was prepared by the reaction of polyaspartic acid, chitosan, and glutaraldehyde [40]. This inhibitor was tested for the inhibition of carbon steel in 3.5% NaCl solution. The inhibition efficiency of 87.6% obtained in the presence of only 20 ppm of PASP/CS indicated the superior [41]. The range of concentrations between 100 and 800 ppm was chosen for the inhibitor. At a concentration of 300 ppm, the maximal inhibitory efficiency was estimated to be 82.3%; however, this estimate drastically dropped at higher concentrations. The rate of corrosion fell significantly after 10 days, but it stayed virtually the same up until 30 days, as seen by the curve that plots the rate of corrosion against time. This served as evidence that the inhibitor may be used for objectives related to extended inhibition.

2.6.6   BIOMATERIAL GRAFTED CHITOSAN

In the field of biomaterials research and engineering, two typical types of biomaterials that are utilized are saccharides and amino acids. It has been found that complexes of chitosan with these components display anti-corrosion characteristics. In order to protect carbon steel against a 0.5 M solution of hydrochloric acid, -cyclodextrin grafted chitosan, also known as -CD–chitosan, was developed [26]. According to the data that were received from the EIS, the Rct value increased from 15 cm² for the blank solution to 265 cm² for 230 ppm of the -CD–chitosan solution, which demonstrates that the proposed inhibitor was successful in reducing the amount of corrosion that occurred. Based on the EIS, weight loss, and polarization data, respectively, the inhibitory efficiency was 91.1%, 94.2%, and 96% when applied to this concentration. The polarization curves suggested that a mixed-type inhibitor was being shown by the -CD–chitosan. In addition to this, it was discovered that the adsorption of the inhibitor on the surface of the steel was in accordance with the Langmuir isotherm. It was hypothesized that the N and O electronegative heteroatoms of chitosan got adsorbed on the anodic sites on the metal surface, most likely by the formation of chelates with the Fe^{2+} that was already present on those sites. In addition, the free OH groups were the ones responsible for the adsorption of -cyclodextrin on the cathodic sites of the metal surface, which involved the H^+ ions that were already there.

2.6.7   TRIAZOLE-MODIFIED CHITOSAN

The chemical modification of chitosan with 4-amino-5-methyl-1,2,4-triazole-3-thiol resulted in the synthesis of novel triazole-modified chitosan, which bears the designation CS-AMT [38]. A solution containing 1.0 M HCl was used to test the effectiveness of the newly synthesized biopolymer as an inhibitor against the corrosion of carbon steel. The data from the EIS were used to determine that a concentration of the inhibitor containing 200 ppm would result in a maximum inhibition efficiency of 97%. According to the data from the polarization, the amount of CS-AMT that was present resulted in a considerable decrease in Icorr. In addition, the inhibitor had an effect on the anodic as well as the cathodic slopes of the Tafel diagram, which suggests that it inhibited both the anodic metal dissolution and the cathodic hydrogen evolution. Ecorr moved in the direction of being negative and had the greatest possible value in that direction when it was measured at a concentration of 200 ppm. This was regarded as the CS-AMT having a stronger influence on the cathodic reaction as compared to that which it had on the anodic reaction. According to the findings of the surface study, the presence of the inhibitor was necessary for the substrate to sustain considerable surface damage. On the other hand, the presence of CS-AMT resulted in a reduction in the amount of damage sustained by the substrate's surface. The creation of a protective layer as a result of the adsorption of inhibitor molecules on the metal surface was thought to be the cause of this improvement [29].

2.6.8 PLANT EXTRACT

Components derived from plants are what make up the majority of the eco-friendly corrosion inhibitors available. The study of green chemistry has resulted in the development of environmentally friendly solutions. Molecules including oxygen and nitrogen as well as nonpolar compounds consisting of aromatic, aliphatic, heterocyclic, and other related components can be found in plant extracts. These chemicals are readily absorbed by the surfaces of metals [42]. In general, each plant extract contains a number of different complex phytochemicals. These phytochemicals are connected with a number of electron-rich centers, which have the potential to act as adsorption centers during interactions between metal inhibitors and other substances. The electron-rich centers consist of polar functional groups, such as hydroxyl ($-OH$), amino ($-NH_2$), ester ($-COOC_2H_5$), amide ($-CONH_2$), acid chloride ($-COCl$), dimethylamino ($-NMe_2$), methoxy ($-OMe$), and ether ($-O-$), in addition to multiple bonds such as $>C=C$, $>C=N-$, $>C=O$, and $-N=O$. Because of their affiliation with high peripheral functions in the form of polar functional groups, most phytochemicals are easily soluble in polar electrolytes, despite the fact that the vast majority of phytochemicals are molecules that have a high degree of complexity. Inorganic salts like KI, which inhibit inhibition protectiveness through synergism, can be used to further improve the protection efficiency of extracts that have a relatively lower protection efficiency against metallic corrosion [43]. This can be done by using extracts that have relatively lower protection efficiency against metallic corrosion [44]. The production and consumption of plant extracts are both on the rise as a direct result of the widespread availability of these substances at a level that is both economically and commercially viable. Many different plant parts, such as the leaves, barks, fruits, peels, and flowers, are utilized to make extracts that are then put to use as corrosion inhibitors [27]. Plants are typically extracted using a two-phase system that consists of an organic and an aqueous phase. This system is called a biphasic system [45].

Due to the widespread adoption of steel and the complexity of today's working environments, there is a pressing need for additional research into corrosion control strategies, particularly those that emphasize environmentally friendly corrosion inhibitors. Inhibition of mild steel corrosion in 0.5 M sulfuric acid (H_2SO_4) by a concentration of 100 ppm of mucilage derived from Okra fruits was demonstrated by mass loss measurements. The effectiveness of an inhibitor grows with both its concentration and the length of time it is immersed (up to maintain gap between 72 hours) [46]. The effectiveness of glycine max leaf extract as a corrosion inhibitor for mild steel in 0.5 M HCl medium was studied, and it was shown to be 91.07% effective at 2 g/L of green corrosion inhibitor concentration and 308 K [47]. The effects of Morus alba pendula leaf extract on carbon steel corrosion were investigated across a spectrum of extract concentrations and temperatures. With the help of potassium iodide, a green corrosion inhibitor was altered. Fourier transform infrared analysis is used to determine functional groups. At 298.15 K, using only the extract resulted in a high inhibitory efficiency of 93%. The extract's inhibitory effectiveness on steel surfaces was 96% when 10 mM KI was added for synergy [48]. When used as an inhibitor, Silybum marianum extract on 304 stainless steel yielded 96% efficiency at a concentration of 1.0 g/L of extract. The extract also shown promising anti-corrosion properties, particularly at elevated temperatures. The effectiveness of the eco-friendly corrosion inhibitor is demonstrated by a decrease in activation energy upon its administration [49]. Carbon steel's resistance to corrosion in an alkaline atmosphere was tested using a range of mulberry leaves inhibitor concentrations. According to the electrochemical study, an inhibitor concentration of 400 mg/L yielded an 88% efficiency [50]. Corrosion testing was conducted with mild steel in an alkaline environment using areca nut husk as the corrosion inhibitor. Under ideal conditions, the system's efficiency reached 91.66%. As the temperature rose, the protective covering on the metal's surface became more clearly visible and the efficiency dropped [51]. Dissolved ions such as chloride, sulfate, bicarbonate, sodium, and magnesium give seawater corrosion its distinctively heterogeneous media composition. The protective effects of rice straw extract against steel corrosion in 3.5% sodium chloride were investigated using gravimetric methods. With a corrosion rate of only 0.013 mm/year, the maximum inhibitor efficiency achieved was 92%. After being submerged in rice straw extract, steel was substantially less likely to acquire corrosion products [52].

2.7 OFFSHORE OIL AND GAS INDUSTRY FUNCTIONAL MATERIALS CHALLENGES, PROSPECTS, AND CONCLUSION

Recent research on clay minerals as functional materials has largely centered on their potential use as a matrix or functional component in high-tech materials. Biocompatible nanocomposites and colloidal soft matter made from clay are an interesting development [53]. Engineering clay minerals as selective and active adsorbents and catalysts are made possible by the presence of multiple, distinct layers and interlayers at the nanometer scale [54]. Notably, the resulting clay-based adsorbents and catalysts can be endowed with electric, optical, photonic, and magnetic functionalities. It is widely believed that clay minerals play a significant role in subterranean petroleum reservoirs, including the creation of kerogen, the catalytic breaking of kerogen into petroleum hydrocarbon, the migration of crude oil, and the ongoing modification of hydrocarbon composition [55]. However, the precise chemical principles underlying the synthesis of crude oil from clay minerals are still a mystery. Therefore, improving our understanding of clay mineral interactions in natural and biological systems is important for both developing catalytic conversion of biomass as a source of sustainable energy and comprehending the geological process of oil creation. Advanced functional materials, such as environmentally responsive functional materials, self-healing materials, and smart materials, play a crucial supporting role in the development of high technology while also continuously advancing related industries, such as petroleum engineering, to the cutting edge of the global technological landscape. Functional materials are high-tech materials used for non-structural applications; they have excellent electrical, magnetic, optical, thermal, acoustic, mechanical, chemical, and biological properties. The advancement of oil and gas technology is inextricably linked to the creation and implementation of state-of-the-art functional materials that are optimized for use in oil fields. Using the most recent discoveries in fusion functional materials, including those that employ organic, inorganic, nanocomposites, and other materials, researchers are constantly improving offshore drilling technology. Some cutting-edge functional materials with prospective uses in the oil sector are sorted out by methods such as expert consultation, patent search analysis, and literature research, thereby generating new inspiration and ideas for improving the growth of offshore drilling technology.

REFERENCES

[1] M. Abdallah, H. E. Megahed, M. A. Radwan, E Abdfattah, M. Abdallah, and E. Abdfattah, "Polyethylene glycol compounds as corrosion inhibitors for aluminium in 0.5M hydrochloric acid solution," *J. Am. Sci.*, vol. 8, no. 11, pp. 1545–1003, 2012.

[2] M. S. Al-Otaibi, A. M. Al-Mayouf, M. Khan, A. A. Mousa, S. A. Al-Mazroa, and H. Z. Alkhathlan, "Corrosion inhibitory action of some plant extracts on the corrosion of mild steel in acidic media," *Arab. J. Chem.*, vol. 7, no. 3, pp. 340–346, 2014, doi: 10.1016/J.ARABJC.2012.01.015.

[3] A. Ghazoui et al., "Comparative study of pyridine and pyrimidine derivatives as corrosion inhibitors of C38 steel in molar HCl," *Int. J. Electrochem. Sci.*, vol. 7, pp. 7080–7097, 2012.

[4] H. Ashassi-Sorkhabi, N. Ghalebsaz-Jeddi, F. Hashemzadeh, and H. Jahani, "Corrosion inhibition of carbon steel in hydrochloric acid by some polyethylene glycols," *Electrochim. Acta*, vol. 51, no. 18, pp. 3848–3854, 2006.

[5] M. Goyal, S. Kumar, I. Bahadur, C. Verma, and E. E. Ebenso, "Organic corrosion inhibitors for industrial cleaning of ferrous and non-ferrous metals in acidic solutions: A review," *J. Mol. Liq.*, vol. 256, pp. 565–573, 2018.

[6] P. Samyn, G. Schoukens, J. Quintelier, and P. De Baets, "Friction, wear and material transfer of sintered polyimides sliding against various steel and diamond-like carbon coated surfaces," *Tribol. Int.*, vol. 39, no. 6, pp. 575–589, 2006, doi: 10.1016/J.TRIBOINT.2005.07.029.

[7] P. B. Raja et al., "Reviews on corrosion inhibitors—A short view," *Chem. Eng. Commun.*, vol. 203, pp. 1145–1156, 2016, doi: 10.1080/00986445.2016.1172485.

[8] C. M. Abreu, M. Izquierdo, P. Merino, X. R. Nóvoa, and C. Pérez, "A new approach to the determination of the cathodic protection period in zinc-rich paints," *Corrosion*, vol. 55, no. 12, pp. 1173–1181, 1999, doi: 10.5006/1.3283955.

[9] F. Mansfeld and C. H. Tsail, "Determination of coating deterioration with EIS: I. Basic relationships," *Corrosion*, vol. 47, no. 12, pp. 958–963, 1991, doi: 10.5006/1.3585209.

[10] G. F. Zhao, W. N. Di, and R. Y. Wang, "Development and potential application of advanced functional materials in the oil field," in *Materials Science Forum*, 2019, vol. 944, pp. 637–642. ACS publication.

[11] Y. Gong, Z. Wang, F. Gao, S. Zhang, and H. Li, "Synthesis of new benzotriazole derivatives containing carbon chains as the corrosion inhibitors for copper in sodium chloride solution," *Ind. Eng. Chem. Res.*, vol. 54, no. 49, pp. 12242–12253, 2015. American Chemical Society.

[12] D. A. Pillard, J. S. Cornell, D. L. DuFresne, and M. T. Hernandez, "Toxicity of benzotriazole and benzotriazole derivatives to three aquatic species," *Water Res.*, vol. 35, no. 2, pp. 557–560, 2001.

[13] A. A. Olajire, "Corrosion inhibition of offshore oil and gas production facilities using organic compound inhibitors—A review," *J. Mol. Liq.*, vol. 248, pp. 775–808, 2017.

[14] L. Cui, Z. Lu, and L. Wang, "Toward low friction in high vacuum for hydrogenated diamondlike carbon by tailoring sliding interface," *ACS Appl. Mater. Interfaces*, vol. 5, no. 13, pp. 5889–5893, 2013, doi: 10.1021/AM401192U/ASSET/IMAGES/AM401192U. SOCIAL.JPEG_V03.

[15] L. Cui, Z. Lu, and L. Wang, "Probing the low-friction mechanism of diamond-like carbon by varying of sliding velocity and vacuum pressure," *Carbon N. Y.*, vol. 66, no. 66, pp. 259–266, 2014, doi: 10.1016/j.carbon.2013.08.065.

[16] V. Rajeswari, D. Kesavan, M. Gopiraman, and P. Viswanathamurthi, "Physicochemical studies of glucose, gellan gum, and hydroxypropyl cellulose—Inhibition of cast iron corrosion," *Carbohydr. Polym.*, vol. 95, no. 1, pp. 288–294, 2013, doi: 10.1016/J. CARBPOL.2013.02.069.

[17] M. M. Solomon, S. A. Umoren, I. I. Udosoro, and A. P. Udoh, "Inhibitive and adsorption behaviour of carboxymethyl cellulose on mild steel corrosion in sulphuric acid solution," *Corros. Sci.*, vol. 52, no. 4, pp. 1317–1325, 2010, doi: 10.1016/J.CORSCI.2009.11.041.

[18] M. G. Stewart and A. Al-Harthy, "Pitting corrosion and structural reliability of corroding RC structures: Experimental data and probabilistic analysis," *Reliab. Eng. Syst. Saf.*, vol. 93, no. 3, pp. 373–382, 2008.

[19] W. O. O. D. Maureen, V. A. Ana Lisa, and V. W. Lorenzo. "Corrosion-related accidents in petroleum refineries: Lessons learned from accidents in EU and OECD countries," 2013.

[20] W. Wu, G. Cheng, H. Hu, and Q. Zhou, "Risk analysis of corrosion failures of equipment in refining and petrochemical plants based on fuzzy set theory," *Eng. Fail. Anal.*, vol. 32, pp. 23–34, 2013.

[21] A. Groysman, "Corrosion in systems for storage and transportation of petroleum products and biofuels." In *Corrosion 2015*. OnePetro, 2015.

[22] . M. R. Simons. "Report of offshore technology conference (OTC) presentation." NACE International oil and gas production, 2008.

[23] A. M. Abdel-Gaber, B. A. Abd-El-Nabey, I. M. Sidahmed, A. M. El-Zayady, and M. Saadawy, "Inhibitive action of some plant extracts on the corrosion of steel in acidic media," *Corros. Sci.*, vol. 48, no. 9, pp. 2765–2779, 2006.

[24] O.-H. Lee et al., "Assessment of phenolics-enriched extract and fractions of olive leaves and their antioxidant activities," *Bioresour. Technol.*, vol. 100, no. 23, pp. 6107–6113, 2009.

[25] S. S. de Assunção Araújo Pereira et al., "Inhibitory action of aqueous garlic peel extract on the corrosion of carbon steel in HCl solution," *Corros. Sci.*, vol. 65, pp. 360–366, 2012.

[26] H. Li, H. Li, Y. Liu, and X. Huang, "Synthesis of polyamine grafted chitosan copolymer and evaluation of its corrosion inhibition performance," *J. Korean Chem. Soc.*, vol. 59, no. 2, pp. 142–147, 2015.

[27] C. Verma, E. E. Ebenso, I. Bahadur, and M. A. Quraishi, "An overview on plant extracts as environmental sustainable and green corrosion inhibitors for metals and alloys in aggressive corrosive media," *J. Mol. Liq.*, vol. 266, pp. 577–590, 2018.

[28] H. Ashassi-Sorkhabi, M. R. Majidi, and K. Seyyedi, "Investigation of inhibition effect of some amino acids against steel corrosion in HCl solution," *Appl. Surf. Sci.*, vol. 225, no. 1–4, pp. 176–185, 2004.

[29] H. Ashassi-Sorkhabi and A. Kazempour, "Chitosan, its derivatives and composites with superior potentials for the corrosion protection of steel alloys: A comprehensive review," *Carbohydr. Polym.*, vol. 237, p. 116110, 2020.

[30] S. A. Umoren, M. J. Banera, T. Alonso-Garcia, C. A. Gervasi, and M. V Mirífico, "Inhibition of mild steel corrosion in HCl solution using chitosan," *Cellulose*, vol. 20, no. 5, pp. 2529–2545, 2013.

[31] T. Rabizadeh and S. Khameneh Asl, "Chitosan as a green inhibitor for mild steel corrosion: Thermodynamic and electrochemical evaluations," *Mater. Corros.*, vol. 70, no. 4, pp. 738–748, 2019.

[32] N. L. Chen, P. P. Kong, H. X. Feng, Y. Y. Wang, and D. Z. Bai, "Corrosion mitigation of chitosan Schiff base for Q235 steel in 1.0 M HCl," *J. Bio-and Tribo-Corrosion*, vol. 5, no. 1, pp. 1–8, 2019.

[33] R. Menaka and S. Subhashini, "Chitosan Schiff base as eco-friendly inhibitor for mild steel corrosion in 1 M HCl," *J. Adhes. Sci. Technol.*, vol. 30, no. 15, pp. 1622–1640, 2016, doi: 10.1080/01694243.2016.1156382.

[34] R. Menaka, R. Geethanjali, and S. Subhashini, "Electrochemical investigation of eco-friendly chitosan Schiff base for corrosion inhibition of mild steel in acid medium," *Mater. Today Proc.*, vol. 5, no. 8, pp. 16617–16625, 2018. doi: 10.1016/j.matpr.2018.06.022.

[35] H. Darmokoesoemo, S. Suyanto, L. S. Anggara, A. N. Amenaghawon, and H. S. Kusuma, "Application of carboxymethyl chitosan-benzaldehyde as anticorrosion agent on steel," *Int. J. Chem. Eng.*, vol. 2018, pp. 1–8, 2018, doi: 10.1155/2018/4397867.

[36] S. M. Elsaeed, E. S. H. El Tamany, H. Ashour, E. G. Zaki, E. A. Khamis, and H. A. El Nagy, "Corrosion and hydrogen evolution rate control for X-65 carbon steel based on chitosan polymeric ionic liquids: Experimental and quantum chemical studies," *RSC Adv.*, vol. 8, no. 66, pp. 37891–37904, 2018, doi: 10.1039/c8ra05444d.

[37] V. Srivastava, D. S. Chauhan, P. G. Joshi, V. Maruthapandian, A. A. Sorour, and M. A. Quraishi, "PEG-functionalized chitosan: A biological macromolecule as a novel corrosion inhibitor," *ChemistrySelect*, vol. 3, no. 7, pp. 1990–1998, 2018, doi: 10.1002/SLCT.201701949.

[38] D. S. Chauhan, M. A. Quraishi, A. A. Sorour, S. K. Saha, and P. Banerjee, "Triazole-modified chitosan: A biomacromolecule as a new environmentally benign corrosion inhibitor for carbon steel in a hydrochloric acid solution," *RSC Adv.*, vol. 9, no. 26, pp. 14990–15003, 2019, doi: 10.1039/c9ra00986h.

[39] K. Wan, P. Feng, B. Hou, and Y. Li, "Enhanced corrosion inhibition properties of carboxymethyl hydroxypropyl chitosan for mild steel in 1.0 M HCl solution," *RSC Adv.*, vol. 6, no. 81, pp. 77515–77524, 2016.

[40] T. Chen, D. Zeng, and S. Zhou, "Study of polyaspartic acid and chitosan complex corrosion inhibition and mechanisms," *Polish J. Environ. Stud.*, vol. 27, no. 4, pp. 1441–1448, 2018.

[41] M. Wu, Y. Li, Z. Li, and H. Baorong, "Corrosion inhibition performance of chitosan and phosphonic chitosan for mild steel in seawater," *J. Chinese Soc. Corros. Prot.*, vol. 30, no. 3, pp. 192–196, 2010.

[42] S. Issaadi, T. Douadi, A. Zouaoui, S. Chafaa, M. A. Khan, and G. Bouet, "Novel thiophene symmetrical Schiff base compounds as corrosion inhibitor for mild steel in acidic media," *Corros. Sci.*, vol. 53, no. 4, pp. 1484–1488, 2011.

[43] G. Bahlakeh, M. Ramezanzadeh, and B. Ramezanzadeh, "Experimental and theoretical studies of the synergistic inhibition effects between the plant leaves extract (PLE) and zinc salt (ZS) in corrosion control of carbon steel in chloride solution," *J. Mol. Liq.*, vol. 248, pp. 854–870, 2017, doi: 10.1016/j.molliq.2017.10.120.

[44] J. Tan, L. Guo, H. Yang, F. Zhang, and Y. El Bakri, "Synergistic effect of potassium iodide and sodium dodecyl sulfonate on the corrosion inhibition of carbon steel in HCl medium: A combined experimental and theoretical investigation," *RSC Adv.*, vol. 10, no. 26, pp. 15163–15170, 2020.

[45] C. Verma, E. E. Ebenso, M. A. Quraishi, and C. M. Hussain, "Recent developments in sustainable corrosion inhibitors: Design, performance and industrial scale applications," *Mater. Adv.*, vol. 2, p. 3806, 2021, doi: 10.1039/d0ma00681e.

[46] S. Banerjee, V. Srivastava, and M. M. Singh, "Chemically modified natural polysac-charide as green corrosion inhibitor for mild steel in acidic medium," *Corros. Sci.*, vol. 59, pp. 35–41, 2012.

[47] D. K. Verma and F. Khan, "Corrosion inhibition of mild steel in hydrochloric acid using extract of glycine max leaves," *Res. Chem. Intermed.*, vol. 42, no. 4, pp. 3489–3506, 2016.

[48] M. Jokar, T. S. Farahani, and B. Ramezanzadeh, "Electrochemical and surface charac-terizations of morus alba pendula leaves extract (MAPLE) as a green corrosion inhibi-tor for steel in 1 M HCl," *J. Taiwan Inst. Chem. Eng.*, vol. 63, pp. 436–452, 2016.

[49] N. Soltani, N. Tavakkoli, M. K. Kashani, A. Mosavizadeh, E. E. Oguzie, and M. R. Jalali, "Silybum marianum extract as a natural source inhibitor for 304 stainless steel corrosion in 1.0 M HCl," *J. Ind. Eng. Chem.*, vol. 20, no. 5, pp. 3217–3227, 2014.

[50] B. Tang, "Evaluation of corrosion inhibition effect of white mulberry leaves extract on HRB500 carbon steel in alkaline chloride solutions," *Int. J. Electrochem. Sci.*, vol. 15, pp. 12524–12533, 2020.

[51] N. Raghavendra and J. I. Bhat, "Chemical components of mature areca nut husk extract as a potential corrosion inhibitor for mild steel and copper in both acid and alkali media," *Chem. Eng. Commun.*, vol. 205, no. 2, pp. 145–160, 2018.

[52] N. K. Othman, S. Yahya, and M. C. Ismail, "Corrosion inhibition of steel in 3.5% NaCl by rice straw extract," *J. Ind. Eng. Chem.*, vol. 70, pp. 299–310, 2019.

[53] C. H. Zhou, L. Z. Zhao, A. Q. Wang, T. H. Chen, and H. P. He, "Current fundamental and applied research into clay minerals in China," *Appl. Clay Sci.*, vol. 119, pp. 3–7, 2016, doi: 10.1016/J.CLAY.2015.07.043.

[54] C. H. Zhou, "An overview on strategies towards clay-based designer catalysts for green and sustainable catalysis," *Appl. Clay Sci.*, vol. 53, no. 2, pp. 87–96, 2011, doi: 10.1016/J. CLAY.2011.04.016.

[55] L. M. Wu, C. H. Zhou, J. Keeling, D. S. Tong, and W. H. Yu, "Towards an understanding of the role of clay minerals in crude oil formation, migration and accumulation," *Earth-Sci. Rev.*, vol. 115, no. 4, pp. 373–386, 2012, doi: 10.1016/j.earscirev.2012.10.001.

3 Application of Functional Ceramics in Oil and Gas Industries
Properties and Current Status

Ajeet Kumar Prajapati, Suraj Aryan, Sukriti Singh, and Deepak Dwivedi
Rajiv Gandhi Institute of Petroleum Technology

Nellya Serikova
DNV France SARL

Bijoy K. Purohit
Loyola Academy Degree and PG College

CONTENTS

ABBREVIATIONS

EOR   Enhanced oil recovery
MAO   The micro-arc oxidation

DOI: 10.1201/9781003242550-3

39

3.1   INTRODUCTION

Oil and gas products are essential for the preservation of industrial development, thus a major concern for many countries. Fuel oil, gasoline, and natural gas are the industry's three highest-volume products and are made up of a complex blend of organic molecules and hydrocarbons. Exploration or upstream, crude oil or natural gas production, downstream, pipeline, and service and supply are the major sectors of the oil and gas industries. Offshore extractions and production environments are subjected to a highly corrosive nature and directly affect productivity. The devices included in this process are mainly related to the transport of fluid (pumps, propellers, impellers, and blowers) and are exposed to the corrosive atmosphere and seawater [1]. Therefore, careful consideration must be given to material selection at every stage of design, system operation, and construction of operational equipment such as valves, elbows, bends, and tees [2]. Chemical deterioration was due to the surface conditions, and inherent properties of the corrosive fluid and metal. Thus, the design and the usage of materials in the petroleum sector have a fundamental requirement to combine high mechanical strength with erosion and corrosion resistance [3].

Carbon steel is the material that is mostly used in both upstream and downstream of the petroleum industry for construction pipeline. Because of its superior mechanical qualities and affordable pricing compared to higher alloy alternatives, carbon steel is therefore an essential commodity [4]. However, pure carbon steel has low corrosion resistance, and it must have its surface protected in such adverse environments. Many oxide ceramics, including alumina, chromium, silica, titania, yttria, and zirconia, are used as surface coating materials to increase resistance to cavitation, fretting, erosion, wear, and corrosion [5]. They are especially beneficial in situations when both corrosion and wear resistance are necessary [6].

Advanced functional ceramics are defined as ceramics that, besides showing structural properties, are designed for special applications requiring additional properties such as electric, magnetic, or optical. They have some unique qualities, including high hardness, high refractoriness to high temperatures, lightweight, low density, lower linear thermal expansion, and higher operating temperatures. Research in these fields has led to the production of highly functional, energy-efficient, and environmentally friendly materials that have proved very beneficial compared to conventional materials used. Functional ceramics are applied to modern technologies which include energy, environment, transportation, biological sciences, and information and communication technology [7]. Equipment for the oil and gas sectors must function well in both corrosive and abrasive harsh environments. Functional ceramics are extensively used in the petroleum sector because they offer good electrical insulation, abrasion resistance, thermal conductivity, and chemical and high-temperature resistance [8].

3.2   CERAMIC MATERIALS FOR OIL AND GAS INDUSTRIES

The oil and gas sector uses a variety of ceramic materials to meet the increasing demands for upstream processing of petroleum and gas. These comprise a broad range of ceramic products made from high-purity alumina, zirconia, silicon nitride, and other materials to produce specialized components that can withstand specific and challenging environments, such as high-temperature/high-pressure conditions, corrosive environments, high mechanical wear environments, and electrical insulation [9].

Alumina ceramics: Alumina oxide (Al_2O_3) is a very dense, extremely strong substance. Alumina ceramic provides resistance to environment related to both corrosion and high temperatures. High volume resistivity and low loss characteristics at high frequencies are two of the electrical characteristics of aluminum oxide. A variety of additives and bonding techniques can be used to change the chemical and thermo-mechanical characteristics of alumina ceramics. Different types of alumina are consequently characterized mainly by purity. Alumina oxide ceramic is used for a variety of things, including refractory anchors, temperature sensors, knife sharpeners, and pump seals.

Zirconia ceramics: Zirconium oxide ceramic, also known as zirconia (ZrO_2), is extremely hard, has great flexibility and strength, and has low thermal conductivity. It has strong corrosion resistance. The material has a greater melting point than alumina and a high resistance to corrosion because of several additives. Oxygen sensors and other technical devices use zirconia. Magnesia partially stabilized zirconia is a great material choice for industrial ceramic parts including bushings and wear sleeves, valve and pump components, and industrial tooling applications.

Silicon nitride ceramics: Silicon nitride ceramic (Si_3N_4) material has the following important characteristics as high strength at a wide range of temperatures, high chemical resistance, high corrosion resistance, high fracture toughness, high shock resistance, high wear resistance, high thermal conductivity, low density, and low thermal expansion than other industrial ceramics. It has applications in the machine wear surfaces, burner nozzles, hydraulic fracturing or fracking, couplings, pipeline, and molten metal processing components. It used to be very expensive to produce a high-density silicon nitride component. Thus, new silicon nitride-based ceramic alloys (SiAlON) that offer improved mechanical performance features while drastically lowering manufacturing costs are used.

Other ceramic materials: Ceramics made of crystalline magnesium silicate called steatite have exceptional mechanical and electrical properties. This steatite ceramic substance is used in thermostats, welding equipment, ceramic heaters, resistors, and igniters. Tungsten-based ceramics include ammonium paratungstate, tungsten, tungsten carbide, blue tungsten oxide, graded carbide powders, yellow tungsten oxide, and cemented tungsten carbide products including carbide wear pads and carbide pellets. Cordierite ceramics (magnesium aluminum silicate) have excellent thermal shock resistance, due to its low coefficients of thermal expansion. These ceramics have mostly applications in kiln furniture, ceramic thermocouples.

3.2.1 CARBON GRAPHITE MATERIALS FOR THE OIL AND GAS INDUSTRIES

Carbon/graphite materials are heterogenous in nature and comprised of a complex mixture of fillers (carbon blacks, natural graphite, petroleum coke, synthetic graphite), thermoplastic binders (tars, petroleum pitch, coal tar pitch, and synthetic resins), and other additives (oxidation inhibitors and film-forming agents). Then, the resultant mixers are molded by either isostatic pressing or die processing. Carbonization is a heat treatment that pyrolyzes the binder at temperatures typically lower than 1,300°C. To carbonize, it is subjected to heat treatments without oxygen and forms carbon graphite. When the binder is carbonized, an amorphous, coked binder is left behind that holds the filler particles together and gives structural integrity in the finished products. A thermal treatment called graphitization uses temperatures as high as 3,100°C. The filler particles and amorphous carbon binder undergo a partial transformation to a more orientated, graphite-like structure during the optional graphitization process, which modifies the crystal structure of the carbonized particles [10]. The characteristics of carbonized and graphited objects must often be modified to provide the best performance.

Impregnations (metals, thermoset resins, thermoplastic resins) are frequently used to modify a material's friction, modulus and strength, porosity, various temperature range, tolerance to operating environment, and electrical resistivity to meet the needs of a particular application. Metal impregnations are frequently employed in seal and bearing applications, to increase the base carbon's strength, tribological performance, and to lowering permeability. To improve the physical qualities of the produced product, the impregnations are used to seal the natural porosity of the carbon/graphite materials for seals and bearing applications [11]. The common heavy metal impregnates are lead, antimony, copper, and babbitt (alloy of antimony, copper, and tin). The usage of heavy metals, which is raising concerns about the environment, human health, and safety, is the deficiency of metal impregnations. Another alternative to heavy metal impregnations involves a phenolic resin impregnation of the base carbon material followed by carbonization resin impregnations. This method fills any remaining porosity in the carbon/graphite material. The phenolic resin degrades to a lesser extent during pyrolysis, and the initial porosity is partially reopened. A final impregnation that makes the material impermeable or several cycles of impregnation and carbonization work well to efficiently close porosity. All carbon materials are defined as carbon/graphite materials with closed porosity attained through these kinds of procedures [12].

However, there are many applications for carbon/graphite materials in intermittent lubrication, including mechanical seals and faces that can survive some of the most extreme dry operating conditions at high speeds and high temperatures. There are many grades of carbon/graphite materials that are employed as seal face materials for dry running applications. These materials are used in applications that are upstream (mostly rotating parts) and downstream (pump in refineries). The choice of material is based on mechanical attributes like strength, temperature limit, thermal conductivity, etc.

3.3 CHALLENGES DUE TO CORROSION ISSUES IN OIL AND GAS INDUSTRIES

Metals are extensively used in almost all sectors of the oil and gas industries. The damaging attack on a material caused by a reaction with its environment is called corrosion. The oil and gas industries experience corrosion because of a direct chemical reaction between a metallic component and various substances. Controlling corrosion involves identifying and understanding the mechanisms of corrosion, using corrosion-resistant materials, altering designs and equipment, utilizing protective systems, and using treatment methods. The usage of corrosion-resistant alloys is widespread across many sectors, particularly those involved in chemical processing [13]. These alloys deliver dependable performance in a variety of industries, including gas, oil, pharmaceuticals, gas, and energy. Excellent resistance to corrosion attack, resistance to stress corrosion cracking, and ease of manufacturing and welding are all facilitated using alloys.

Metals including stainless steel, chrome, iron, titanium, cobalt, nickel, and molybdenum are all used to make corrosion-resistant alloys. Such metals, when combined, can increase corrosion resistance more than other elements, including carbon steel. High-temperature corrosion is the degradation of a metallic substance at temperatures and pressures of more than 400°C. Several processes for corrosion include carburization, chlorination, nitration, oxidation, and sulfidation. The refinery industry of today is heavily reliant on practical knowledge of the variables influencing sulfidic corrosion [14]. This effect relates to the attack on materials resulting from the hydrogen sulfide formed by the thermal degradation of organic sulfur molecules [15]. Sulfidic corrosion is controlled by the composition of crude oil, which includes the amount and type of organic sulfide compounds as well as the presence of other corrosive substances such as naphthenic acids, hydrochloric acid, and salts. Additionally, it depends on the exposed materials, such as carbon or stainless steel, and system process variables, such as exposure time, temperature, and transfer line flow velocity [16]. The type and chemical content of the metal and alloy, as well as the surface state, all affect the erosion/corrosion process. The processing fluid's fast flow, which may also contain solid particles, has the potential to sweep away the protective film covering the metal surface. Use of highly corrosion-resistant materials, infrequent cathodic protection, use of appropriate coatings, alteration of environment (such as settling and filtration to remove solid particles or lowering of temperature), and appropriate shape or geometry design are just a few preventive measures that can be used [17,18]. It is important to note that the use of functional ceramic coatings can reduce corrosion problems in the oil and gas industries.

3.3.1 CERAMIC COATINGS IN OIL AND GAS INDUSTRIES

Coating techniques are most popular for the prevention of corrosion in petroleum industries, and the coating types include inorganic zinc coatings, epoxy coatings, and ceramic coatings. Inorganic zinc coatings are quite efficient for protecting against rust, salt, and the impacts of the weather. To prevent equipment and mechanical parts from corroding, they are used in chemical industries and refineries. Zinc-rich primers can be used in combination with strong polyester coatings in a variety of colors to produce a corrosion-resistant surface. Epoxy coatings, which are placed in thicknesses ranging from 4 to 6 mm, are particularly good at preventing corrosion. Epoxy is a highly cost-effective corrosion prevention option with the added benefit of preventing chemical erosion [8]. When compared to organic coatings such as epoxy and polyurethane, ceramic coatings have greater thermal and UV protection properties. However, the abrasion resistance is not as excellent. Organic coatings, particularly polyurethane-based coatings, have the highest mechanical resilience. However, polyurethane coatings need a skilled hand and sophisticated equipment, significantly increasing the cost. Ceramic coatings have the distinct benefit of being fully transparent, making them suitable for use on glasses and windows. They will not hinder visibility over time as they fade off. Ceramic coatings have greater chemical resistance than acrylic- or epoxy-based coatings. When exposed to direct UV radiation, epoxy-based coatings tend to tear faster [19].

Ceramic coatings are utilized extensively for heat protection, biological coatings, and hard coatings. Increased lifetime, corrosion prevention, stability on high-temperature components, reduced friction, protection against acidic/thermal corrosion, and improved surface attractiveness are all benefits of ceramic coatings. Drawbacks of ceramic coatings include their brittle nature and difficulty in repair, the possibility of de-bonding during expansion and shrinkage, the ease with which corrosion can form at cracks, they weigh more than organic coatings, and applying them requires additional tools, materials, and labor [20].

Ceramic coatings are available in a variety of forms, including dry-film lubricants, plasma spray coating, sputter coating, thermal spray coating, and several wet electrochemical and chemical coatings. Depending on the use and coating method, ceramic films can range in thickness from 50 nm to a few micrometers. The best technique for analyzing mechanical properties is nanoindentation. If the coating is thin, however, the measurement may combine the mechanical properties of the substrate and the coating. The main purpose of ceramic coating is to strengthen the resistance of carbon materials against oxidation at high temperatures [21]. Thermo-mechanical superior new nanoscale ceramic coatings like boron nitride, cerium oxide, Si_3N_4, and silicon carbide have recently been investigated in metal and alloy coatings to generate potential high-temperature structural materials [22,23]. However, finding a combination of oxide/carbon/silica glass/refractory carbide/refractory oxide that will satisfy these materials' physical compatibility, or match their thermal expansion coefficients at each boundary, is challenging in reality [24].

The most popular and enticing way of depositing various materials is thermal spraying, which is extensively used in the chemical, oil and gas, oil, and aerospace industries. The primary thermal spraying methods for coating purpose such as detonation cannon, electric arc, flame, plasma, and high-velocity oxygen fuel [25]. In essence, the plasma spray method involves a heat-softened or molten substance onto a surface to create a coating. The powdered material is introduced into a plasma flame at a very high temperature, where it is quickly heated and accelerated to a high velocity. The heated substance strikes the surface of the substrate and swiftly cools to produce a coating. The process parameters can be controlled in order to provide low porosity and strong binding strength of the coatings as well as to reduce the oxidation of sprayed particles [4]. Due to the ceramic oxide material's high melting point and the benefit of plasma-sprayed coating over other coating techniques, the high-quality plasma-sprayed coating can be produced. Due to this, functionally enhancing the surface by depositing Al_2O_3–40 wt% TiO_2 using the plasma spray approach is effective [26]. This kind of alloy effectively prevents corrosion, eliminating the need for costlier maintenance and repair.

Al_2O_3-TiO_2 composite powder-based ceramics are used as a promising material. Excellent resistance to abrasion, heat, and corrosion are all features of aluminum titanate ceramic composite. Furthermore, it is said to have a low frictional coefficient and a high fracture toughness [27]. The micro-arc oxidation (MAO) process in an alkaline solution with nanoadditive NaF or Al_2O_3 was employed to produce novel ceramic coatings on rare-earth-containing magnesium alloy. Most of the coating's component is a MgO crystal phase. The crystal phase of the ceramic coating is virtually unaffected by the extra doping.

The coatings' topography is altered by the additives. In comparison to Al_2O_3 nanoparticles, NaF doping creates MAO coatings with smaller, more uniform nodules, creating a coating that is both functional and protective. Utilizing nanoadditives significantly improves the corrosion resistance of MAO coatings. The magnesium substrate has stronger corrosion resistance due to fluoride doping of the MAO coating than aluminate nanoparticles [28].

3.3.1.1 Challenges with Ceramic Coatings

The primary issues with ceramic coating applications have been noted to be their demanding application conditions and high processing costs [29].

Application difficulty

The ceramic coating needs to be a clean surface and should also be in perfect condition, with no scratches or flaws, to avoid unpleasant results. Rust, scratches, and other paint flaws will remain under the protective layer, but due to the bright ceramic finishing, they will be significantly more noticeable. Again, these coatings should not be applied in moist, poorly ventilated environments. This could lead to bubbling, an unfavorable process resulting in an oxide concentration. Therefore, a dry environment is also necessary for applying the ceramic coating [30].

Inappropriate for outdoor use

Although one of the main advantages of ceramic coatings is UV protection, these coating layers shouldn't be applied in direct sunshine. Sunlight speeds up the surface drying of the coating, resulting in defects and deteriorating the paint. Another reason to avoid utilizing these techniques outside is the wind, which carries dust and other minute particles that will damage the paint layer [31].

High cost

The ceramic coating production process is complicated, and skilled laborers were required due to the challenges involved in applying the coating, which increased the total cost of the project.

3.3.2   FUNCTIONAL CERAMIC/POLYMER NANOCOMPOSITES FOR THE OIL AND GAS INDUSTRIES

Multiphase composite materials with ceramic and polymer matrix components have been used in a range of applications recently. Despite their high specific strength and flexibility, polymers have several disadvantages in industrial applications. Ceramics are extremely stiff, strong, and stable at high temperatures, but they are also fragile and need to be improved upon for usage in industry. Therefore, it is much desired to enhance the characteristics of composites made of ceramic and polymer [32].

Additionally, nanotechnology has been one of the most significant technologies used to enhance the properties of composite materials. Due to their small size, high specific surface area, and small volume of pores, nanoparticles have distinct physical and chemical properties. Nanocomposite materials can be categorized into three groups: ceramic nanocomposites, metal nanocomposites, and polymer nanocomposites. They can therefore be used as fillers in composite materials, and their dispersion in ceramic and polymer matrices will create nanocomposite materials with distinct properties. Numerous studies have been conducted on ceramic/polymer nanocomposites due to their enormous potential for usage in a wide range of applications, including the oil and gas industries, packaging, membrane, adsorption, catalysis, etc. [32]. These materials have superior absorptive, barrier, coating, electrical, mechanical, physical, and permeability properties, making them more heat- and flame-resistant than pure ceramic/polymer materials. The type of matrices and fillers employed has a significant impact on the properties of nanocomposites. As fillers, various nanomaterials are included in ceramic and polymer matrices, such as carbon nanotubes, graphene, nanoclay, silica, and reinforcing fibers. Graphene is one of the best candidates for a nanofiller and has advantages as an additive [33]. Graphene has received a lot of attention due to its unique properties, such as its large specific surface area, mechanical toughness, and electrical conductivity, and requires a lower loading fraction than other nanomaterials.

3.4 OUTLOOK

3.4.1 CERAMIC SENSORS FOR OIL AND GAS SECTORS

Exploration for oil and gas in deeper areas is raising severe safety issues throughout the world to improve oil and gas recovery. Monitoring of numerous essential factors, such as high pressure, high temperature, and chemicals, in real time for longer distances performs average in these harsh environmental circumstances.

Sensing technology should be able to monitor wellbore integrity and safety in addition to better oil and gas recovery. As a result, there is a rising need for upgraded sensors with improved measurement capabilities as well as sensors that produce accurate data for oil and gas production. To raise the apparent recoverable assets, new enhanced oil recovery (EOR) techniques are being used all over the world, and this intensive focus on enhancing oil recovery makes provision for the adoption of fiber-optic monitoring [34]. The use of fiber-optic sensing devices for production and pipeline monitoring has increased because of increased digitization in the oil and gas sectors. Even though this wonderful technology has been available for a long time, experts are continuously looking into relatively fresh ways to make it tolerate relatively greater temperatures and considerable pressures with little disturbance.

3.4.2 WASTEWATER TREATMENT FOR OIL AND GAS SECTORS

Purification and reuse of water from secondary effluents is the solution to the issues of water scarcity and poor water quality. The majority of secondary effluents from wastewater treatment plants are made up of many organic pollutants, suspended particles, and microorganisms [35]. For eliminating such pollutants from secondary effluents, separation procedures like activated carbon adsorption and membrane-based filtering have significant advantages over conventional treatment technologies including coagulation, flocculation, advanced oxidation processes, and ion exchange [36–39].

Porous ceramic membranes have most widely used polymeric membranes in terms of higher flux, better separation qualities, greater fouling resistance, longer working lives, no requirement for chemical additions, use of less energy, compact in design, taking less space, and more convenient to use [40–44]. Among various ceramic membranes, silicon carbide (SiC) membranes have distinguished themselves as good porous materials due to all of the aforementioned qualities with great mechanical strength and superior hydrothermal stability, exceptionally good heat and chemical resistance [45–48]. However, the high cost of SiC membranes due to the high-temperature production processes and their multi-step preparation process is the significant barrier to their widespread usage in wastewater treatment. By developing efficient techniques, it is necessary to reduce the cost of SiC membranes by considering the use of economical and sustainable materials to get around these challenges and satisfy industrial demand.

3.4.3 ANTI-FOULING CERAMIC MATERIALS FOR OIL AND GAS SECTORS

Heat-exchanger fouling occurs mostly because of frequent heating and cooling operations. It affects the oil and gas industries by causing a loss of thermal duty, an increase in pressure drops, equipment shutoff for cleaning, which will result in decreased production and elevated operating costs [49]. If the temperature falls below 32°C–38°C, paraffin wax may crystallize at the heat-exchanger plate wall. When heating crude oil, particle fouling occurs due to presence of suspended materials (clay, debris, or rust), crystallization soluble salts, and constituent chemical fouling (asphaltene precipitation). Again, steam boilers and condensers were the main targets of fouling issues. Scale is one of the most common deposits being formed when hard water is heated (or cooled) in heat transfer equipment (i.e., heat exchangers, condensers, evaporators, cooling towers, boilers, and pipe walls). The type of scale depends on the mineral content of the available water, with the most common form of scale being calcium carbonate. According to Breckenridge et al. [39], a clean boiler dissipated 13% more steam than one that was contaminated. To overcome the fouling problem, a ceramic coating was developed and applied to carbon steel with a simple and inexpensive coating method to deduce the corrosion effect and increases the life of equipment, but it is worth concluding that the developed ceramic coating has great potential to be applied in real boiler and heat-exchanger systems to improve their overall thermal efficiency [50].

3.5 CONCLUSION

Oil and gas extraction and processing depend heavily on advanced ceramic materials. Functional ceramics is futuristic materials for a given application in a highly corrosive environment and related to high temperature and pressure in the oil and gas industries. Ceramics can be made from a wide range of materials, and there is an equally wide range of additives that can be used to control various attributes including porosity, strength, and thermal integrity, giving them the flexibility to be used in several situations. Being formed of cheap raw materials, these applications in oil and gas are economically viable as compared with conventional metals or alloys. The use of nanotechnology and fiber-optic sensing (fiber Bragg grating sensors and distributed fiber-optic sensors) techniques in recent years, to find a workable sensing solution, has considerably benefited the oil and gas industries. Nowadays, the oil and gas sector has shown a lot of interest in the improvement of composite characteristics using graphene as a nanofiller to polymer/ceramic matrices. Polymer/ceramic graphene nanocomposites are a strong contender for use as adsorbents, catalysts, coating, and anticorrosion materials, as well as in the separation of water and oil and the elimination of organic pollutants and oil spots from the environment.

REFERENCES

[1] E. E. Cordes et al., "Environmental impacts of the deep-water oil and gas industry: A review to guide management strategies," *Front. Environ. Sci.*, vol. 4, no. SEP, pp. 1–26, 2016, doi: 10.3389/fenvs.2016.00058.

[2] L. T. Popoola, A. S. Grema, G. K. Latinwo, B. Gutti, and A. S. Balogun, "Corrosion problems during oil and gas production and its mitigation," *Int. J. Ind. Chem.*, vol. 4, no. 1, pp. 1–15, 2013, doi: 10.1186/2228-5547-4-35.

[3] A. H. Al-Moubaraki and I. B. Obot, "Corrosion challenges in petroleum refinery operations: Sources, mechanisms, mitigation, and future outlook," *J. Saudi Chem. Soc.*, vol. 25, no. 12, p. 101370, 2021, doi: 10.1016/j.jscs.2021.101370.

[4] Y. T. Al-Janabi, "An overview of corrosion in oil and gas industry: Upstream, midstream, and downstream sectors," *Corros. Inhib. Oil Gas Ind.*, Part-1, pp. 3–39, 2020, doi: 10.1002/9783527822140.ch1.

[5] S. K. Mobbassar Hassan and A. M. Abdullah, "Corrosion of general oil-field grade steel in CO_2 environment - An update in the light of current understanding," *Int. J. Electrochem. Sci.*, vol. 12, no. 5, pp. 4277–4290, 2017, doi: 10.20964/2017.05.12.

[6] M. A. Zavareh, A. A. D. M. Sarhan, B. B. A. Razak, and W. J. Basirun, "Plasma thermal spray of ceramic oxide coating on carbon steel with enhanced wear and corrosion resistance for oil and gas applications," *Ceram. Int.*, vol. 40, no. 9 PART A, pp. 14267–14277, 2014, doi: 10.1016/j.ceramint.2014.06.017.

[7] P. Greil, "Advanced engineering ceramics," *Adv. Eng. Mat.*, vol. 4, no. 5, pp. 247–254, 2002.

[8] M. Askari, M. Aliofkhazraei, R. Jafari, P. Hamghalam, and A. Hajizadeh, "Downhole corrosion inhibitors for oil and gas production – A review," *Appl. Surf. Sci. Adv.*, vol. 6, p. 100128, 2021, doi: 10.1016/j.apsadv.2021.100128.

[9] M. A. Materials, *"Innovative solutions for severe service environments in oil & gas applications,"* 2022.

[10] D. H. Everett, *Industrial carbon and graphite*, vol. 181, no. 4610, 1958.

[11] A. Yuxiu and Q. Weijia, "Applications and prospects of graphene in oilfield," *Arch. Pet. Environ. Biotechnol.*, vol. 2, no. 3, 2017, doi: 10.29011/2574-7614.100119.

[12] M. Sang, Y. Meng, S. Wang, Z. Long, "Graphene/cardanol modified phenolic resin for the development of carbon fiber paper-based composites," *RSC Adv.*, vol. 8, no. 43, pp. 24464–24469, 2018. doi: 10.1039/c8ra02699h. PMID: 35539179; PMCID: PMC9082169.

[13] M. Nuri Rahuma, "Corrosion in oil and gas industry: A perspective on corrosion inhibitors," *J. Mater. Sci. Eng.*, vol. 03, no. 03, p. 4172, 2014, doi: 10.4172/2169-0022.1000e110.

[14] R. B. Rebak, "Sulfidic corrosion in refineries - A review," *Corrosion Rev.*, vol. 29, no. 3–4, pp. 123–133, 2011, doi: 10.1515/CORRREV.2011.021.

[15] M. A. Quraishi, D. S. Chauhan, and F. A. Ansari, "Development of environmentally benign corrosion inhibitors for organic acid environments for oil-gas industry," *J. Mol. Liq.*, vol. 329, p. 115514, 2021, doi: 10.1016/j.molliq.2021.115514.

[16] M. Shahid, "Corrosion protection with eco-friendly inhibitors," *Adv. Nat. Sci. Nanosci. Nanotechnol.*, vol. 2, no. 4, pp. 1–7, 2011, doi: 10.1088/2043-6262/2/4/043001.

[17] C. Mele, F. Lionetto, and B. Bozzini, "An erosion-corrosion investigation of coated steel for applications in the oil and gas field, based on bipolar electrochemistry," *Coatings*, vol. 10, no. 2, pp. 1–11, 2020, doi: 10.3390/coatings10020092.

[18] Y. Kawahara, "An overview on corrosion-resistant coating technologies in biomass/ waste-to-energy plants in recent decades," *Coatings*, vol. 6, no. 3, pp. 1–24, 2016, doi: 10.3390/coatings6030034.

[19] Y. X. Wang and S. Zhang, "Toward hard yet tough ceramic coatings," *Surf. Coatings Technol.*, vol. 258, pp. 1–16, 2014, doi: 10.1016/j.surfcoat.2014.07.007.

[20] H. H. Jasim, A. A. Jasim, and A. H. Al-Hilfi, "Development of a new type of epoxy coating for crude oil storage tanks," *Am. Int. J. Res. Sci.*, vol. 1 & 2, no. January 2013, pp. 13–143, 2013 [Online]. Available: http://www.iasir.net.

[21] A. Forero, D. A. Giacometti, M. S. Alencar, R. P. Silva, and A. Labes, "Nanocoatings applied to corrosion protection at the oil and gas industry trends," *Proc. Annu. Offshore Technol. Conf.*, vol. 2, pp. 1263–1278, 2013, doi: 10.4043/24425-ms.

[22] S. Dabees, S. Mirzaei, P. Kaspar, V. Holcman, and D. Sobola, "Characterization and evaluation of engineered coating techniques for different cutting tools—Review," *Materials (Basel)*, vol. 15, no. 16, pp. 1–40, 2022, doi: 10.3390/ma15165633.

[23] M. H. Bocanegra-Bernal and B. Matovic, "Mechanical properties of silicon nitride-based ceramics and its use in structural applications at high temperatures," *Mater. Sci. Eng. A*, vol. 527, no. 6, pp. 1314–1338, 2010, doi: 10.1016/j.msea.2009.09.064.

[24] K. G. Nickel, P. Quirmbach, and J. Pötschke, "High temperature corrosion of ceramics and refractory materials," *Shreir's Corros.*, vol. 1, no. December, pp. 668–690, 2010, doi: 10.1016/B978-044452787-5.00079-2.

[25] E. Rodríguez et al., "*We are IntechOpen, the world ' s leading publisher of Open Access books built by scientists, for scientists TOP 1%*," Intech, vol. 32, no. tourism, pp. 137–144, 1989 [Online]. Available: https://www.intechopen.com/books/advanced-biometric-technologies/liveness-detection-in-biometrics.

[26] Y. T. Al-Janabi, "An overview of corrosion in oil and gas industry," *Corros. Inhib. Oil Gas Ind.*, Part-I, pp. 1–39, 2020, doi: 10.1002/9783527822140.ch1.

[27] S. Akbar, P. Dutta, and C. Lee, "High-temperature ceramic gas sensors: A review," *Int. J. Appl. Ceram. Technol.*, vol. 3, no. 4, pp. 302–311, 2006, doi: 10.1111/j.1744-7402.2006.02084.x.

[28] A. P. I. Popoola and O. S. I. Fayomi, "An investigation of the properties of Zn coated mild steel," *Int. J. Electrochem. Sci.*, vol. 7, no. 7, pp. 6555–6570, 2012.

[29] G. S. Rohrer et al., "Challenges in ceramic science: A report from the workshop on emerging research areas in ceramic science," *J. Am. Ceram. Soc.*, vol. 95, no. 12, pp. 3699–3712, 2012, doi: 10.1111/jace.12033.

[30] M. Sathish, N. Radhika, and B. Saleh, "A critical review on functionally graded coatings: Methods, properties, and challenges," *Compos. Part B Eng.*, vol. 225, no. April, p. 109278, 2021, doi: 10.1016/j.compositesb.2021.109278.

[31] B. Saleh et al., "30 Years of functionally graded materials: An overview of manufacturing methods, applications and future challenges," *Compos. Part B Eng.*, vol. 201, no. August, p. 108376, 2020, doi: 10.1016/j.compositesb.2020.108376.

[32] Y. K. Mashkov, V. A. Egorova, O. V. Chemisenko, and O. V. Maliy, "Polymer nanocomposites development and research for petrochemical and oil and gas production equipment," *Procedia Eng.*, vol. 152, no. 3812, pp. 545–550, 2016, doi: 10.1016/j.proeng.2016.07.653.

[33] A. H. Hartog, "Distributed sensors in the oil and gas industry," *Opt. Fibre Sensors*, Part-6, pp. 151–191, 2020, doi: 10.1002/9781119534730.ch6.

[34] N. V. Krishna Prasad et al., "Ceramic sensors: A mini-review of their applications," *Front. Mater.*, vol. 7, pp. 1–26, 2020, doi: 10.3389/fmats.2020.593342.

[35] L. Fan, T. Nguyen, F. A. Roddick, and J. L. Harris, "Low-pressure membrane filtration of secondary effluent in water reuse: Pre-treatment for fouling reduction," *J. Memb. Sci.*, vol. 320, no. 1–2, pp. 135–142, 2008, doi: 10.1016/j.memsci.2008.03.058.

[36] C. Reith and B. Birkenhead, "Membranes enabling the affordable and cost effective reuse of wastewater as an alternative water source," *Desalination*, vol. 117, no. 1–3, pp. 203–209, 1998, doi: 10.1016/S0011-9164(98)00097-6.

[37] J. L. Acero, F. J. Benitez, A. I. Leal, F. J. Real, and F. Teva, "Membrane filtration technologies applied to municipal secondary effluents for potential reuse," *J. Hazard. Mater.*, vol. 177, no. 1–3, pp. 390–398, 2010, doi: 10.1016/j.jhazmat.2009.12.045.

[38] A. Farsi, S. H. Jensen, P. Roslev, V. Boffa, and M. L. Christensen, "Inorganic membranes for the recovery of effluent from municipal wastewater treatment plants," *Ind. Eng. Chem. Res.*, vol. 54, no. 13, pp. 3462–3472, 2015, doi: 10.1021/acs.iecr.5b00064.

[39] L. P. Brechenridge, "Effects of scale on the evaporation in locomotive boilers, railroad gazette," *Effects Scale Locomot. Boil. Railr. Gaz.*, vol. 31, P-60, 1899.

[40] S. G. Lehman and L. Liu, "Application of ceramic membranes with pre-ozonation for treatment of secondary wastewater effluent," *Water Res.*, vol. 43, no. 7, pp. 2020–2028, 2009, doi: 10.1016/j.watres.2009.02.003.

[41] P. S. Goh and A. F. Ismail, "A review on inorganic membranes for desalination and wastewater treatment," *Desalination*, vol. 434, no. April, pp. 60–80, 2018, doi: 10.1016/j.desal.2017.07.023.

[42] Z. He, Z. Lyu, Q. Gu, L. Zhang, and J. Wang, "Ceramic-based membranes for water and wastewater treatment," *Colloids Surfaces A Physicochem. Eng. Asp.*, vol. 578, no. May, p. 123513, 2019, doi: 10.1016/j.colsurfa.2019.05.074.

[43] S. R. H. Abadi, M. R. Sebzari, M. Hemati, F. Rekabdar, and T. Mohammadi, "Ceramic membrane performance in microfiltration of oily wastewater," *Desalination*, vol. 265, no. 1–3, pp. 222–228, 2011, doi: 10.1016/j.desal.2010.07.055.

[44] H. Zhu, X. Wen, and X. Huang, "Characterization of membrane fouling in a microfiltration ceramic membrane system treating secondary effluent," *Desalination*, vol. 284, pp. 324–331, 2012, doi: 10.1016/j.desal.2011.09.019.

[45] R. J. Ciora et al., "Preparation and reactive applications of nanoporous silicon carbide membranes," *Chem. Eng. Sci.*, vol. 59, no. 22–23, pp. 4957–4965, 2004, doi: 10.1016/j.ces.2004.07.015.

[46] B. Hofs, J. Ogier, D. Vries, E. F. Beerendonk, and E. R. Cornelissen, "Comparison of ceramic and polymeric membrane permeability and fouling using surface water," *Sep. Purif. Technol.*, vol. 79, no. 3, pp. 365–374, 2011, doi: 10.1016/j.seppur.2011.03.025.

[47] D. Das, N. Kayal, G. A. Marsola, L. A. Damasceno, and M. D. de M. Innocentini, "Permeability behavior of silicon carbide-based membrane and performance study for oily wastewater treatment," *Int. J. Appl. Ceram. Technol.*, vol. 17, no. 3, pp. 893–906, 2020.

[48] Z. Li, K. Kusakabe, and S. Morooka, "Preparation of thermostable amorphous Si-C-O membrane and its application to gas separation at elevated temperature," *J. Memb. Sci.*, vol. 118, no. 2, pp. 159–168, 1996, doi: 10.1016/0376-7388(96)00086-5.

[49] D. J. Kukulka, "An evaluation of heat transfer surface materials used in fouling applications," *Heat Transf. Eng.*, vol. 26, no. 5, pp. 42–46, 2005, doi: 10.1080/01457630590927327.

[50] M. D. Nguyen et al., "Anti-fouling ceramic coating for improving the energy efficiency of steel boiler systems," *Coatings*, vol. 8, no. 10, pp. 1–13, 2018, doi: 10.3390/COATINGS8100353.

4 Understanding the Adsorption Behaviour of Corrosion Inhibitors on Metal–Water and Air–Water Interfaces from Molecular Simulations

Himanshu Singh and Sumit Sharma
Ohio University

CONTENTS

ABBREVIATIONS

AFM   Atomic force microscopy
bda   Benzyl dimethylammonium inhibitors
MD   Molecular dynamics
WHAM   Weighted histogram analysis method

DOI: 10.1201/9781003242550-4

4.1 INTRODUCTION

In oil industry, it is estimated that internal corrosion is responsible for ~55% of the total oil pipeline failures [1]. Internal corrosion is caused by the presence of water in oil-and-gas stream. It is economically not feasible to employ corrosion-resistant alloys in the manufacturing of thousands of miles of oil pipelines. The use of corrosion inhibitors has been an efficient alternative in controlling the internal corrosion. Corrosion inhibitors are organic surfactant molecules which are injected in oil-and-gas pipelines in the ppm-level concentration. These molecules adsorb onto the metal surface and form a hydrophobic film, which acts as a barrier to the ingression of water molecules [2]. The use of corrosion inhibitor molecules is also very cost-effective, and therefore, there has always been a search for a robust inhibitor molecule. Determining adsorption mechanism and equilibrium morphology of inhibitor molecules has remained a long-time challenge because of the experimental difficulties in perceiving the behavior of the molecules at the atomic scale. The inhibitor molecules tend to adsorb to the metal–water interface and to aggregate and form micelles above the critical micelle concentration (CMC) in the bulk aqueous medium. It is, therefore, expected that the molecules behave according to their relative preference between forming a micelle and adsorbing onto the metal surface. The micelles formed in the bulk aqueous medium may also adsorb onto the metal surface. The adsorbed molecules may then self-assemble to attain an equilibrium morphology on the surface. Because of the presence of oil-and-gas in the pipeline, some inhibitor molecules are also expected to adsorb onto the oil/gas–water interfaces. For corrosion protection, it is desired that the inhibitors should have a good aqueous solubility so that their bulk concentration is sufficient for adsorption onto the metal–water interface rather than being accumulated at the oil/gas–water interfaces. Several possibilities can, therefore, be surmised in which the inhibitor molecules may behave and organize on the interfaces. Various experiments are performed in the past to understand the kinetics of adsorption of the inhibitors on the surfaces; however, there are some limitations. In the quartz crystal microbalance (QCM), the mass change upon adsorption of the inhibitors falls close to the detection limit [3]. Moreover, the adsorbed mass may also capture the mass of the water molecules that surround the adsorbed polar head groups [4]. Determination of equilibrium adsorbed morphologies of the inhibitors is attempted via atomic force microscopy (AFM) experiments. AFM studies on polar surfaces have estimated inhibitor aggregates in the shapes of planar [5], spherical [6], and cylindrical micelles [7,8]. These morphologies are understood to be governed by hydrophobic interactions [9–13], molecular geometry [14], and Coulombic interactions between the polar head, surface, and the counter-ions [6,7,9]. It is to be noted that while the height resolution of AFM is good (~1 Å), the lateral resolution (~10–50 Å) is inefficient for making accurate atomic scale measurements [15].

Molecular dynamics (MD) simulations are useful in studying the behavior of molecules at atomistic scale. The structures of inhibitor molecules studied in this work are shown in Figure 4.1. These molecules are widely used as corrosion inhibitors in the oil-and-gas industry [16–18]. The cationic and anionic inhibitor molecules carry a charge of +1e and –1e, respectively. To perform MD simulations of inhibitor molecules, the molecules are initially placed in a medium of explicit water molecules.

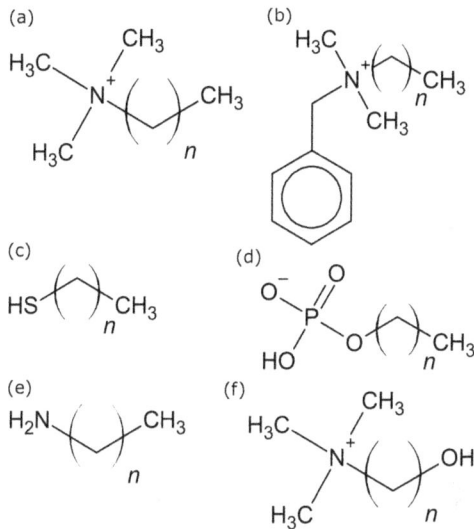

FIGURE 4.1   Structure of (a) cationic trimethylammonium-based molecules (termed as *quat; n*=9 and 15 for quat-10 and quat-16, respectively), (b) cationic benzyl dimethylam-monium-based molecules (termed as *bda; n*=3 and 11 for bda-4 and bda-12, respectively), (c) charge-neutral thiol-based molecule (termed as *decanethiol; n*=9 for decanethiol), (d) anionic phosphate monoester-based molecule (termed as *pe; n*=11 for pe-12), (e) charge-neutral amine-based molecule (termed as *amine; n*=9 for amine-10), and (f) cationic tri-methylammonium-OH molecules (termed as *quat-OH; n*=9 and 15 for quat-10-OH and quat-16-OH, respectively). quat-OH molecules have a similar structure as of quat molecules with the terminal -CH$_3$ group replaced by an -OH group.

Simulations are then performed by integrating classical equations of motion for all the species in the simulation box. Each atom/ion moves according to the forces acting on it from the surrounding species. The forces are the result of interaction potentials which are mainly van der Waals' potential between non-ionic species and Coulombic potential between the ionic species. The interaction potential parameters are obtained from the force fields. Researchers have previously developed the force fields of the species by matching their thermodynamic properties from simulations with the properties obtained from density functional theory (DFT) calculations and/or experiments.

Figure 4.2a shows a setup of an MD system in which the bda-12 molecules are randomly inserted in water. The box is periodic in x- and y-directions and non-peri-odic in the z-direction. In the non-periodic z-direction, a metal lattice is placed at the bottom end and an athermal reflective wall is placed at the top end. The metal lattice is comprised of 2,040 gold atoms placed in six layers in (111) Miller index planes of face-centered cubic (FCC) structure. The dimension of the lattice in x- and y-direc-tions is 49.04 Å × 49.97 Å. The interaction potential parameters of the inhibitor mol-ecules are taken from the general amber force field (GAFF), a popular force field for organic molecules [19]. Partial charges on each atom of the inhibitor molecules are calculated from density functional theory (DFT) using B3LYP hybrid functional

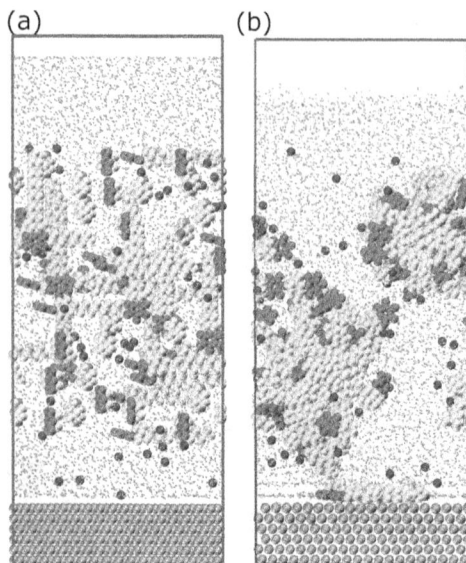

FIGURE 4.2 (a) An initial configuration of the simulation system with bda-12 molecules randomly inserted near the metal–water interface, (b) final configuration of bda-12 molecules obtained from an unbiased simulation run of 100 ns started with the initial configuration in (a). The red beads represent the aromatic rings, the yellow beads represent the alkyl tails of the inhibitor molecules, the blue beads represent bromide ions, and the cyan beads represent water molecules. The gold atoms are shown as orange beads. (Adapted with permission from Singh and Sharma [27]. Copyright (2022), American Chemical Society.)

theory with 6–31G (d, p) basis set in implicit water using Gaussian 16 [20]. Water molecules are modeled using extended simple point charge (SPC/E) model [21]. Counterions are introduced in the system for charged molecules, unless specified otherwise. Bromide ions are used as the counter-ions for cationic molecules, and sodium ions are used as the counter-ions for anionic molecules. The interaction parameters of counterions are taken from Joung–Cheatham's model [22]. Force field for the gold atoms of the metal lattice is obtained from the interface force field developed by Heinz and co-workers [23]. The cutoff for Lennard–Jones and short-range Coulombic potentials is chosen as 10 Å. Interaction of thiol with gold surface is obtained from a previous study [24]. All the simulations are performed in the canonical ensemble in which the number of atoms, volume, and temperature of the system (300 K) are fixed and maintained via Nose–Hoover thermostat. A vacuum gap of ~10 Å is kept above the water so that system pressure is maintained at the saturation pressure corresponding to $T = 300$ K [25]. The MD simulations package, Large-Scale Atomic/Molecular Massively Parallel Simulator (LAMMPS), is used to perform all the MD simulations [26].

An MD simulation of the bda-12 molecules is performed for 100 nanoseconds (ns). Figure 4.2a and b shows the initial and final configurations of the molecules in the system. It is observed that most molecules aggregate to form micelles in the bulk aqueous phase and only a few molecules adsorb by lying flat on the metal surface. It is important to mention that a usual MD simulation may not be able to achieve the equilibrium configuration of molecules in a short time span of hundreds of nanoseconds.

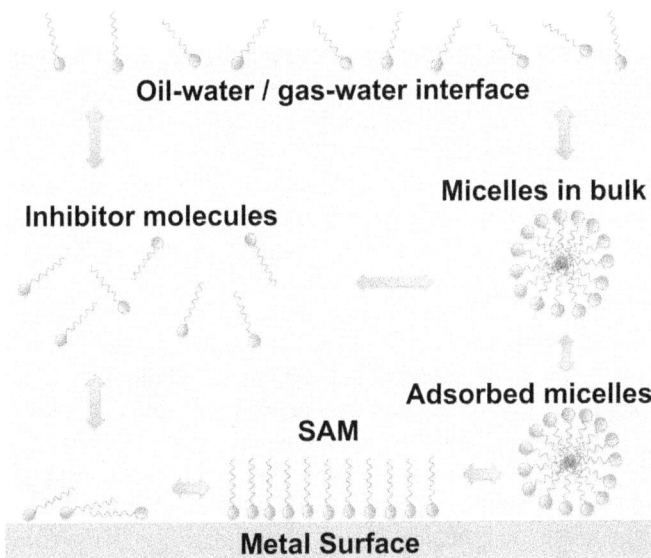

FIGURE 4.3 A cartoon depicting possible states of inhibitor molecules near interfaces. In bulk aqueous phase, inhibitor molecules can be present in unaggregated state and micellar state. At interfaces, inhibitors may attain various morphologies such as lying-down configuration, self-assembled monolayer (SAM), and micellar configuration.

It is experimentally observed that inhibitor molecules take several hours to organize at the metal surfaces [28]. This behavior might be attributed to kinetic barriers that the molecules face in high-density adsorption morphologies. Computationally, it is not feasible to simulate a full atomistic system more than hundreds of nanoseconds. An alternate approach is to perform advanced MD simulations for determining the free energy of different states of molecules. The state with the minimum free energy corresponds to the equilibrium state of the molecules in a system. A cartoon of some possible states of the inhibitor molecules near the interfaces is shown in Figure 4.3. Inhibitor molecules are present in bulk aqueous phase, at metal–water interface and at oil–water/gas–water interfaces.

4.2 FREE ENERGIES IN DIFFERENT STATES

In this chapter, the free energy profiles connecting different states shown in Figure 4.3 are obtained. Free energy profiles are often described as function of an order parameter. An order parameter is a variable that connects the thermodynamic states of interest. For example, if one is interested in calculating the adsorption free energy profile, a reasonable order parameter is the distance of the center-of-mass of the molecule or a micelle from the adsorbing surface, z. One of the methods to obtain the free energy profiles is the umbrella sampling method [29]. In umbrella sampling method, a biased potential is applied to the center-of-mass of the molecules/micelles given by,

$$U_{biased} = \frac{1}{2}k\left(\xi - \xi_o\right)^2 \tag{4.1}$$

where ξ is the order parameter which defines the current state of a molecules/ micelles in the system and ξ_o is the set value of the order parameter where the molecules/micelles are biased. The constant k is the force constant, which in this study is mainly taken as 40 $k_B T/\text{Å}^2$ for obtaining free energy profiles on metal–water interfaces, and 5 $k_B T/\text{Å}^2$ for free energy profiles across air–water interfaces. To study the adsorption free energy profiles of the molecules/micelles on metal–water interface, ξ is considered as the distance of center-of-mass of the molecules/micelles from the metal surface, z. The set positions, z_o, are spaced at an interval of 0.5 Å to enable a good sampling of the molecules/micelles across the distance from the interfaces. The biased probability distributions obtained from the umbrella sampling simulations are then combined using weighted histogram analysis method (WHAM) [30], which removes the bias and generates the free energy profile with respect to z. The umbrella sampling is performed using the implementation of COLVARS package in LAMMPS [31]. In each umbrella sampling simulation, the system is equilibrated for the first 40 ns, and the next 40 ns is used for free energy calculations.

The free energy profiles for adsorption of the inhibitor molecule and micelle onto metal–water interface are generated for the quat-10 and quat-16 molecules. To study the aqueous solubility of the inhibitor molecules, the free energy profiles are generated across the air–water interface for quat-10, quat-16, and amine-10 molecules, the simulation details of which are discussed later. To determine the equilibrium adsorption morphology of inhibitors on the metal surface, a modified umbrella sampling-based methodology is introduced [27,32]. The details of the functional form of its biased potential are discussed further. Using this methodology, the equilibrium adsorption morphologies of bda-4, bda-12, pe-12, and decanethiol molecules on metal–water interfaces are predicted. All the free energy profiles reported herein are obtained from our previous publications [27,33–35].

4.2.1 Adsorption Free Energy in Infinite Dilution

Free energy profiles for adsorption of quat-10 and quat-16 molecules to metal–water interface in infinite dilution are calculated [34]. The free energy profiles of the molecules as a function of the distance of their center-of-mass, z, are shown in Figure 4.4. The free energies are minimum when the molecules are adsorbed on the metal surface. A strong adsorption free energy of ~35 $k_B T$ indicates that a single molecule is stable in the adsorbed state and unstable in the bulk aqueous state. Upon adsorption, the molecule lies flat on the surface as shown in the inset. It should also be noted that there is no free energy barrier to adsorption of the molecules.

4.2.2 Adsorption Free Energy of Inhibitor Micelles

From the MD simulations of inhibitor molecules in the bulk aqueous phase (time span ~ 20 ns), it is observed that the molecules aggregate and form spherical micelles [36]. The quat-10 and quat-16 inhibitors form micelles of size 18 and 19 molecules, respectively [34,36]. The quat micelles are shown in Figure 4.5a and b. One interesting question is to estimate whether the micelles also adsorb onto the metal surface the way unaggregated molecules adsorb. The free energy profiles of adsorption of the

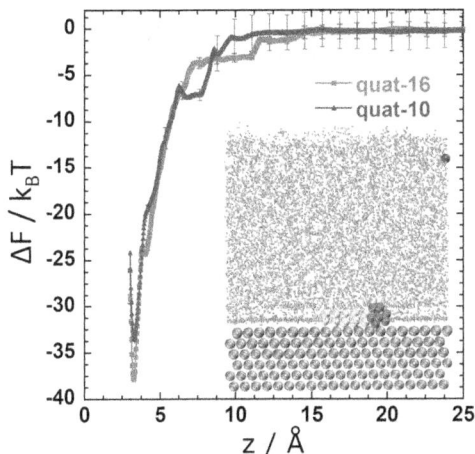

FIGURE 4.4 Free energy profiles of quat molecules as a function of their distance, z from the metal–water interface. The molecules undergo barrier-less adsorption to the metal surface. The inset figure shows that upon adsorption, the molecule lies flat on the surface. Error bars are obtained from three independent estimates of the free energy profile. (Adapted with permission from Singh and Sharma [34]. Copyright (2020), Taylor & Francis.)

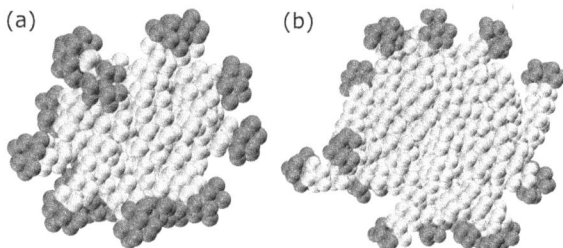

FIGURE 4.5 (a) quat-10 micelle comprising of 18 quat-10 molecules, (b) quat-16 micelle comprising of 19 quat-16 molecules. The polar head groups (shown in red) are exposed toward the aqueous phase, and the hydrophobic alkyl tails (shown in yellow) form the core of the spherical micelles. (Adapted with permission from Singh and Sharma [34]. Copyright (2020), Taylor & Francis.)

quat micelles are therefore determined with respect to the distance of their center-of-mass, z, from the surface and shown in Figure 4.6a and b [34]. It is worthwhile to note that unlike the inhibitor molecules in infinite dilution, the cationic micelles experience a long-range free energy barrier to adsorption. The barrier to adsorption begins at $z \approx 50 - 60\,\text{Å}$ from the metal surface and goes up to ~ 5 k_BT at its peak at $z \approx 22\,\text{Å}$. Beyond $z \approx 22\,\text{Å}$, the free energy drops sharply, and the micelles strongly adsorb to the metal surface. The reason of the long-range free energy barrier is that, in the bulk aqueous phase, the cationic micelles are surrounded by a solvation shell of the counterions (bromides), and the counter-ions are further solvated by the water molecules. The free energy barrier is, therefore, associated with the loss of this solvation shell as the micelle approaches the metal surface. A similar study is previously performed with the micelles of uncharged inhibitor molecules in which the counter-ions are not present. In the absence of the counter-ions, the uncharged micelles are not surrounded by a solvation shell and therefore do not experience a free energy barrier to adsorption [34].

The free energies for adsorption of the cationic micelles are minimum at the metal surface, which indicates that the micelles are stable in the adsorbed state. The true free energy minimum is not attained and not shown in Figure 4.6a and b. It is because close to the metal surface $z \approx 22$ Å, the micelles no longer remain intact, but keep deforming and disintegrating. It is therefore difficult to compute the free energies below $z \approx 22$ Å. The inset images in Figure 4.6a and b refer to the equilibrium state of the micelles. At the metal surface, the micelles take ~200 ns to disintegrate completely [33]. Upon disintegration, the constituent molecules of the micelles lie flat on the surface. The lying-down configuration enhances their interaction with the metal surface.

The above studies explain the kinetics of adsorption of inhibitor molecules and micelles on the metal–water interfaces. The inhibitor molecules and micelles adsorb strongly to the metal surface. While the unaggregated molecules adsorb without experiencing any free energy barrier, the cationic inhibitor micelles experience a long-range free energy barrier to adsorption. Though the adsorption of the quat micelles only explain the equilibrium morphology of 18/19 number of quat molecules, it is experimentally observed that the inhibitor molecules form a much denser adsorption morphology on the metal surface [5–8]. The further discussion is about finding the equilibrium adsorption morphology of inhibitors when they are present at sufficiently high concentrations near the metal–water interface.

The equilibrium adsorption morphology of inhibitor molecules is the one that has the lowest free energy among the other adsorption morphologies. A new order

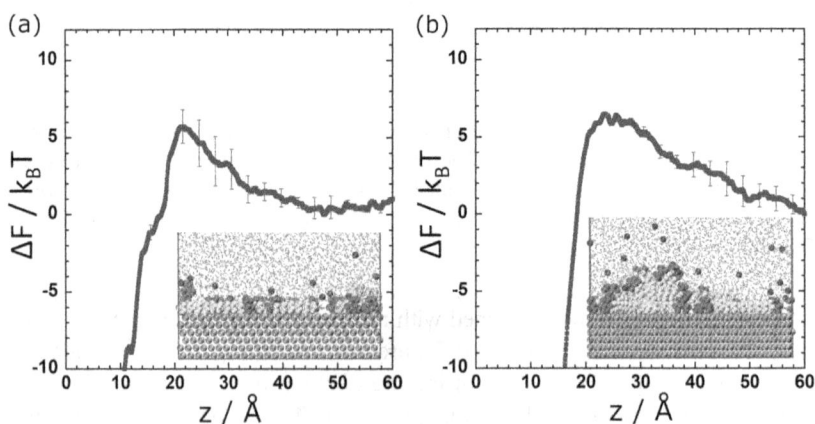

FIGURE 4.6 Free energy profiles of (a) quat-10 micelle, and (b) quat-16 micelle as a function of their distance, z from the metal surface. A long-range free energy barrier is observed from $z \approx 50$–60 Å to $z \approx 22$ Å for the adsorption of micelles. The inset images show the configuration of the micelles at their free energy minimum location. Error bars in the profiles are obtained from three independent estimates of the free energy profiles. (Adapted with permission from Singh and Sharma [34]. Copyright (2020), Taylor & Francis.)

parameter, $\xi(z, z_o, n, m)$, is defined which connects various adsorption morphologies. The order parameter is devised to be a measure of the number of the inhibitor molecules on the surface. Therefore, the order parameter is referred as the *adsorption number* function and is defined as [27],

$$\xi(z, z_o, n, m) = \sum_{i=1}^{N} \frac{1 - (z_i/z_o)^n}{1 - (z_i/z_o)^m} \tag{4.2}$$

where N represents the total number of inhibitor molecules in the system; z_i is the distance of polar head of the ith inhibitor molecule from the top layer of the metal lattice; z represents the collection of z_i of all the N inhibitor molecules; z_0 is the cutoff distance that indicates the inflection point of the function; and n and m are integers $(n < m)$ that control the long-range behavior and stiffness of the function. The adsorption number function is $\xi(z, z_o, n, m) \rightarrow N$ as the molecules come close enough to the surface $(z_i \rightarrow 0)$, and $\xi(z, z_o, n, m) \rightarrow 0$ as the molecules move away from the surface $(z_i \gg z_o)$. Therefore, the adsorption amount is well captured by $\xi(z, z_o, n, m)$, which is a continuous and differentiable function in the range of $(0, N)$. It is advantageous to use the differentiable function as opposed to using the integral number of adsorbed molecules so that the umbrella sampling potential can be described based on $\xi(z, z_o, n, m)$ which does not change stepwise with respect to z, and therefore, the resulting biased forces are well defined. For this study, the constant parameters are chosen as $z_0 = 18$ Å, $n = 2$, and $m = 4$. From now on, the adsorption number function is simply referred as ξ. The dependence of ξ on z for a single molecule $(N = 1)$ is shown in Figure 4.7.

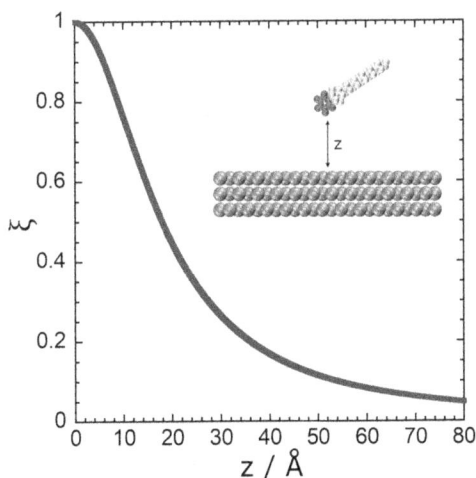

FIGURE 4.7 The adsorption number function, $\xi(z, z_o, n, m)$, plotted as a function of z for $N = 1$. The values of other parameters are $z_0 = 18$ Å, $n = 2$, and $m = 4$. The inset image shows the distance of the polar head group of a molecule from the metal surface, z. (Adapted with permission from Singh and Sharma [27]. Copyright (2022), American Chemical Society.)

The simulations of bda-4 and bda-12 are performed with 50 inhibitor molecules and ~7,000 water molecules. The initial configuration of the bda-12 molecules is shown in Figure 4.2a. The decanethiol simulations are performed with 200 inhibitor molecules and ~25,000 water molecules. A larger system is considered for decanethiol as a significant more adsorption is observed in this system. The adsorption behavior of an equimolar mixture of 50 bda-12 and 50 pe-12 molecules is studied with ~12,400 water molecules. The number of water molecules is chosen according to the number of inhibitor molecules to maintain the same concentration of inhibitors in all the simulation systems.

4.3 ADSORPTION MORPHOLOGY OF BDA-12 MOLECULES

The free energy profile of bda-12 molecules is shown in Figure 4.8a. The free energy minimum is at $\xi = 27.5$, which corresponds to the equilibrium adsorption morphology of the molecules. The side and top views of the equilibrium morphology are shown in Figure 4.8b and in the inset of Figure 4.8a, respectively. At equilibrium, the molecules adsorb in such a way that some molecules lie flat on the surface and a hemimicelle sits on top of lying-down molecules. A denser morphology in which all molecules of the simulation system are adsorbed $(\xi = 50)$ is unstable at the surface [27]. The top view of the equilibrium morphology shows that the metal surface is nearly fully covered by the bda-12 molecules. Due to longer alkyl tails, bda-12 molecules have stronger hydrophobic interactions among themselves in comparison with bda-4 molecules. Therefore, the gold surface is better packed by the adsorbed bda-12

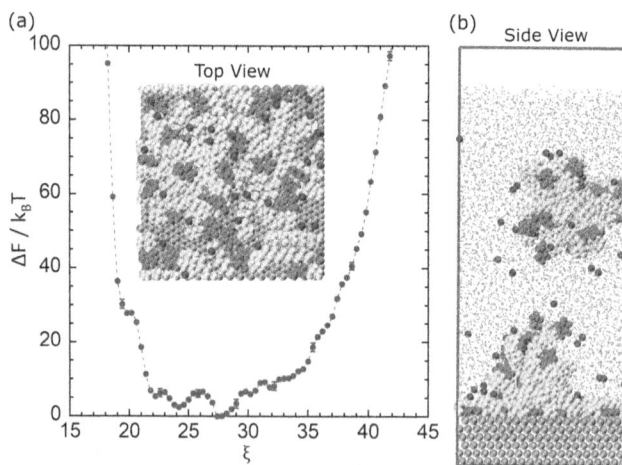

FIGURE 4.8 (a) Free energy profile of bda-12 molecules as a function of the adsorption number function, ξ near the metal–water interface, (b) side view of the equilibrium configuration at $\xi = 27.5$. The inset in (a) shows the top view of the equilibrium adsorption morphology. Error bars in (a) are obtained from three independent estimates of the free energy profile. (Adapted with permission from Singh and Sharma [27]. Copyright (2022), American Chemical Society.)

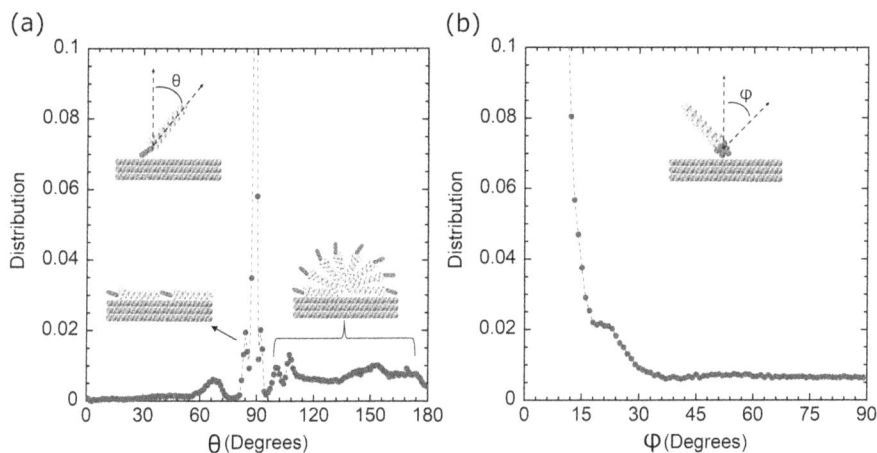

FIGURE 4.9 Distribution of orientation of (a) alkyl tails and (b) aromatic rings of adsorbed bda-12 molecules on the metal surface in the equilibrium adsorption morphology. In (a), θ is the angle between the surface normal and the vectors connecting the polar head groups to the terminal methyl group of the alkyl tails. In (b), φ is the angle between the surface normal and the vectors normal to the aromatic rings. A molecule is considered as adsorbed when its center-of-mass lies within one molecular length (~23 Å) from the metal surface. There are ~30 bda-12 molecules adsorbed on the metal surface, with ~16 molecules lying flat onto the surface and ~14 molecules arranged in the hemispherical micelle. The height of the adsorbed hemispherical micelle is around one molecular length. (Adapted with permission from Singh and Sharma [27]. Copyright (2022), American Chemical Society.)

molecules. The inhibitor molecules in the bulk phase are stable in a micellar state, due to the strongly favorable micellization free energy [36].

Figure 4.9a shows the distribution of orientation of alkyl tails of adsorbed bda-12 molecules with the surface normal in the equilibrium adsorption morphology. The cartoons shown in the inset explain the various features observed in the distribution. The peak at $\theta \approx 90°$ indicates that ~16 molecules lie flat onto the surface. Another broad distribution in between $90° < \theta < 180°$ shows that ~14 molecules form a hemispherical micelle on top of the lying-down molecules. This arrangement of the molecules also matches with the configuration of the molecules shown previously in Figure 4.8b. The distribution of orientation of the aromatic rings of the adsorbed molecules is shown in Figure 4.9b. In Figure 4.9b, the peak below $\varphi < 15°$ corresponds to the aromatic rings that lie flat onto the surface. These aromatic rings correspond to the lying-down molecules on the surface. The uniform distribution in between $30° < \varphi < 90°$ corresponds to the aromatic rings of the molecules of the hemispherical micelle.

4.4 ADSORPTION MORPHOLOGY OF BDA-4 MOLECULES

Using the same umbrella sampling methodology based on ξ, the free energy profile of bda-4 molecules is calculated. Figure 4.10a and b shows the free energy profile and the equilibrium adsorption morphology of bda-4 molecules. The free energy minimum is at $\xi = 24$ Å. A total of ~19 bda-4 molecules adsorb on the surface as compared

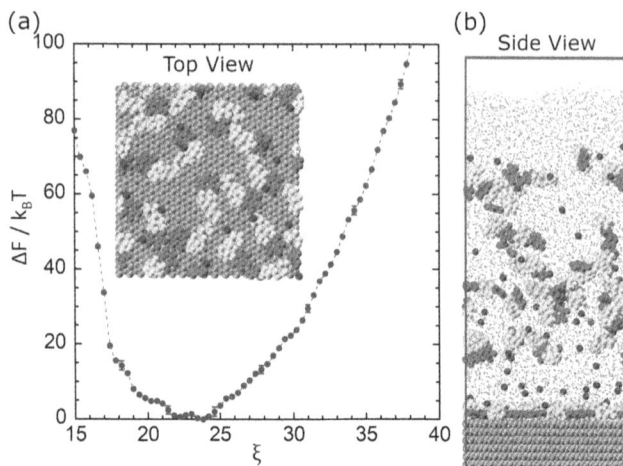

FIGURE 4.10 (a) Free energy profile of bda-4 molecules as a function of the adsorption number function, ξ near the metal–water interface, (b) side view of the equilibrium configuration at $\xi = 24$. The inset in (a) shows the top view of the equilibrium adsorption morphology. Error bars in (a) are obtained from three independent free energy profiles. (Adapted with permission from Singh and Sharma [27]. Copyright (2022), American Chemical Society.)

to ~30 molecules in case of bda-12. Unlike the bda-12 molecules, bda-4 molecules do not aggregate to form micelles because they have smaller hydrophobic length. In the adsorbed state, the molecules lie flat on the surface. The lying-down configuration is also reported previously via SFG spectroscopy measurements [37]. The bda-4 molecules are unable to achieve full surface coverage. The sparse adsorption of bda-4 molecules is attributed to the accumulation of positive charge on the surface, which makes the adsorption unfavorable. It is important to mention that charge accumulates even in the case of adsorption of bda-12 molecules, but their favorable hydrophobic interactions overcompensate for the unfavorable charge accumulation, thereby promoting full coverage of the molecules on the metal surface. A similar behavior is observed in electrochemical corrosion tests where bda-12 molecules are found to show much better inhibition efficiency than bda-4 molecules [38,39].

4.5 ADSORPTION MORPHOLOGY OF DECANETHIOL MOLECULES

In the previous sections, adsorption morphologies of cationic molecules with two different tail lengths are determined. While bda-12 molecules attain nearly full surface coverage, bda-4 molecules are unable to cover the metal surface entirely. In this section, adsorption of decanethiol molecules, which are uncharged, is studied. Figure 4.11a shows the free energy profile of decanethiol molecules. A dramatic change in the free energy minimum location $\left(\xi = 161 \right)$ is observed. A significantly larger number of molecules (~196) adsorb as compared to bda-12 molecules (~30). The configuration of the equilibrium adsorbed morphology is shown in Figure 4.11b

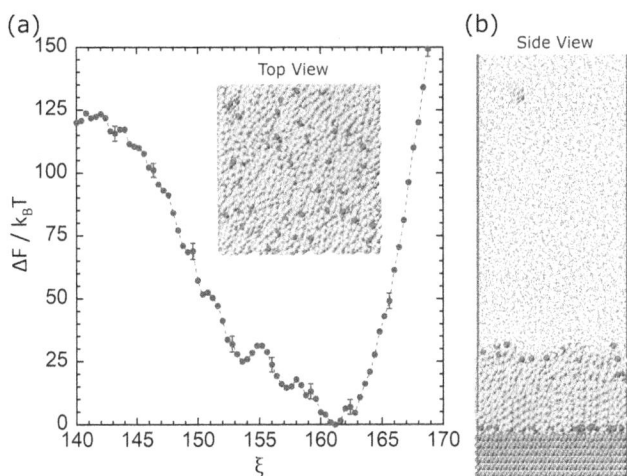

FIGURE 4.11   (a) Free energy profile of decanethiol molecules as a function of the adsorption number function, ξ near the metal–water interface, (b) side view of the equilibrium configuration at $\xi = 161$. The inset in (a) shows the top view of the equilibrium adsorption morphology. Error bars in (a) are obtained from three independent free energy profiles. The red beads represent the sulfur atoms of the molecules, and the yellow beads represent the alkyl tails of the molecules. (Adapted with permission from Singh and Sharma [27]. Copyright (2022), American Chemical Society.)

and the inset of Figure 4.11a. Decanethiol molecules adsorb in a bilayer morphology. The molecules in the first layer are aligned in a standing-up orientation on the surface. The molecules in the second layer are, however, horizontally oriented. A bilayer adsorption morphology is also observed in a previous electrochemical impedance spectroscopy measurement [40]. The dense packing of decanethiol molecules, as depicted in Figure 4.11b, is explained by the lack of charge accumulation and favorable hydrophobic interactions between the alkyl tails. The strong thiol–gold interaction also promotes the dense packing of the molecules [24]. It is also observed experimentally that the corrosion rate with use of decanethiol molecules is far smaller than with the use of bda molecules [39,40].

4.6   ADSORPTION MORPHOLOGY OF AN EQUIMOLAR MIXTURE OF BDA-12 AND PE-12 MOLECULES

A better packing efficiency of uncharged (decanethiol) molecules over charged (bda-12) molecules indicates that accumulation of charged headgroups is unfavorable at the surface. Moreover, a comparison of adsorption morphologies of bda-12 molecules with bda-4 molecules suggests that hydrophobic length promotes adsorption. In this section, adsorption morphology is studied upon mixing bda-12 molecules with pe-12 molecules in 1:1 ratio. Figure 4.12a shows the free energy profile of adsorption morphologies of the mixture of the molecules. Interestingly, the number of adsorbed molecules at free energy minimum $\left(\xi = 41\right)$ is ~43, indicating that a larger number of

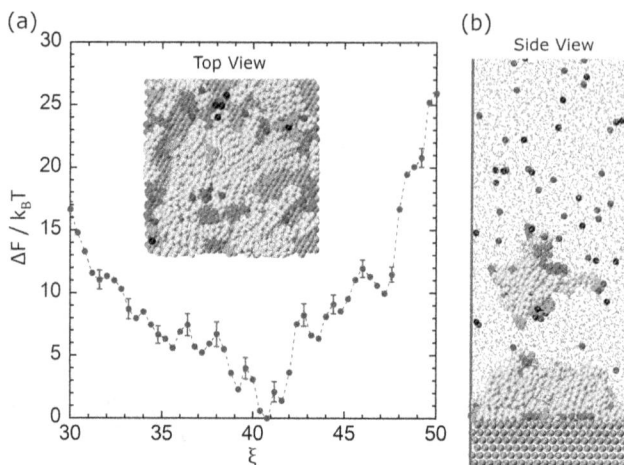

FIGURE 4.12 (a) Free energy profile of a mixture of bda-12 molecules and pe-12 molecules as a function of the adsorption number function, ξ near the metal–water interface, (b) side view of the equilibrium configuration at $\xi = 41$. The inset in (a) shows the top view of the equilibrium adsorption morphology. Error bars in (a) are obtained from three independent free energy profiles. The red beads represent the aromatic rings of the bda-12 molecules, the green beads represent the head groups of the pe-12 molecules, the yellow beads represent the alkyl tails of both the molecules, the blue beads represent the bromide counter-ions, and the black beads represent the sodium counter-ions. (Adapted with permission from Singh and Sharma [27]. Copyright (2022), American Chemical Society.)

molecules adsorb on the surface as compared to just the bda-12 molecules (~30). This suggests that the charged molecules, when mixed with oppositely charged molecules, show synergistic effect in adsorption.

4.7 FREE ENERGY OF INHIBITORS ACROSS AIR–WATER INTERFACES

In previous sections, adsorption behavior of inhibitor molecules on metal–water interface is explored. Another desirable characteristic of corrosion inhibitors is their high aqueous solubility. In an oil-and-gas pipeline, inhibitors may get accumulated at the oil–water or gas–water interfaces because of their amphiphilicity [41], and become partially unavailable to adsorb on metal–water interfaces. It is, therefore, important to study the solubility of the inhibitors in water. A more water-soluble inhibitor will have higher concentration in the aqueous phase and thus will have higher adsorption amount at the metal–water interface. The free energy profiles of inhibitor molecules across the air–water interface are determined. The simulation system for studying the behavior of inhibitor molecules across air–water interface is shown in Figure 4.13. The simulation box is periodic in x- and y-directions but non-periodic in the z-direction. The dimension of the box is 52.61 Å × 52.61 Å × 100 Å which consists of a ~50 Å water column being held with an attractive surface at one end (Z = 0 Å), and a vapor space of ~50 Å above the water column. The number of water molecules in

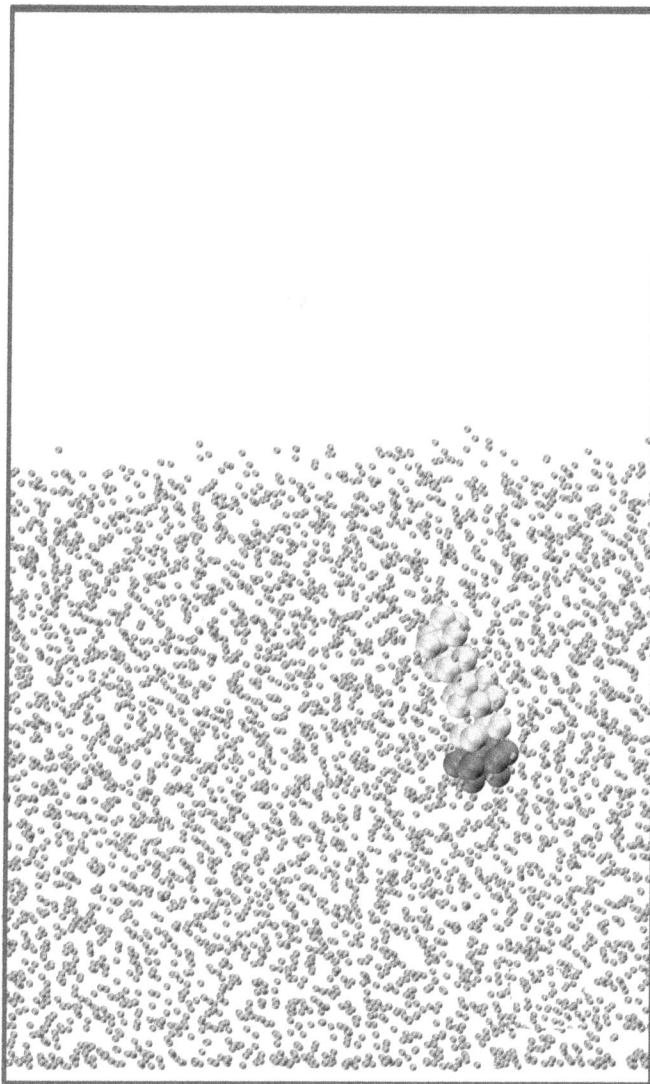

FIGURE 4.13 Simulation system for obtaining the free energy profiles of an inhibitor molecule across air–water interface. The red beads represent the head group, and the yellow beads represent the alkyl tail of the quat-10 molecule. Counter-ion is not inserted in the simulation system; therefore, the simulations of the quat molecules are performed in the charged system. (Adapted with permission from Singh and Sharma [35]. Copyright (2022), American Chemical Society.)

the system is 4,382. The surface at $Z = 0\,\text{Å}$ has an attractive interaction only with the water molecules but no interaction with the inhibitor molecule. There is an athermal surface at $Z = 100\,\text{Å}$ which is non-interacting and reflective to all the species of the system. The vapor phase is chosen beyond the water column because it is non-polar, and therefore a good proxy for an oil, or a gas phase. The vapor phase is simply represented by an empty space above the water column, so that a few water molecules may intermittently enter this space. For ease of representation, the vapor phase is henceforth referred to as the air phase. The free energy profiles are obtained for the inhibitors quat-10, quat-16, amine-10, and the alkanes C-10 and C-16 using the umbrella sampling method. The order parameter for the free energy profiles is the distance of the center-of-mass of the molecules from air–water interface, z. In each umbrella sampling simulation, the system is equilibrated for the first 20 ns, and the next 20 ns is used for free energy calculations. The simulations of the charged molecules, such as quat, are performed in absence of their counter-ion. It is found that in the presence of a counter-ion, the free energy profile obtained from bringing the quat molecule from water-to-air differs from the profile obtained from bringing the quat molecule from air-to-water [35]. This hysteresis in the free energy profiles does not occur when the simulations are performed in the absence of the counter-ion, that is, in the charged system [35]. To simulate a charged system, a compensating background charge is intrinsically applied by the MD simulation package, LAMMPS so that the system energy remains finite.

Figure 4.14a shows the free energy profiles of quat-10, quat-16, and amine-10 molecules along with those of the alkanes C-10 and C-16 [35,42]. The inhibitors and the alkanes have the free energy minimum at the air–water interface. The free energies of cationic quat molecules are smaller in the water than in air, whereas those of the amine molecule and alkanes are smaller in the air than in water. This suggests that hydration of the cationic inhibitor molecules is favorable, but that of charge-neutral molecules is unfavorable. It is observed that the free energy profiles of the quat

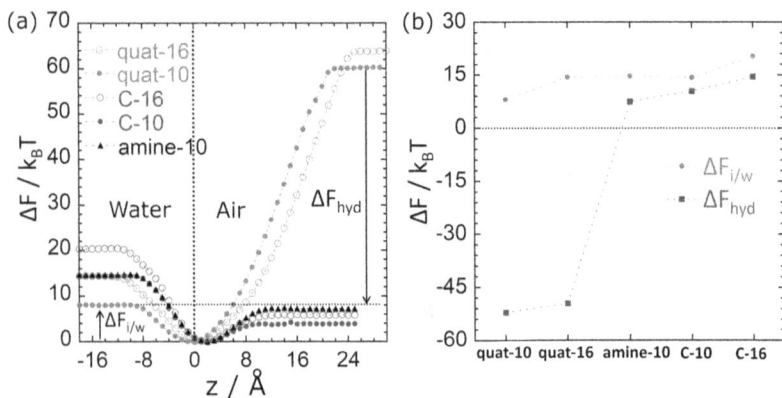

FIGURE 4.14 (a) Free energy profiles of quat-16, quat-10, C-16, C-10, and amine-10 molecules across the air–water interface. The air–water interface is located at $z = 0$ Å. The interface is considered as the position where the average density of water is 0.5 g/cc.[35] The region $z < 0$ Å represents water, and the region $z > 0$ Å represents air. The hydration free energy, ΔF_{hyd}, and the transfer free energy, $\Delta F_{i/w}$, of quat-10 molecule are indicated by the arrows. (b) The ΔF_{hyd} and $\Delta F_{i/w}$ values of the inhibitor and alkane molecules obtained from (a). All the molecules have the free energy minimum at the interface. The cationic quat molecules have a favorable ΔF_{hyd}, and the uncharged molecules have an unfavorable ΔF_{hyd}. (Adapted with permission from Singh and Sharma [35]. Copyright (2022), American Chemical Society.)

molecules rise sharply beyond as a convex function, that is, $\Delta F''(z) > 0 \forall (z) > 0 \, \text{Å}$. This increase in the free energy is not seen in case of the charge-neutral amine molecule or the alkanes. The reason is that in the air, the charged head group of the quat molecules points downward and tries to remain solvated in the water. This pulls up a water finger above the air–water interface, thereby increasing the interfacial area. The sharp rise in the free energy profiles is because of this increase in the interfacial energy. The snapshots shown in Figure 4.15a–c illustrate the increase in the interfacial area when the quat-16 molecule goes beyond $z > 12.5 \, \text{Å}$. The water finger grows until $z \approx 24 \, \text{Å}$, beyond which it breaks (Figure 4.15d). Correspondingly, the free energy also increases until $z \approx 24 \, \text{Å}$ and then stops increasing beyond $z > 24 \, \text{Å}$. The amine and the alkane molecules do not have a charged head group and therefore no water finger is formed when they are moved up in the air.

Figure 4.14b shows the free energy of hydration (termed as ΔF_{hyd}) and the free energy of transfer from the interface to water (termed as $\Delta F_{i/w}$). The ΔF_{hyd} for quat-10 and quat-16 are $-52.1 \, k_B T$ and $-49.5 \, k_B T$, respectively, and for C-10 and C-16 are $10.4 \, k_B T$ and $14.5 \, k_B T$, respectively. From these values, it is understood that the quat head group contributes to decreasing the hydration free energy by $-63 \, k_B T$. Similarly, from the $\Delta F_{i/w}$ values, it can be noted that the quat head group aids in reducing the transfer free energies by $6 \, k_B T$. The $\Delta F_{i/w}$ for the C-10 and amine-10 molecules are the same, indicating that the uncharged amine group plays no effect in the reduction of transfer free energy.

The comparison of the hydration free energies of the various inhibitor molecules reveals that the polar group plays an important role in enhancing the solvation tendency of the molecules. An attempt is made to enhance aqueous solubility of inhibitor molecules by adding a polar moiety to terminal position of hydrophobic length of the molecules. The idea is that a small polar hydroxyl group would enhance the water solubility of the molecules, while the rest of the hydrophobic alkyl groups would retain

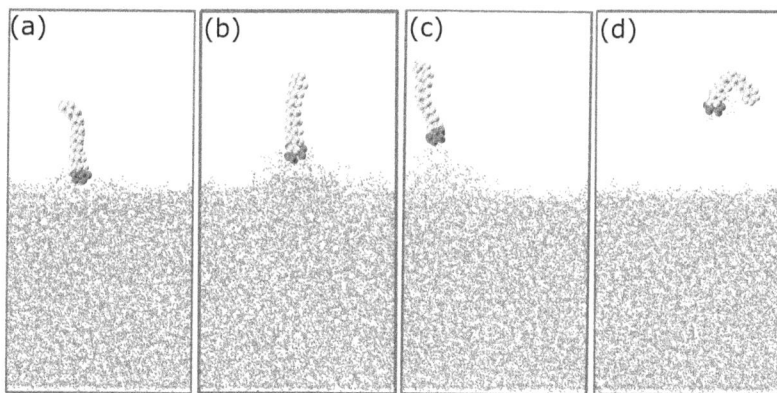

FIGURE 4.15 Snapshots of the quat-16 molecule at different locations from the air–water interface: (a) $z = 12.5 \, \text{Å}$, (b) $z = 16 \, \text{Å}$, (c) $z = 24 \, \text{Å}$, and (d) $z = 26 \, \text{Å}$. The formation of the water finger is apparent in between $12.5 \, \text{Å} \lesssim z \lesssim 24 \, \text{Å}$, beyond which the water finger breaks and the quat molecule stays in the air with a few water molecules surrounding its polar head group. Counter-ion is not present in the system. (Adapted with permission from Singh and Sharma [35]. Copyright (2022), American Chemical Society.)

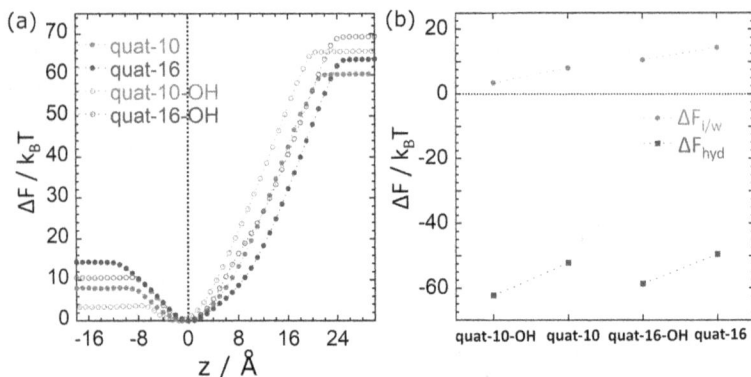

FIGURE 4.16 (a) Free energy profiles of quat-10-OH and quat-16-OH molecules along with those of quat-10 and quat-16 molecules across the air–water interface. (b) The ΔF_{hyd} and $\Delta F_{i/w}$ values of the inhibitor and alkane molecules obtained from (a). A significant decrease in the ΔF_{hyd} and $\Delta F_{i/w}$ values (~10 $k_B T$ and ~4 $k_B T$, respectively) is observed upon introduction of the -OH groups in the quat molecules. (Adapted with permission from Singh and Sharma [35]. Copyright (2022), American Chemical Society.)

the required hydrophobicity for corrosion inhibition. Figure 4.1f shows the structure of the quat molecule upon replacing its terminal methyl group with a hydroxyl group (-OH) [41]. The quat molecules with the presence of the -OH groups are referred to as the quat-OH molecules. The free energy profiles of quat-10-OH and quat-16-OH molecules along with those of the quat molecules are shown in Figure 4.16a.

As speculated, a significant reduction in the ΔF_{hyd} and $\Delta F_{i/w}$ values are observed (Figure 4.16b). The values of quat-10 and quat-16 molecules are reduced by 10.2 $k_B T$ and 9.2 $k_B T$, respectively, upon introduction of the -OH groups. An appreciable drop of 4.6 $k_B T$ and 3.8 $k_B T$ in the $\Delta F_{i/w}$ values is also observed.

We have previously seen that the micellization tendency of the inhibitor molecules is not desired. The quat molecules experience a long-range free energy barrier to adsorption on the metal surface. This is because in the bulk aqueous phase, they are surrounded by a corona of counter-ions and water molecules. The solvation shell poses a barrier for the micelles to adsorb onto the metal surface. The unaggregated quat molecules, however, adsorb without overcoming any free energy barrier. For corrosion protection, it is beneficial if the micellization of the charged inhibitor molecules can be prevented. Though we have seen before that the small cationic molecules such as bda-4 do not form micelles [27,32], they show a poor adsorption tendency on the metal surface. The newly modeled quat-10-OH molecules, which retain a long hydrophobic length, are therefore examined for their micellization tendency.

4.8 MICELLIZATION TENDENCY OF QUAT-10-OH MOLECULES

An MD simulation is performed in which the quat-10-OH molecules are initially placed close to each other. Figure 4.17a shows the initial configuration of the 18 number of quat-10-OH molecules in a three-dimensional periodic bulk aqueous system.

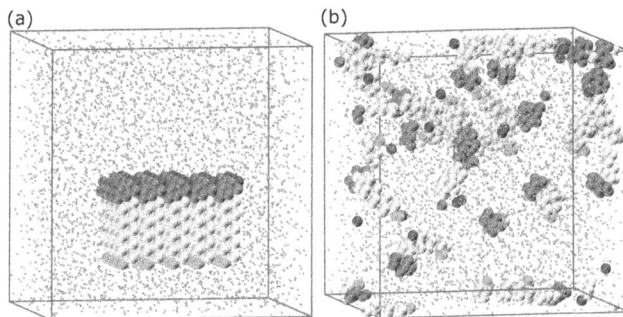

FIGURE 4.17 (a) Initial configuration of the 18 quat-10-OH molecules placed close to each other. The red beads represent the polar head groups, the yellow beads represent the hydrophobic alkyl tails, the green beads refer to the -OH groups present at the terminal position of the alkyl tails, the blue beads represent the bromide ions, and the cyan dots represent the water molecules. (b) Configuration of the quat-10-OH molecules after an MD simulation of 50 ns. The quat-10-OH molecules disintegrate and do not prefer to aggregate any further. (Adapted with permission from Singh and Sharma [35]. Copyright (2022), American Chemical Society.)

The counter-ions (bromides) are also placed closed to the charged head groups to minimize the charge accumulation. The simulation is then performed for 50 ns in the isobaric–isothermal ensemble (fixed number of species N, temperature $T = 300\,K$, and pressure $P = 1\,atm$). Figure 4.17b shows the final configuration of the system after 50 ns. It is apparent that the closely packed molecules have disintegrated and no longer remain in the aggregated state. The kinetics of disintegration of the aggregate in terms of the *aggregation number* is shown in Figure 4.18a. Aggregation number is defined as the number of molecules present in the largest aggregate in the system. At $t = 0\,ns$, the aggregation number is 18 because initially, the largest aggregate is comprised of 18 molecules. In a duration of 10 ns, the aggregate has completely disintegrated resulting in an aggregation number of ~2. No further increase in the aggregation number is observed for the next 40 ns. This confirms that the quat-10-OH molecules do not aggregate and form micelles as the quat-10 molecules do [33]. It is worth mentioning here that the quat-10-OH molecules do not stay in an aggregate even when they are closely placed together. The quat-10 molecules, on the other hand, quickly aggregate to form micelles even when they are sparsely placed in the aqueous medium [33,34].

The reason why the quat-10-OH molecules are unstable in an aggregated state is that the -OH groups, being polar, tend to remain solvated with the water molecules. Figure 4.18b shows the radial distribution function (RDF) between the -OH groups and the water oxygens. A large peak at $r = 2.6\,Å$ indicates that the -OH groups possibly form hydrogen bonds with the surrounding water molecules. The RDF between the nitrogen of the head groups and water molecules is also shown for comparison. A smaller peak as compared to that of the -OH groups further proves that the hydroxyl groups prefer to remain highly solvated, because of which the quat-10-OH molecules do not form micelles.

(a)

(b)

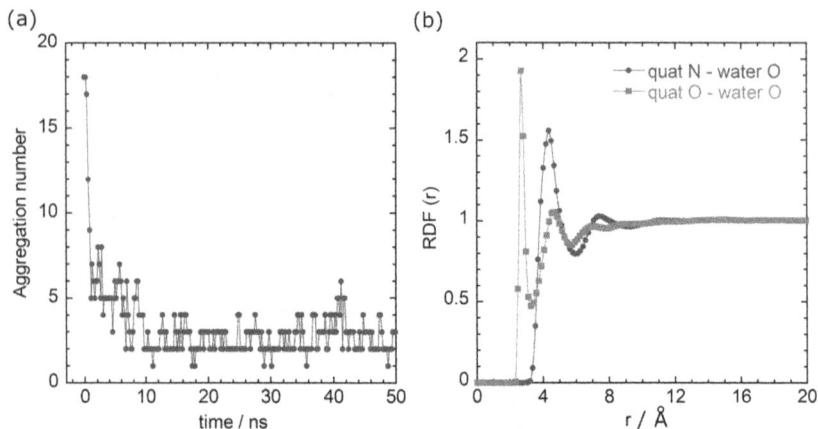

FIGURE 4.18   (a) Aggregation number of the quat-10-OH molecules as a function of time. Aggregation number of the largest aggregate is calculated by finding the size of all the aggregates of the molecules using the algorithm of density-based spatial clustering of applications with noise (DBSCAN). The drop in the aggregation number from 18 to ~2 in 10 ns indicates that the quat-10-OH molecules do not prefer to stay in an aggregate. (b) Radial distribution function (RDF) of the -OH groups and nitrogen of the head groups with respect to water oxygens. A large peak at $r = 2.6$ Å shows that the -OH groups are highly solvated in water. (Adapted with permission from Singh and Sharma [35]. Copyright (2022), American Chemical Society.)

4.9   CONCLUSION

In this work, the behavior of corrosion inhibitor molecules near metal–water and air–water interfaces is studied via advanced MD simulations. In infinite dilution, the inhibitor molecules adsorb strongly to the metal surface without experiencing any free energy barrier. The micelles of cationic inhibitors, however, experience a long-range free energy barrier to adsorption on the metal surface. The micelles are stable in the adsorbed state rather than in the bulk aqueous phase. It takes ~200 ns for a micelle to completely disintegrate on the metal surface. Upon disintegration, the constituent molecules of the micelles lie flat on the surface. A new free energy methodology is developed to determine the equilibrium adsorption morphologies of the inhibitors on metal surface. It is found that the hydrophobic interactions promote a dense adsorption morphology, whereas the accumulation of the charged head groups opposes aggregation of the molecules on the metal surface. Decanethiol, being a charge-neutral inhibitor molecule with a long alkyl tail, displays a closely packed dense adsorption morphology. The cationic bda-12 molecules show a better surface coverage than bda-4 molecules. The mixture of cationic bda-12 and anionic pe-12 molecules shows a synergistic effect in adsorption. The free energies of hydration reveal that the addition of a hydroxyl group to the terminal position of the alkyl tails significantly enhances the aqueous solubility of the inhibitor molecules. The presence of the hydroxyl group also reduces the micellization tendency of the quat-10 molecules, which is advantageous for barrier-less adsorption of the quat-10-OH molecules onto the metal surface.

ACKNOWLEDGMENTS

This work is supported by the NSF CAREER grant 2046095. The authors thank the researchers at the Institute for Corrosion and Multiphase Technology (ICMT), Ohio University, for useful discussions. This research used resources of the Oak Ridge Leadership Computing Facility at the Oak Ridge National Laboratory, which is supported by the Office of Science of the U.S. Department of Energy under Contract No. DE-AC05-00OR22725. Computational resources for this work were also provided by the Ohio Supercomputer Center (project number PAA0031), and National Science Foundation XSEDE grant number DMR190005.

REFERENCES

[1] Regulator, A. E., *Report 2013-B: Pipeline Performance in Alberta, 1990–2012.* Basel: Alberta Energy Regulator, **2013**.

[2] Malik, M. A.; Hashim, M. A.; Nabi, F.; Al-Thabaiti, S. A.; Khan, Z., Anti-Corrosion Ability of Surfactants: A Review. *Int. J. Electrochem. Sci* **2011**, *6* (6), 1927–1948.

[3] Caruso, F.; Serizawa, T.; Furlong, D. N.; Okahata, Y., Quartz Crystal Microbalance and Surface Plasmon Resonance Study of Surfactant Adsorption onto Gold and Chromium Oxide Surfaces. *Langmuir* **1995**, *11* (5), 1546–1552.

[4] Stålgren, J.; Eriksson, J.; Boschkova, K., A Comparative Study of Surfactant Adsorption on Model Surfaces Using the Quartz Crystal Microbalance and the Ellipsometer. *J. Colloid Interf. Sci.* **2002**, *253* (1), 190–195.

[5] Xiong, Y.; Brown, B.; Kinsella, B.; Nešić, S.; Pailleret, A., Atomic Force Microscopy Study of the Adsorption of Surfactant Corrosion Inhibitor Films. *Corrosion* **2014**, *70* (3), 247–260.

[6] Subramanian, V.; Ducker, W. A., Counterion Effects on Adsorbed Micellar Shape: Experimental Study of the Role of Polarizability and Charge. *Langmuir* **2000**, *16* (10), 4447–4454.

[7] Jaschke, M.; Butt, H.-J.; Gaub, H.; Manne, S., Surfactant Aggregates at a Metal Surface. *Langmuir* **1997**, *13* (6), 1381–1384.

[8] Sharma, B.; Basu, S.; Sharma, M., Characterization of Adsorbed Ionic Surfactants on a Mica Substrate. *Langmuir* **1996**, *12* (26), 6506–6512.

[9] Zhang, R.; Somasundaran, P., Advances in Adsorption of Surfactants and Their Mixtures at Solid/Solution Interfaces. *Adv. Colloid Interface Sci.* **2006**, *123*, 213–229.

[10] Tariq, M.; Serro, A.; Colaço, R.; Saramago, B.; Lopes, J. C.; Rebelo, L. P. N., Effect of Alkyl Chain Length on the Adsorption and Frictional Behaviour of 1-Alkyl-3-Methylimidazolium Chloride Ionic Liquid Surfactants on Gold Surfaces. *Colloid Surface A* **2011**, *377* (1–3), 361–366.

[11] Sharma, S.; Ko, X.; Kurapati, Y.; Singh, H.; Nešić, S., Adsorption Behavior of Organic Corrosion Inhibitors on Metal Surfaces—Some New Insights From Molecular Simulations. *Corrosion* **2019**, *75* (1), 90–105.

[12] Jiménez-Ángeles, F.; Khoshnood, A.; Firoozabadi, A., Molecular Dynamics Simulation of the Adsorption and Aggregation of Ionic Surfactants at Liquid–Solid Interfaces. *J. Phys. Chem. C* **2017**, *121* (46), 25908–25920.

[13] Sharma, S.; Singh, H.; Ko, X., A Quantitatively Accurate Theory to Predict Adsorbed Configurations of Linear Surfactants on Polar Surfaces. *J. Phys. Chem. B* **2019**, *123* (34), 7464–7470.

[14] Ko, X.; Sharma, S., Adsorption and Self-Assembly of Surfactants on Metal–Water Interfaces. *J. Phys. Chem. B* **2017**, *121* (45), 10364–10370.

[15] Sudarsan, V., Materials for Hostile Chemical Environments. In A.K. Tyagi and S. Banerjee (Eds.), *Materials under Extreme Conditions*. Amsterdam: Elsevier, **2017**, pp. 129–158.

[16] Hegazy, M.; Abdallah, M.; Awad, M.; Rezk, M., Three Novel Di-Quaternary Ammonium Salts as Corrosion Inhibitors for API X65 Steel Pipeline in Acidic Solution. Part I: Experimental Results. *Corros. Sci.* **2014**, *81*, 54–64.

[17] Jovancicevic, V.; Ahn, Y. S.; Dougherty, J. A.; Alink, B., CO_2 Corrosion Inhibition by Sulfur-Containing Organic Compounds. In *Corrosion 2000* (pp. 1–18). Houston: OnePetro, **2000**.

[18] Ren, S.; He, Y.; Belarbi, Z.; Wang, X.; Young, D.; Singer, M.; Mohamed-Saïd, M.; Camperos, S., Methodology for Corrosion Inhibitor Characterization Applied to Phosphate Ester and Tetrahydropyrimidinium Model Compounds. In *Corrosion 2021* (pp. 1–17). Houston: OnePetro, **2021**.

[19] Wang, J.; Wolf, R. M.; Caldwell, J. W.; Kollman, P. A.; Case, D. A., Development and Testing of a General Amber Force Field. *J. Comput. Chem.* **2004**, *25* (9), 1157–1174.

[20] Frisch, M.; Trucks, G.; Schlegel, H.; Scuseria, G.; Robb, M.; Cheeseman, J.; Scalmani, G.; Barone, V.; Petersson, G.; Nakatsuji, H., *Gaussian 16 Revision C. 01. 2016*. Vol. 421, Gaussian Inc., Wallingford, CT, **2016**.

[21] Berendsen, H.; Grigera, J.; Straatsma, T., The Missing Term in Effective Pair Potentials. *J. Phys. Chem-US* **1987**, *91* (24), 6269–6271.

[22] Joung, I. S.; Cheatham III, T. E., Determination of Alkali and Halide Monovalent Ion Parameters for Use in Explicitly Solvated Biomolecular Simulations. *J. Phys. Chem. B* **2008**, *112* (30), 9020–9041.

[23] Heinz, H.; Vaia, R.; Farmer, B.; Naik, R., Accurate Simulation of Surfaces and Interfaces of Face-Centered Cubic Metals Using 12– 6 and 9– 6 Lennard-Jones Potentials. *J. Phys. Chem. C* **2008**, *112* (44), 17281–17290.

[24] Coppage, R.; Slocik, J. M.; Ramezani-Dakhel, H.; Bedford, N. M.; Heinz, H.; Naik, R. R.; Knecht, M. R., Exploiting Localized Surface Binding Effects to Enhance the Catalytic Reactivity of Peptide-Capped Nanoparticles. *J. Am. Chem. Soc.* **2013**, *135* (30), 11048–11054.

[25] Patel, A. J.; Varilly, P.; Chandler, D., Fluctuations of Water Near Extended Hydrophobic and Hydrophilic Surfaces. *J. Phys. Chem. B* **2010**, *114* (4), 1632–1637.

[26] Plimpton, S., Fast Parallel Algorithms for Short-Range Molecular Dynamics. *J. Comput. Phys.* **1995**, *117* (1), 1–19.

[27] Singh, H.; Sharma, S., Determination of Equilibrium Adsorbed Morphologies of Surfactants at Metal-Water Interfaces Using a Modified Umbrella Sampling-Based Methodology. *J. Chem. Theory Comput.* **2022**, *18* (4), 2513–2520.

[28] Ding, Y.; Brown, B.; Young, D.; Singer, M. The Effect of Temperature and Critical Micelle Concentrations (CMC) on the Inhibition Performance of a Quaternary Ammonium-Type Corrosion Inhibitor. In *Corrosion 2020* (pp. 1–14). Houston: OnePetro, **2020**.

[29] Torrie, G. M.; Valleau, J. P., Nonphysical Sampling Distributions in Monte Carlo Free-Energy Estimation: Umbrella Sampling. *J. Comput. Phys.* **1977**, *23* (2), 187–199.

[30] Kumar, S.; Rosenberg, J. M.; Bouzida, D.; Swendsen, R. H.; Kollman, P. A., The Weighted Histogram Analysis Method for Free-Energy Calculations on Biomolecules. I. The Method. *J. Comput. Chem.* **1992**, *13* (8), 1011–1021.

[31] Fiorin, G.; Klein, M. L.; Hénin, J., Using Collective Variables to Drive Molecular Dynamics Simulations. *Mol. Phys.* **2013**, *111* (22–23), 3345–3362.

[32] Singh, H.; Sharma, S. Understanding the Adsorption Morphologies of Quaternary Ammonium-Based Corrosion Inhibitors at Metal-Water Interfaces via Molecular Simulations. In *Corrosion 2021* (pp. 1–8). Houston: OnePetro, **2021**.

[33] Singh, H.; Sharma, S., Disintegration of Surfactant Micelles at Metal–Water Interfaces Promotes Their Strong Adsorption. *J. Phys. Chem. B* **2020**, *124* (11), 2262–2267.

[34] Singh, H.; Sharma, S., Free Energy Profiles of Adsorption of Surfactant Micelles at Metal-Water Interfaces. *Mol. Simulat.* **2021**, *47* (5), 420–427.

[35] Singh, H.; Sharma, S. Understanding the Hydration Thermodynamics of Cationic Quaternary Ammonium and Charge-Neutral Amine Surfactants. *J Phys Chem B.* **2022**, *126* (47), 9810–9820. doi: 10.1021/acs.jpcb.2c03562. Epub 2022 Nov 17. PMID: 36395484.

[36] Singh, H.; Kurapati, Y.; Sharma, S. Aggregation and Adsorption Behavior of Organic Corrosion Inhibitors Studied Using Molecular Simulations. In *Corrosion 2019* (pp. 1–12). Houston: OnePetro, **2019**.

[37] Khan, M. R.; Singh, H.; Sharma, S.; Asetre Cimatu, K. L., Direct Observation of Adsorption Morphologies of Cationic Surfactants at the Gold Metal–Liquid Interface. *J. Phys. Chem. Lett.* **2020**, *11* (22), 9901–9906.

[38] Olivo, J.; Young, D.; Brown, B.; Nesic, S. Effect of Corrosion Inhibitor Alkyl Tail Length on the Electrochemical Process Underlying CO_2 Corrosion of Mild Steel. In *Corrosion 2018* (pp. 1–15). Houston: OnePetro, **2018**.

[39] Moradighadi, N.; Lewis, S.; Olivo, J. D.; Young, D.; Brown, B.; Nešić, S., Determining Critical Micelle Concentration of Organic Corrosion Inhibitors and its Effectiveness in Corrosion Mitigation. *Corrosion* **2021**, *77* (3), 266–275.

[40] Belarbi, Z.; Dominguez Olivo, J.; Farelas, F.; Singer, M.; Young, D.; Nešić, S., Decanethiol as a Corrosion Inhibitor for Carbon Steels Exposed to Aqueous CO_2. *Corrosion* **2019**, *75* (10), 1246–1254.

[41] Singh, H.; Sharma, S. Designing Corrosion Inhibitors with High Aqueous Solubility and Low Tendency Towards Micellization: A Molecular Dynamics Study. In *Corrosion 2020* (pp. 1–11). Houston: OnePetro, **2020**.

[42] Singh, H.; Sharma, S., Hydration of Linear Alkanes is Governed by the Small Length-Scale Hydrophobic Effect. *J. Chem. Theory Comput.* **2022**, *18* (6), 3805–3813.

5 Application of Modern Functional Materials in Petroleum Exploration and Process Development

Ganesh Kumar, Yogendra Kumar, and Jitendra S. Sangwai
Indian Institute of Technology Madras

Deepak Dwivedi
Rajiv Gandhi Institute of Petroleum Technology

CONTENTS

ABBREVIATIONS

EIA   Energy Information Administration
EOR   Enhanced oil recovery
IFT   Interfacial tension

DOI: 10.1201/9781003242550-5

5.1 INTRODUCTION

Energy Information Administration (EIA) projects the global energy demand to rise by 50% by 2050, owing to economic expansion, population growth, and industrial advancements, particularly in developing countries [1]. In order to meet this global energy demand, crude oil is currently the most frequently used energy source, accounting for 33% of global consumption [2]. However, many of the oilfields have already matured globally and reached a plateau period, resulting in a decline in oil production [3]. The inefficiency in discovering new oilfields maximizes oil recovery from matured, already existing, and decommissioned oilfields. After primary and secondary oil recovery, a significant quantity of crude oil is still trapped inside the reservoir [4]. In order to recover a fraction of this trapped oil, the enhanced oil recovery (EOR) technique is used. The EOR techniques include thermal (hot water flooding, steam flooding, and in situ combustion), miscible gas injection (CO_2), microbial, and chemical (alkaline, surfactant, and polymer) EOR [5]. New EOR techniques for extracting residual crude oil trapped in reservoirs must be developed economically to meet the energy demand for the next several decades. On the contrary, nanoparticles have demonstrated their potential in biofuel, corrosion resistance, bio-surfactant, mass transfer, and catalytic applications [6–8]. Recently, nanotechnology has been proposed as one of the most promising EOR technologies because it can easily enter the pore channels and alter reservoir properties, improving oil recovery. The pictorial depiction of oil recovery from pores of oil-bearing sediments using nanoparticles and their proposed EOR mechanisms is given in Figure 5.1.

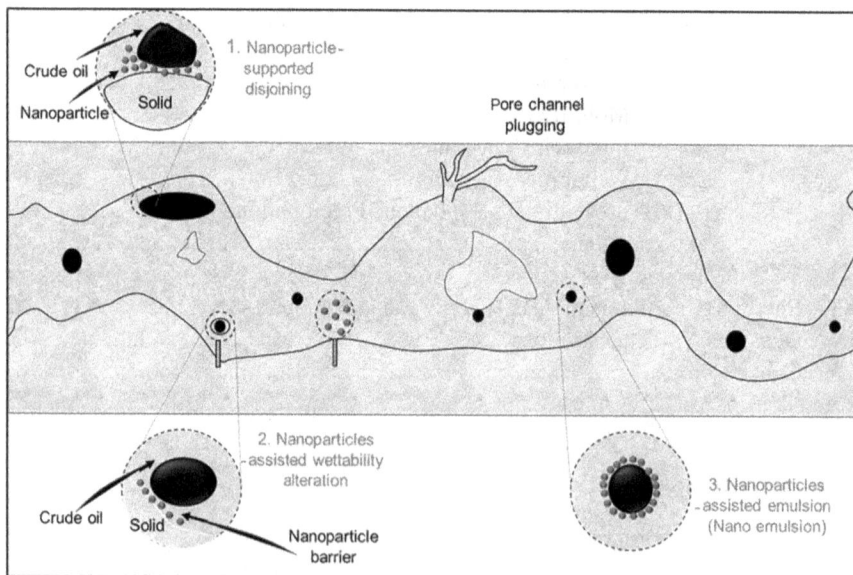

FIGURE 5.1 Nanoparticle-assisted mechanism of enhanced oil recovery in oil-bearing sediments.

Nanofluid-enabled wettability alteration, nanoemulsion formation, and disjoining pressure contribute effectively to EOR, which are discussed in the later sections of this chapter.

The application of modern functional materials in petroleum exploration and development is expanding. It expands from the surface of the land to the depths of the ocean, from high- to low-grade petroleum resources, and from partial structural traps to the whole basin [9]. Nanotechnology has recently gotten much interest in the petroleum and petrochemical sectors, especially in liquid–liquid separation, EOR, drilling fluids, process intensification, and chemical kinetics [10,11]. Although nanoparticle applications in the oil and gas industries are still in their early stages, they have already demonstrated significant potential as new materials, such as additives for drilling fluid and cementing material, completion fluids, and injection fluids. Nanoparticles that are smaller than the pore throat diameter might quickly pass through rock pore channels and access the trapped crude oil [12]. Moreover, nanoparticles with appropriate wettability can be utilized as an alternate stabilizer to surfactant for forming the Pickering emulsions. When nanoparticles reach the crude oil bank, they are irreversibly adsorbed at the water–oil interface, resulting in the formation of a very stable emulsion [13–15]. Nanoparticles are proposed for use as a novel emulsifier in oil reservoirs because they may be able to displace more leftover oil and decrease the surfactant adsorption on the reservoir rock surface, which led to improved oil recovery [16–18]. However, the biggest challenge for this application is the adverse operating environment found in oil reservoirs, which includes high temperatures and salinity.

5.2   EOR MECHANISMS

Understanding the mechanics behind the utilization of nanoparticles is crucial for their potential application as novel EOR agents. Various EOR mechanisms have been uncovered by recent research on the interactions between nanofluids, oil, and rocks. The EOR mechanisms of nanoparticles, on the other hand, still lack a complete understanding. Nanoparticles can be applied in EOR processes in three ways: nanofluids, nanocatalysts, and nanoemulsions. The EOR mechanisms are discussed below.

5.2.1   Nanofluids

The term nanofluid is used to refer to a base fluid containing nanoparticles in a colloidal suspension. The main nanofluid EOR mechanisms include pore channel plugging, disjoining pressure, interfacial tension (IFT) reduction, mobility ratio, wettability alteration, and asphaltene precipitation prevention.

5.2.1.1   Pore Channel Plugging

The most common causes of plugging the pore channels are log jamming and mechanical entrapment. It occurs when injected particles have a larger diameter than the pore channels through which they flow, resulting in mechanical entrapment. However, the size of pore channels is generally, on a microscale, thousands of times larger than nanoparticles. As a result, nanoparticles are capable of penetrating pore channels and passing through the channels without being mechanically entrapped [5]. However, metal nanoparticles of large size might block reservoir pore channels [19]. These nanoparticles must be evaluated before being injected into reservoirs to make sure that improving oil recovery is feasible.

Log jamming is another cause of pore channel plugging. It occurs due to the blockage of pore channels whose size is larger than each nanoparticle. As nanofluids move from pore to throat, their velocity increases because of differential pressure and the narrowing of the flow area. As a result, smaller water molecules move more quickly compared to nanoparticles, which causes the nanoparticles to aggregate near the entry of the pore throat. However, in some circumstances, pore throat plugging brought on by the log jamming helps enhance the effectiveness of nanofluid flooding. As nanoparticles aggregate, they block the pore throat, which raises the pressure in the neighboring pores and releases the trapped oil. Following the release of trapped oil, the surrounding pressure decreases, the blockage gradually disappears, and nanoparticles begin to move with water [5]. This phenomenon is known as temporary log jamming.

5.2.1.2   Disjoining Pressure

As nanoparticles in the nanofluid self-assemble in a structured and ordered manner, a wedge-shaped film is formed between the oil droplet and the rock surface. The film detaches the oil droplets from the rock's surface, increasing the amount of oil that can be retrieved. The film is generated as a result of pressure known as structural disjoining pressure (Figure 5.2). The structural disjoining pressure generation is osmotic. This pressure is primarily driven by electrostatic repulsion and Brownian motion between nanoparticles [5]. An imbalance in the interfacial forces between the rock, aqueous, and oil phases is responsible for the pressure due to the ability of the nanofluid to spread across the rock's surface [20]. The force behind the spreading of nanoparticles is the structural disjoining pressure gradient ($\Delta\Upsilon$), also known as the film tension gradient in the wedge direction from the bulk fluid [21]. The film tension gradient is larger at the wedge's vertex because of the nanoparticles' structuring inside the wedge-shaped confinement (Figure 5.2). Nanofluids spread at the tip of the wedge when the film tension rises toward the wedge's vertex, enhancing the dynamic spreading characteristic of nanofluids. Furthermore, when the film thickness reduces, the spreading coefficient rises exponentially [22]. The structural disjoining pressure rises when the film thickness decreases close to the wedge's vertex, improving oil recovery.

5.2.1.3   Mobility Ratio

Mobility ratio can be defined as the displacing fluid (injection fluid) mobility divided by the displaced fluid (oil) mobility. If the displacing fluid mobility is high, a viscous fingering effect occurs, which results in poor sweep efficiency. Hence, the mobility of the displacing fluid must be controlled during the EOR process to achieve better sweep efficiency. It can be expressed by Eq. (5.1):

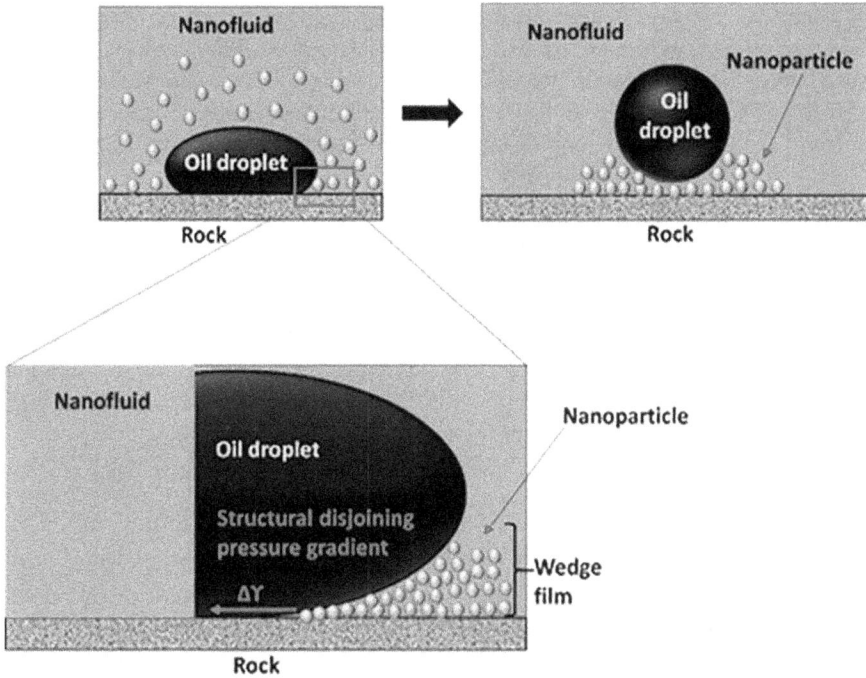

FIGURE 5.2 Nanoparticles in the nanofluid self-assemble in a structured and ordered manner, forming a wedge film resulting in the structural disjoining pressure.

$$M = \frac{\lambda_{displacing\ fluid}}{\lambda_{displaced\ fluid}} = \frac{\lambda_i}{\lambda_o} = \frac{\dfrac{k_{ri}}{\mu_i}}{\dfrac{k_{ro}}{\mu_o}} = \frac{k_{ri}\ \mu_o}{k_{ro}\ \mu_i} \tag{5.1}$$

where M represents the mobility ratio; λ_i and λ_o represent the mobility of the displacing (injection) fluids and displaced fluid (oil), respectively; k_{ri} and k_{ro} represent the relative permeability of the injection fluids and oil, respectively; and μ_i and μ_o represent the viscosity of the injection fluids and oil, respectively. When ($M < 1$), it indicates a favorable mobility ratio, which means the mobility of the displaced (oil) fluid is higher than the displacing (injection) fluid. The opposite is true for an unfavorable mobility ratio ($M > 1$). A favorable mobility ratio means the displaced fluid (oil) can flow through the reservoir rock more quickly than the displacing (injection) fluid.

5.2.1.4 Interfacial Tension Reduction

IFT is a key factor in determining the distribution and movement of fluids across the porous media for EOR. IFT reduction eases oil extraction from the reservoir pores by lowering capillary pressure in the pore channel of the reservoir. If the IFT between the water–oil interface is high, the immiscibility of the two fluids prevents the oil displacement from the reservoir. IFT reduction between the fluid phases is observed when hydrophilic nanoparticles are added to the pore system, which helps enhance oil recovery [23]. The IFT value gives a clue about the miscibility of reservoir fluid; the lower the IFT value, the higher the fluid's miscibility.

Similar to surfactants, nanoparticles can be positioned at the interface between oil and water phases to reduce IFT. Surfactants get adsorbed at oil–water interfaces because of their amphiphilic nature. However, neutral-wet nanoparticles get adsorbed at the water–oil interface as they need less energy in that state [24]. In these circumstances, nanoparticles help modify the free surface energy by eliminating the interface between two fluids and substituting it with the fluid particle interface. To lower the energy, either the IFT must drop (adding surfactant) or the area (integration area) must decrease (adding nanoparticles to the fluid system). Hence, the addition of nanoparticles in a fluid system helps reduce its IFT.

5.2.1.5 Wettability Alteration

Wettability alteration of the reservoir rock's surface is a key factor during oil recovery. It influences the relative permeability, fluid saturation, and capillary pressure during EOR [24]. Regardless of the geological composition, most oil reservoirs are typically mixed wet in nature, which means they are neither totally water-wet nor entirely oil-wet. During EOR, shifting the reservoir's wettability to water-wet is usually desirable since it improves water imbibition [25]. The most popular method for studying rock wettability is contact angle measurement. Contact angles are commonly measured from the denser phase during three-phase contact angle measurements. Anderson states that a solid substrate is water-wet if the contact angle (θ) is ($0 \leq \theta < 75°$), intermediate-wet when θ is ($75° \leq \theta \leq 105°$), and oil-wet when θ is ($105° < \theta \leq 180°$), as shown in Figure 5.3 [26]. As the reservoir rock's wettability shifts toward water-wet, the adhesive force that holds oil on the rock surface decreases, allowing crude oil to flow more readily and enhancing oil recovery [27]. Recent research studies have demonstrated that dispersing nanoparticles in different fluids may modify the contact angle, shifting the reservoir rock's wettability toward strongly water-wet conditions.

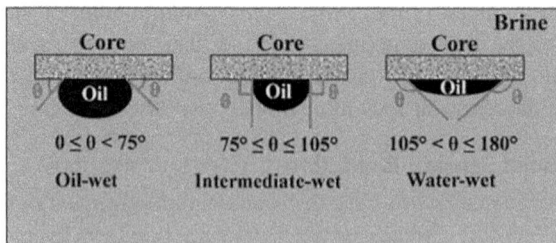

FIGURE 5.3 Schematic diagram depicting the wettability conditions of rock in a core substrate-oil–brine/nanofluid system.

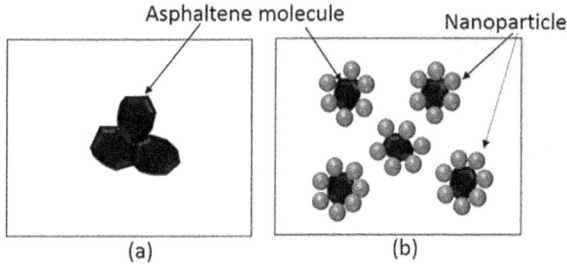

FIGURE 5.4 (a) Flocculation of asphaltene without nanoparticles and (b) prevention of asphaltene precipitation with nanoparticles.

5.2.1.6 Prevention of Asphaltene Precipitation

Asphaltene precipitation was observed in some EOR methods (e.g., CO_2 flooding) because of reservoir conditions that cause a change in the wettability, reduction in formation permeability, and block transportation pipes [28]. Hence, it is important to avoid asphaltene precipitation during the EOR to improve oil recovery. One of the prominent solutions to prevent asphaltene precipitation is by using nanoparticles. Figure 5.4a and b shows that nanoparticles get adsorbed on the asphaltene surface, which greatly reduces asphaltene flocculation in reservoir pores [29].

5.2.2 NANOEMULSIONS

The emulsion stabilized by nanoparticles with droplet diameter less than 100 nm is commonly known as nanoemulsions. These emulsions prove to be more effective at overcoming challenges compared to other conventional emulsions. These emulsions remain stable and can withstand harsh reservoir conditions compared to conventional microemulsions [30–32]. Additionally, these emulsions possess better viscosity and can reduce the viscous fingering effect by controlling the mobility ratio during EOR applications [5]. Therefore, during EOR, nanoemulsions can be used rather than relatively large polymers that retain and get adsorbed on the reservoir rock, which may block and damage the formation [27,33,34]. Moreover, nanoemulsion droplets are smaller than the diameter of pore throats, allowing them to easily pass through with less retention, resulting in higher oil recovery [35].

5.2.3 NANOCATALYSTS

Nanocatalysts are nano-sized metal particles that act as a catalyst during the steam injection process for heavy oils. Nanocatalysts offer several advantages over traditional catalysts. Using nanocatalysts, bitumen can be converted into lighter products via aquathermolysis for upgrading heavy oil [36]. A key benefit of aquathermolysis is that it reduces the viscosity of heavy oils [37]. The primary explanation for this phenomenon is that nanocatalysts exhibit greater reactivity than micron-sized catalysts due to their increased specific surface area. This rise of nanocatalysts' surface area leads to greater oil contact with the catalysts and improves their interaction. The utilization of nanocatalysts (e.g., iron or nickel) facilitates the breaking of carbon–sulfur bonds within asphaltenes, resulting in an increase in aromatic and saturated heavy oil components [38]. It may be expressed as:

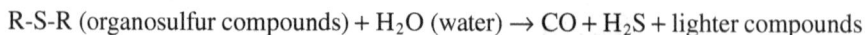

R-S-R (organosulfur compounds) + H_2O (water) $\rightarrow$ CO + H_2S + lighter compounds

As carbon monoxide is produced, it interacts with water to form hydrogen during the water–gas shift reaction process. These processes take place at temperatures between 200°C and 300°C. These hydrogen molecules attack the oil's unsaturated and unstable molecules to create lighter, saturated molecules through hydrogenolysis [39].

5.3 ENERGY EXTRACTION FROM GAS HYDRATE USING NANOPARTICLE

Gas hydrates are considered the most significant energy source available on earth, and nearly 50% of carbon is trapped inside clathrate hydrates. The energy extraction from these hydrate sources is a cumbersome process. Depressurization, thermal stimulation, and additives are pretty popular to extract methane from these hydrates. The use of nanomaterials for the destruction of hydrate cages is the front-runner owing to its higher process efficacy and negligible environmental hazards. In order to destroy gas hydrates efficiently, it is important to use nanoparticles with uniformly arranged spikes, which are spherical in shape [40,41]. For an efficient destruction process of the clathrate cell containing an encaged methane molecule, the length, distance between them, and shape of spherical nanoparticle spikes must be optimal (rectilinear, curved, thickened, etc.). Electrochemistry makes it quite easy to form nanostructures that look like natural biological objects, such as sea urchins. Their primary construction material for innovative structures is polystyrene [40]. It is the polystyrene microsphere that provides a three-dimensional surface for zinc oxide. Spikes extend in all directions from hollow, spherical nanostructures. A hydrodynamic jet of seawater pre-saturated with spherical nanoparticles will gradually destroy the upper layer of gas hydrate accumulations. Spherical particles crash into crystalline cells of gas hydrates, causing them to destroy and release methane molecules. By doing so, methane and its homologs are formed in a solution, which is extracted forcibly on the surface and then by gas lift [42]. Nanostructures such as MOF, ZIF, functionalized silica, and magnetite are common nanoparticles used that have tunable characteristics [43] and can be utilize in methane extraction from hydrate reserves.

5.4 OCEANIC CO_2 SEQUESTRATION

Apart from energy extraction potential as discussed in the previous section, oceans also have a huge scope of CO_2 sequestration in subsea sediments and deep oceans. Anthropogenic CO_2 can be sequestered on the sea floor in hydrate-forming zones [44]. However, diffusion and kinetic barriers make the sequestration process cumbersome. Nanoparticles with tunable characteristics along with additives have immense potential to solve aforesaid issues. Nanoparticle incorporation at the time of gas induction demonstrates positive affection of nucleation kinetics, gas consumption, and reduction of induction time. Sulfate-functionalized, amine-functionalized, and passivated nanoparticles have been well received by the scientific community in the last decade. Any breakthrough in this process will open up new avenues for large-scale CO_2 sequestration in oceanic conditions. Alternatively, direct air capture is another sector where functionalized nanoparticles have immense potential especially for the promotion of capture efficiency [45].

5.5 CONCLUSION

Nanoparticle tunability plays a vital role in EOR, wettability alteration, and IFT reduction. The role of nanoparticle morphology seemingly contributed to the enhancement of effectiveness in EOR, methane extraction, and CO_2 sequestration through hydrates. The enhanced kinetic mechanism in the presence of nanoparticles is owing to Brownian motion, diffusiophoresis, and thermophoresis. The perturbation around nanoparticles enhances the mixing of phases; hence, small changes in morphology have drastic effects on their process efficacy. Functionalized nanoparticles reduce IFT, thus facilitating wettability alteration during oil extraction from reservoirs. Moreover, nanoparticles also reduce the viscosity of crude oil, promoting its flowing characteristics. It can easily pass through the pore channels without being mechanically entrapped. It also helps prevent asphaltene adsorption on the reservoir rock surface and reduces the flocculation of asphaltene in the reservoir pores. Nanoemulsions possess better viscosity and can reduce the viscous fingering effect by controlling the mobility ratio during EOR applications. Nanocatalysts help reduce the viscosity of crude oil by upgrading heavy oil by converting bitumen into lighter products via aquathermolysis reactions.

REFERENCES

[1] IEA. *World Energy Outlook 2021*. Paris: 2021. https://www.iea.org/reports/world-energy-outlook-2021.

[2] Druetta P, Raffa P, Picchioni F. Chemical enhanced oil recovery and the role of chemical product design. *Appl Energy* 2019;252:113480. https://doi.org/10.1016/j.apenergy.2019.113480.

[3] Babadagli T. Development of mature oil fields - A review. *J Pet Sci Eng* 2007;57:221–46. https://doi.org/10.1016/j.petrol.2006.10.006.

[4] Egbogah EO. EOR target oil and techniques of its estimation. *J Pet Sci Eng* 1994;10:337–49. https://doi.org/10.1016/0920-4105(94)90024-8.

[5] Sun X, Zhang Y, Chen G, Gai Z. Application of nanoparticles in enhanced oil recovery: A critical review of recent progress. *Energies* 2017;10:345. https://doi.org/10.3390/en10030345.

[6] Bhadouria AS, Kumar A, Raj D, Verma A, Singh S, Tripathi P, et al. Corrosion mitigation in oil reservoirs during CO_2 injection using nanomaterials. in: *Nanotechnology for CO_2 Utilization in Oilfield Applications*. in *Nanotechnology for CO_2 Utilization in Oilfield Applications* 2022, pp. 127–46. https://doi.org/10.1016/B978-0-323-90540-4.00014-4.

[7] Sakthipriya N, Kumar G, Agrawal A, Doble M, Sangwai JS. Impact of biosurfactants, surfactin, and Rhamnolipid produced from Bacillus subtilis and Pseudomonas aeruginosa, on the enhanced recovery of crude oil and its comparison with commercial surfactants. *Energy Fuels* 2021;35:9883–93. https://doi.org/10.1021/acs.energyfuels.1c00679.

[8] Kumar Y, Yogeshwar P, Bajpai S, Jaiswal P, Yadav S, Pathak DP, et al. Nanomaterials: Stimulants for biofuels and renewables, yield and energy optimization. *Mater Adv* 2021;2:5318–43. https://doi.org/10.1039/D1MA00538C.

[9] Zhao GF, Di WN, Wang RY. Development and potential application of advanced functional materials in the oil field. *Mater Sci Forum* 2019;944:637–42. https://doi.org/10.4028/www.scientific.net/MSF.944.637.

[10] Jaiswal P, Kumar Y, Shukla R, Nigam KDP, Panda D, Guha Biswas K. Covalently immobilized nickel nanoparticles reinforce augmentation of mass transfer in milli-channels for two-phase flow systems. *Ind Eng Chem Res* 2022;61:3672–84. https://doi.org/10.1021/acs.iecr.1c04419.

[11] Kumar G, Kakati A, Mani E, Sangwai JS. Nanoparticle stabilized solvent-based emulsion for enhanced heavy oil recovery. *SPE Canada Heavy Oil Technical Conference*, Alberta, Canada: SPE, 2018. https://doi.org/10.2118/189774-MS.

[12] McAuliffe CD. Oil-in-water emulsions and their flow properties in porous media. *J Pet Technol* 1973;25:727–33. https://doi.org/10.2118/4369-PA.

[13] Chevalier Y, Bolzinger M-A. Emulsions stabilized with solid nanoparticles: Pickering emulsions. *Colloids Surfaces A Physicochem Eng Asp* 2013;439:23–34. https://doi.org/10.1016/j.colsurfa.2013.02.054.

[14] Kumar G, Kakati A, Mani E, Sangwai JS. Stability of nanoparticle stabilized oil-in-water Pickering emulsion under high pressure and high temperature conditions: comparison with surfactant stabilized oil-in-water emulsion. *J Dispers Sci Technol* 2021;42:1204–17. https://doi.org/10.1080/01932691.2020.1730888.

[15] Tambe DE, Sharma MM. The effect of colloidal particles on fluid-fluid interfacial properties and emulsion stability. *Adv Colloid Interface Sci* 1994;52:1–63. https://doi.org/10.1016/0001-8686(94)80039-1.

[16] Almahfood M, Bai B. The synergistic effects of nanoparticle-surfactant nanofluids in EOR applications. *J Pet Sci Eng* 2018;171:196–210. https://doi.org/10.1016/j.petrol.2018.07.030.

[17] Kumar G, Behera US, Mani E, Sangwai JS. Engineering the wettability alteration of sandstone using surfactant-assisted functional silica nanofluids in low-salinity seawater for enhanced oil recovery. *ACS Eng Au* 2022;2:421–35. https://doi.org/10.1021/acsengineeringau.2c00007.

[18] Zhou Y, Wu X, Zhong X, Sun W, Pu H, Zhao JX. Surfactant-augmented functional silica nanoparticle based nanofluid for enhanced oil recovery at high temperature and salinity. *ACS Appl Mater Interfaces* 2019;11:45763–75. https://doi.org/10.1021/acsami.9b16960.

[19] Hashemi R, Nassar NN, Pereira Almao P. Enhanced heavy oil recovery by in situ prepared ultradispersed multimetallic nanoparticles: A study of hot fluid flooding for Athabasca bitumen recovery. *Energy Fuels* 2013;27:2194–201. https://doi.org/10.1021/ef3020537.

[20] Chengara A, Nikolov AD, Wasan DT, Trokhymchuk A, Henderson D. Spreading of nanofluids driven by the structural disjoining pressure gradient. *J Colloid Interface Sci* 2004;280:192–201. https://doi.org/10.1016/j.jcis.2004.07.005.

[21] Wasan DT, Nikolov AD. Spreading of nanofluids on solids. *Nature* 2003;423:156–9. https://doi.org/10.1038/nature01591.

[22] Hendraningrat L, Li S, Torsæter O. A coreflood investigation of nanofluid enhanced oil recovery. *J Pet Sci Eng* 2013;111:128–38. https://doi.org/10.1016/j.petrol.2013.07.003.

[23] Xu D, Bai B, Wu H, Hou J, Meng Z, Sun R, et al. Mechanisms of imbibition enhanced oil recovery in low permeability reservoirs: Effect of IFT reduction and wettability alteration. *Fuel* 2019;244:110–9. https://doi.org/10.1016/j.fuel.2019.01.118.

[24] Davoodi S, Al-Shargabi M, Wood DA, Rukavishnikov VS, Minaev KM. Experimental and field applications of nanotechnology for enhanced oil recovery purposes: A review. *Fuel* 2022;324:124669. https://doi.org/10.1016/j.fuel.2022.124669.

[25] Al-Anssari S, Barifcani A, Wang S, Maxim L, Iglauer S. Wettability alteration of oil-wet carbonate by silica nanofluid. *J Colloid Interface Sci* 2016;461:435–42. https://doi.org/10.1016/j.jcis.2015.09.051.

[26] Anderson W. Wettability literature survey - Part 2: Wettability measurement. *J Pet Technol* 1986;38:1246–62. https://doi.org/10.2118/13933-PA.

[27] Perazzo A, Tomaiuolo G, Preziosi V, Guido S. Emulsions in porous media: From single droplet behavior to applications for oil recovery. *Adv Colloid Interface Sci* 2018;256:305–25. https://doi.org/10.1016/j.cis.2018.03.002.

[28] Abu Tarboush BJ, Husein MM. Adsorption of asphaltenes from heavy oil onto in situ prepared NiO nanoparticles. *J Colloid Interface Sci* 2012;378:64–9. https://doi.org/10.1016/j.jcis.2012.04.016.

[29] Jadhav RM, Kumar G, Balasubramanian N, Sangwai JS. Synergistic effect of nickel nanoparticles with tetralin on the rheology and upgradation of extra heavy oil. *Fuel* 2022;308:122035. https://doi.org/10.1016/J.FUEL.2021.122035.

[30] Umar AA, Saaid IBM, Sulaimon AA, Pilus RBM. A review of petroleum emulsions and recent progress on water-in-crude oil emulsions stabilized by natural surfactants and solids. *J Pet Sci Eng* 2018;165:673–90. https://doi.org/10.1016/j.petrol.2018.03.014.

[31] Zhou Y, Yin D, Chen W, Liu B, Zhang X. A comprehensive review of emulsion and its field application for enhanced oil recovery. *Energy Sci Eng* 2019;7:1046–58. https://doi.org/10.1002/ese3.354.

[32] Kumar G, Mani E, Sangwai JS. Impact of surface-modified silica nanoparticle and surfactant on the stability and rheology of oil-in-water pickering and surfactant-stabilized emulsions under high-pressure and high-temperature. *J Mol Liq* 2023;379:121620. https://doi.org/10.1016/j.molliq.2023.121620.

[33] Binks BP, Philip J, Rodrigues JA. Inversion of silica-stabilized emulsions induced by particle concentration. *Langmuir* 2005;21:3296–302. https://doi.org/10.1021/la046915z.

[34] Druetta P, Picchioni F. Polymer and nanoparticles flooding as a new method for enhanced oil recovery. *J Pet Sci Eng* 2019;177:479–95. https://doi.org/10.1016/j.petrol.2019.02.070.

[35] McAuliffe CD. Crude-oil-water emulsions to improve fluid flow in an oil reservoir. *J Pet Technol* 1973;25:721–6. https://doi.org/10.2118/4370-PA.

[36] Hashemi R, Nassar NN, Pereira Almao P. Nanoparticle technology for heavy oil in-situ upgrading and recovery enhancement: Opportunities and challenges. *Appl Energy* 2014;133:374–387. https://doi.org/10.1016/j.apenergy.2014.07.069.

[37] Bera A, Belhaj H. Application of nanotechnology by means of nanoparticles and nano-dispersions in oil recovery - A comprehensive review. *J Nat Gas Sci Eng* 2016;34:1284–309. https://doi.org/10.1016/j.jngse.2016.08.023.

[38] Rezvani H, Riazi M, Tabaei M, Kazemzadeh Y, Sharifi M. Experimental investigation of interfacial properties in the EOR mechanisms by the novel synthesized Fe3O4@Chitosan nanocomposites. *Colloids Surfaces A Physicochem Eng Asp* 2018;544:15–27. https://doi.org/10.1016/j.colsurfa.2018.02.012.

[39] Hamedi Shokrlu Y, Babadagli T. In-situ upgrading of heavy oil/bitumen during steam injection by use of metal nanoparticles: A study on in-situ catalysis and catalyst transportation. *SPE Reserv Eval Eng* 2013;16:333–44. https://doi.org/10.2118/146661-PA.

[40] Bondarenko V, Svietkina O, Sai K. Effect of mechanoactivated chemical additives on the process of gas hydrate formation. *Eastern-European J Enterp Technol* 2018;1:17–26. https://doi.org/10.15587/1729-4061.2018.123885.

[41] Bondarenko V, Svietkina O, Sai K. Study of the formation mechanism of gas hydrates of methane in the presence of surface-active substances. *Eastern-European J Enterp Technol* 2017;5:48–55. https://doi.org/10.15587/1729-4061.2017.112313.

[42] Vorob'yev AE. Prospects of nanotechnologies of developing gaseous-hydrate resources of the Russian Arctic shelf. *Vestn MGTU* 2016;19:70–81. https://doi.org/10.21443/1560-9278-2016-1/1-70-81.

[43] Kumar Y, Sinha ASK, Nigam KDP, Dwivedi D, Sangwai J. Functionalized nanoparticles: Tailoring properties through surface energetics and coordination chemistry for advanced biomedical applications. *Nanoscale* 2023;15:6075–6104. https://doi.org/10.1039/D2NR07163K.

[44] Kumar Y, Sangwai J. Environmentally sustainable large-scale CO_2 sequestration through hydrates in offshore basins: Ab initio comprehensive analysis of subsea parameters and economic perspective. *Energy Fuels* 2023. https://doi.org/10.1021/acs.energyfuels.3c00581.

[45] Chowdhury S, Kumar Y, Shrivastava S, Patel S, Sangwai J. A review on the recent scientific and commercial progresses on the direct air capture technology to manage atmospheric CO_2 concentrations and future perspectives. *Energy Fuels*, 2023. https://doi.org/10.1021/acs.energyfuels.2c03971

6 Application of Self-Cleaning Materials in the Oil and Gas Industries

*Prerna Tripathi, Yogendra Yadawa, Anshika Verma,
Pranav Kumar, A.S.K. Sinha, and Deepak Dwivedi*
Rajiv Gandhi Institute of Petroleum Technology

Anteneh Mesfin Yeneneh and Khadija Al-Balushi
National University of Science and Technology

CONTENTS

DOI: 10.1201/9781003242550-6

ABBREVIATIONS

CVD   Chemical vapor deposition
PDMS   Polydimethyl siloxane
SNPs   Silica nanoparticles

6.1   INTRODUCTION

Self-cleaning is one of those many technologies that have been inspired by nature. While considering nature's diverse features, one is left in astonishment by the surfaces present in nature exhibiting self-cleaning property, for instance, the plant leaves such as lotus [1], legs of strider [2], and butterfly wings [3]. The outstanding water-repellent qualities of many natural items inspired scientists to create materials with great water repellency. Self-cleaning (SC) technology relies on the formation of spherical water droplets, each of which could remove dirt particles from the surface. The major classification of self-cleaning coatings is into hydrophobic and hydrophilic coatings [4], both utilizing water to clean themselves. In the case of a hydrophilic surface, water over the surface takes away the impurities such as dirt while the water drops glide and drift over the surface thereby cleaning them. One point to note here is that the hydrophilic coatings having metal oxides which possess the inherent property of chemical decomposition of complex deposits through a mechanism incorporating sunlight [5]. Among the many uses of a superhydrophobic/hydrophilic surface, self-cleaning stands out. To give a surface practical self-cleaning property, it is essential to design a highly hydrophobic/hydrophilic substance as the coating. The primary benefits of employing such coatings are the decrease in maintenance costs, the improvement in durability, the prevention of snow or ice adhesion, and the protection against environmental contamination. Recent tragic pipeline explosions in the United States have been blamed on corrosion and old pipe foundations. The Pipeline and Hazardous Materials Safety Administration has produced a study showing that there were approximately 6,298 documented accidents between 2010 and 2019 that had an economic impact as significant financial investments were made to solve corrosion-related concerns [6]. Cathodic protection, exterior protective coatings, corrosion inhibitors and biocides, and interior coating treatments are some of the approaches that have been developed to halt internal and external corrosion in pipelines [7,8]. However, preventative measures are typically costly and time-consuming. To combat corrosion in pipelines, it is important to develop comprehensive, long-term policies for their operation and maintenance [6]. Diverse coating treatments have been developed by specialists around the world to protect low-grade carbon steel pipelines against corrosion and fouling in real-world marine environments [9,10]. Since the superhydrophobic coatings drastically lessen the water/solid contact area, they have become more popular as corrosion inhibitors in the maritime industry [11]. They are successful because they reduce the corrosive species in water's reactivity with steel. Pipeline corrosion and fouling can be avoided at a low cost with the use of superhydrophobic coatings, which are simple to install [11]. Additionally, superhydrophobic polymer coatings for pipelines in marine circumstances provide additional benefits including self-cleaning, anti-icing, oil/water separation, and reduced viscous drag [12,13]. In this study, we introduce the concept of superhydrophobic surfaces that form in nature and then proceed to investigate the underlying theories that dictate wetting and non-wetting behavior on solid surfaces. We then take a look at some methods for fabricating these coatings from scratch. Finally, the production and application of superhydrophobic coatings in real-world maritime environments are described in detail.

6.2 WETTABILITY AND MODELS FOR SUPERHYDROPHOBIC SURFACES

In the last few decades, scientists have studied the physical interactions between solids and liquids at great length. Wettability refers to a surface's ability to allow water to spread across it due to surface tension. Wettability is influenced by the amount of contact at the solid surface. Figure 6.1 demonstrates this concept. At the interface between a solid and a liquid, the contact angle is the angle formed by the plane of the solid surface and the tangent to the liquid surface (solid, liquid, and air) [14]. Surface wetting performance can be broken down into four groups defined by the water contact angle (WCA): hydrophilic, hydrophobic, superhydrophilic, and superhydrophobic. Hydrophilic and hydrophobic have their WCA within the range of $10° < \theta < 90°$ and $90° < \theta < 150°$, respectively [15]. Superhydrophilic and superhydrophobic domains are of considerable interest owing to the utmost surface wetting traits having WCA differing in the stretch of $0° < \theta < 10°$ and $150° < \theta < 180°$, respectively. There are four physical models that show the connection between the solid's energy or tension, the equilibrium contact angle, and other solid parameters [14]. Some examples of physical models are the Young's model for ideal solid surfaces, the Wenzel model that considers surface roughness, the Cassie model for liquid contact with a heterogeneous solid surface, and the Cassie-Baxter model for liquid resting atop a textured solid surface with air trapped underneath it.

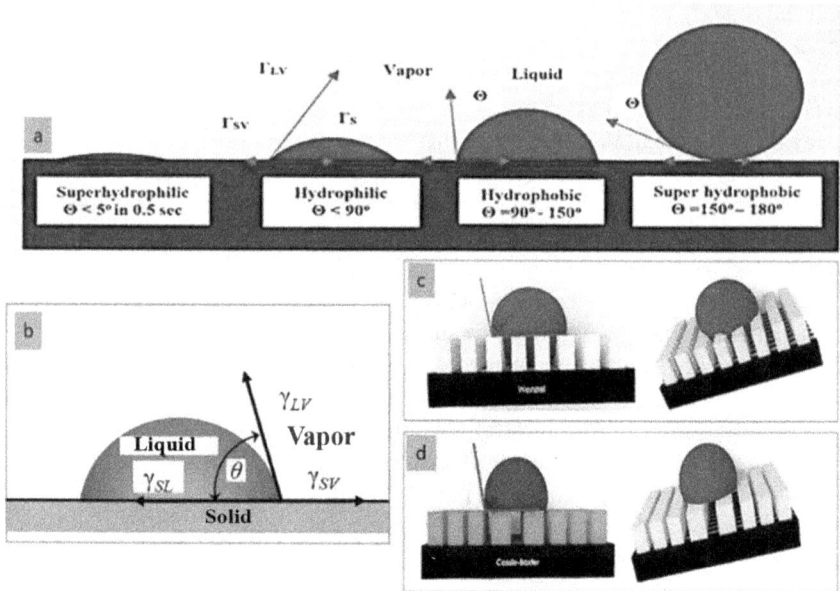

FIGURE 6.1 (a) Hydrophilic, superhydrophilic, hydrophobic, and superhydrophobic surfaces are shown schematically in 2D and 3D views. (Reproduced with permission from [16].) (b) Young's model. (Reproduced with permission from [17].) (c, d) Wetting regime for Wenzel model and Cassie-Baxter model. (Reproduced with permission from [16].)

6.2.1 Young's Model

The wetting models that emerge because of the equilibrium of contact forces along the triple line all begin with Young's equation as their point of departure. It is a term that is used to define the angle at which a liquid meets a solid surface that is isotropic, uniform, stiff, and smooth. According to Young's equation, the only thing that determines the static contact angle is whether the thermodynamic equilibrium of the interfacial tension between the solid, liquid, and vapor interfaces has been reached. It is assumed that in this model, the fluids that surround the solid surface do not react with it [18]. Through his research, Young was able to establish a relationship between the various surface tensions and the contact angles represented in Eq. (6.1).

$$cos\ \theta = \frac{\gamma_{SV} - \gamma_{SL}}{\gamma_{LV}} \qquad (6.1)$$

where θ is the contact angle, and γ_{SV}, γ_{SL}, and γ_{LV} are the interfacial tension of the solid/vapor, solid/liquid, and liquid/vapor interfaces, respectively. Static contact angles of around 120° can be achieved on a smooth surface with the chemical group that has the lowest surface energy, but this is not high enough to qualify as superhydrophobic [19,20]. Therefore, surface roughness is the second essential factor in creating a superhydrophobic surface. It is important to combine increased surface roughness with decreased surface energy to get a larger contact angle and superhydrophobicity. Many scholars, including Wenzel and Cassie-Baxter in particular, have worked to determine the relationship between surface tension, roughness, and contact angle [21].

6.2.2 CASSIE MODEL

When a drop of liquid is deposited on a solid surface, it will either settle into a drop shape within its confines or spread chaotically throughout the substrate [22]. To form a droplet, more energy is required that is available at the solid/liquid or solid/vapor contact (as per Eq. 6.2):

$$\gamma_{SV} - \gamma_{SL} > \gamma_{LV} \qquad (6.2)$$

The contact angle exists at the point when the above imbalance remains unsatisfied; therefore, the liquid droplet stays finite in size. The equilibrated contact angle is estimated between a fluid and a solid surface, which is independent of gravity and is given by (as per Eq. 6.1)

$$\cos \theta = \frac{\gamma_{SV} - \gamma_{SL}}{\gamma_{LV}} \qquad (6.1)$$

As stated above, the cosine of the contact angle gives the ratio between the energy gained by forming a solid/liquid interface and the energy used in forming a liquid/air interface per unit area. The surface condition has a crucial role in determining solid/liquid interfacial tension. A rougher surface increases the energy required to produce a solid/liquid interface, and this is what the contact angle measures (SV-SL).

$$\cos \theta' = \frac{\sigma(\gamma_{SV} - \gamma_{SL})}{\gamma_{LV}} \qquad (6.3)$$

The angle θ' is termed an "apparent" contact angle, and θ is the "real" angle.

6.2.3 WENZEL MODEL

The Wenzel model is a modified version of Young's model that is used to specify the wettability of surfaces that have a uniformly rough texture. According to the findings of Wenzel's research on the wettability of porous surfaces, the roughness of a surface has a substantial impact on the contact angle that exists on that surface. The Wenzel theory is based on two key tenets: The first is that liquid can penetrate abrasive grooves, and the second is that an expansion of the liquid-spreading interfacial area can take place [23,24]. According to Wenzel's homogeneous wetting state at thermodynamic equilibrium [25], it was hypothesized that the apparent contact angle and the surface roughness factor have a linear connection with one another.

$$\cos \theta_w = \frac{r(\gamma_{SV} - \gamma_{SL})}{\gamma_{LV}} = r\cos \theta_{eq} \qquad (6.4)$$

Here, θ_w denotes the angle at which liquid droplets contact a surface that is relatively rough and marks the angle at which such droplets make contact with a surface that is relatively smooth. Both the contact angle of Young and the surface roughness factor are shown to be dependent on the surface roughness factor, denoted by the letter r and defined as the ratio of the actual surface area to the predicted surface area. In light of this, the Wenzel equation indicates that surface roughness does, in fact, influence surface wettability and predicts that when $\theta < 90°$ it enhances and while $\theta > 90°$ lessens it. This is because of the fact that surface roughness does affect surface wettability. However, if the angle is greater than 90°, air bubbles will settle within these prominent grooves and prevent water droplets from entering the hydrophobic pillars, so rendering the surface extremely hydrophobic [26].

6.2.4 CASSIE-BAXTER MODEL

As proposed by Young and Wenzel, under the homogeneous wetting condition, a single drop of liquid completely fills the deep grooves. To prevent liquid from penetrating pores, air (or another fluid) becomes trapped in the rough grooves under the liquid in a wetting state known as Cassie-Baxter [27]. The Cassie-Baxter equation can only be used in practice if the liquid is in physical contact with the solid surface at the top of the projections [28,29].

$$cos\ \theta_{CB} = f_s cos\ \theta - (1 - f_s) = f_s cos\ \theta + f_s - 1 \qquad (6.5)$$

where θ, θ_{CB}, and f_s are Young's contact angle, the ratio of the solid/liquid interface area to the entire area of the liquid/air and solid/liquid interfaces in the projected region, and the Cassie-Baxter contact angle. The roughness factor, r_f, of the wetted surface area will also affect the Cassie-Baxter equation [30,31]:

$$cos\ \theta_{CB} = r_f f_s cos\ \theta + f_s - 1 \qquad (6.6)$$

where $f_s = 1$ and $r_f = r$. The Wenzel equation is derived from the Cassie-Baxter equation. According to the Cassie-Baxter model, the air is more likely to get trapped in the grooves as surface roughness rises. In turn, f_s drops, and the contact angle rises [21]. According to these theories, surface roughness influences the static contact angle, and it increases the hydrophobicity and hydrophilicity. The dominant regimes are Wenzel's and Cassie-Baxter's models for low and high roughness, respectively.

6.3 NATURE'S PATTERNS OF SELF-CLEANING

For scientists and researchers, nature serves as a school where they are inspired to create new materials to meet societal demands. Many animals have acquired distinctive wettability traits via billions of years of species development. Using nature's patterns to create new surfaces or enhance the ones that already exist is referred to as "biomimetic design." Biomimetic architecture involves taking important physical and chemical information from nature and using it to create a variety of advanced synthetic materials that are nevertheless helpful in real-world applications. As shown in Figure 2, this section summarizes the most recent advancements in bio-inspired distinctive wettability, including the self-cleaning lotus effect, flower petals, creepy crawlies, gecko feet, mosquito eyes, etc. The production of hydrophobic and superhydrophobic coatings has been discussed, along with the materials and processes used to achieve these coatings.

6.3.1 LOTUS EFFECT

Many plants' leaves display extraordinary water repellence. The lotus leaf stands out among them as an excellent example and has inspired many scientists to develop unique surface materials for use in industry. Lotus leaves exhibit WCA exceeding 161° and a lower sliding angle of approximately 2° [32]. Raindrops and dust effectively roll down the surface of the leaf due to the lower sliding angle. A remarkable superhydrophobic self-cleaning activity is provided by the isotropic dual (micro-nano) scale roughness papillae and naturally hydrophobic epicuticular wax on the surface [33,34]. Known as the "lotus effect," this effect is now well known. Researchers are carefully examining the magical effect of lotus leaves due to their ability to resist scrubbing. Several researchers have investigated biomimicking this lotus effect to create artificial superhydrophobic surfaces whose WCA is greater than 150°. Using the leaf as a negative template, Sun et al. generated a dual (nano and micro) scale roughness on the underlying polymer by imitating the delicate lotus leaf topography pattern onto a polydimethyl siloxane (PDMS) surface. Since surface texture and chemical composition both affect superhydrophobicity, the WCA on flat PDMS unexpectedly increased from 110° to 160° [35].

6.3.2 RICE LEAVES

Rice leaves, like lotus leaves, are incredibly water-repellent, and the interesting anisotropic wettability of rice leaves is an unusual property. It has a superhydrophobicity rating of 157.2° and papillae that are 5–8 m in diameter and run parallel to the leaf edge [36]. This is because the surface roughness occurs on both the micro- and nanoscales. The arrangement of the papillae parallel to the leaf edge surface makes it easier for water droplets to roll off the leaf edge. Using a rice leaf as a negative template, Gao et al. utilized a two-step replication approach to create a superhydrophobic surface. A rice leaf was used as a negative template to recreate the topography of PDMS film on poly (N-isopropyl acrylamide) (PNIPAAm) film. The authors claim that a thermally sensitive, anisotropically wettable film exists, with a WCA of 119° at 50°C and 77° at 20°C in the direction parallel to the grooves. In perpendicular direction, the WCA was measured to be 87° at 50°C and 49° at 20°C [37].

6.3.3 BUTTERFLY WINGS AND PEACOCK FEATHER

Nature also contains a few insects' legs that are naturally free of filth, such as arthropods, pigeon feathers, and butterfly wings. The epidermal scale (40–80 µm) and the micro-relief of raised ridges covering each wing scale are the two periodic features that make up water droplets (1,200–1,500 nm in width) which are pushed outward by the adhesive forces exerted by these periodic patterns [38]. The iridescent, brilliant, multiscale patterns on the butterfly's wings provide it superhydrophobicity and high chemical sensing abilities. There are two types of wing scales: ground scales (responsible for the wing's structural color) and cover scales (responsible for the wing's superhydrophobicity and self-cleaning properties). The wettability of butterfly wings served as an inspiration for Zheng et al. [39]. It was discovered that a liquid droplet can readily roll off the body in a radial outward direction, with the body acting as the central axis. However, it clings to the skin in a radially inward fashion. Their findings helped explain the behavior of water droplets on butterfly wings, which is essential for the design of low-maintenance coatings. The tails of a peacock's feather are where its iridescent hues and unique eye patterns really shine. Peacock feathers are not only highly iridescent but also remarkably hydrophobic. The number of periods and lattice constant in the photonic crystal structure are responsible for the color shifts.

6.3.4 Water Strider Legs and Insect Compound Eyes

The water strider's legs enable it to move swiftly through the water. Researchers have attempted to replicate the water strider's legs, which they say are coated in needle-shaped micrometer-scale setae and have a surface inclination of 20°. A great many helical nanogrooves, each one capable of holding a single air bubble, are packed into each macrosetae [40]. In addition to these features, it was found that the water strider's legs were hydrophobic. If the water strider's body is above the surface, its legs may swim. The immense support given by its leg allows the water strider to bend even in the face of turbulence in flowing water. Its ability to flush away an amount of water 300 times more than the size of its leg exemplifies its hydrophobic properties. Unattractive light reflection is a feature of the surfaces of some insects [41,42]. Attractive physiological optics with high sensitivity and anti-reflection, for example, emerge from the existence of multiscale structure in the eyes of moths, butterflies, and flies. Ommatidia are a group of small eyes that make up the insect's compound eye. Many different types of artificial compound eyes [43] have been developed to mimic natural eye structures with anti-reflective properties. These arrays are very anti-reflective in addition to being hydrophobic. Recently, silica substrates have been used in the creation of ultra-hydrophilic surfaces with anti-reflective and anti-fogging characteristics [44].

6.4 MATERIALS AND MECHANISM TO PRODUCE HYDROPHOBIC AND SUPERHYDROPHOBIC COATINGS

There are two general kinds of methods that can be used to create hydrophobic and superhydrophobic surfaces: (1) altering a rough surface with a low-surface-energy material, and (2) creating a rough surface from a low-surface-energy material.

6.4.1 Making a Rough Surface and Modifying the Surface with a Material of Low Surface Energy

6.4.1.1 Wet Chemical Reaction and Hydrothermal Reaction

Wet chemical processes were used to create nanostructures with controlled dimensionality and morphology, such as nanowires and nanoparticles [45,46]. On metal substrates including copper, aluminum, and steel, this technique has been frequently utilized to create biomimetic superhydrophobic surfaces. A surface roughness technique was applied by Zhang et al. [47] to etch polycrystalline metals by using an acidic or basic solution of fluoroalkyl silane. The fluoroalkyl silane treatment resulted in superhydrophobicity on the etched surfaces. Using a wet chemical method, mono alkyl phosphonic acid reacts with nickel to produce flower-like microstructures that form a continuous slipcover, resulting in superhydrophobic surfaces on nickel substrates [48]. By using micro-arc oxidation as a pre-treatment and chemically altering the surface with PDMSVT with spin coating on a magnesium alloy, Hao et al. were able to create a biomimetic superhydrophobic surface [49]. To efficiently fabricate useful materials with various patterns and morphologies, the hydrothermal approach was recently developed. It is economical because the initial combination doesn't need to be subjected to any extra calcination, grinding, or milling. Controlling the synthesis temperature, precursor concentrations, etc., allows one to alter the size and shape of the nanoparticles that are produced. By first oxidizing zinc metal and then applying n-octadecyl thiol to the surface, Hou et al. [50] synthesized a superhydrophobic ZnO nanorod layer on a zinc substrate. These two methods both save time and may be scaled. These techniques' adaptability and ease of use enable them to create morphologies that are sensible in both size and shape.

6.4.1.2 Electrochemical Deposition

Electrodeposition and chemical deposition techniques have made considerable contributions to the creation of cutting-edge superhydrophobic surfaces and gadgets over the past decade. Physical deposition and physical vapor deposition were crucial in the creation of the self-cleaning superhydrophobic surface. The electrochemical deposition inducing long-chain fatty acids to build a hierarchical copper mesh on the micro- and nanoscale that exhibited both superhydrophobicity and superoleophilicity was employed by Jiang et al. [51]. Khorsand et al. using an electrodeposition method fabricated a hydrophobic Ni film that featured an embedded micro/nanonetwork. The superhydrophobic surface resisted corrosion even after being submerged in a 3.5 wt% sodium chloride solution for many days [52]. Superhydrophobic and bendable 3D porous copper sheets were fabricated by employing hydrogen bubbles as a dynamic template for metal electrodeposition. Since the porous copper films expanded in the voids between the hydrogen bubbles during electrodeposition, their pore width and wall thickness could be controlled by altering the electrolyte concentration [53].

6.4.1.3 Lithography

Lithography is frequently used to produce micro- and nanopatterns. These days, a wide variety of lithographic techniques are used, including photolithography, X-ray lithography, electron beam lithography, soft lithography, nanosphere lithography, and others. Han et al. [54] developed and optimized superhydrophobic tungsten hierarchical surfaces with CAs greater than 150° and SAs less than 20° that survived 70 scrape cycles, 28 minutes of solid particle impact, or 500 tape stripping cycles. This describes how superhydrophobic surfaces can achieve excellent durability for practical applications. Martine et al. [55] employed the use of plasma etching and electron beam lithography to create a variety of nanopits and nanopillars on the material's surface. Octadecyltrichlorosilane was used to hydrophobize this surface, which resulted in superhydrophobic characteristics with a WCA as high as 164°.

6.4.1.4   Electrospinning Technique

For the creation of fine nanofibers, electrospinning is a popular technique. In order to create a rough enough surface to induce superhydrophobicity, many research organizations adopt this technique. Ma et al. [56] used electrospinning and chemical vapor deposition (CVD) technologies to construct superhydrophobic surfaces. They established that superhydrophobicity is a direct consequence of electrospinning substances. This was shown by the fact that electrospinning results in superhydrophobicity. Within the context of this method, hydrophobic perfluoroalkyl ethyl methacrylate was polymerized (PPFEMA) applied to electrospun polycaprolactone by CVD. With this method, a WCA of roughly 175° was obtained.

6.4.1.5   Etching and CVD

To create functional surfaces with various morphologies, polymers have been widely employed in plasma etching and CVD techniques. By adopting a micro-condensation procedure with plasma chemical patterns, Garrod et al. [57] recreated the surface of the stenocara beetle's back after analyzing it. The microtextures, which were created on Si surfaces, showed extremely hydrophobic behavior with a WCA of roughly 174°. A unique technology combining two dry processing processes was used by Teshima et al. [58] to create a transparent superhydrophobic surface. In this technique, tetramethyl silane was used as the precursor, and selective oxygen plasma etching was used to first create nanotexture on a poly (ethylene terephthalate) (PET) substrate. The surface created using this method displayed a WCA larger than 150°.

6.4.1.6   Sol–Gel Method and Polymerization Reaction

The sol–gel method is a versatile technique for producing superhydrophobic surfaces on a wide range of solid substrates. Utilizing hexamethylenetetramine and ethylene glycol, a powerful bidentate chelating agent to Cu^{2+} and Fe^{2+} ions with high stability constant, Huang et al. [59] built bio-inspired superhydrophobic surfaces on copper alloys. With the help of colloidal monolayers of polystyrene and the sol-dipping technique, Duan et al. [60] were able to create films with ordered pore arrays made of indium oxide. The film's superhydrophobic properties are found to be modulable through the manipulation of pore size.

6.4.1.7   Self-Assembly and Layer-by-Layer (LBL) Methods

The successive adsorption of a substrate in solutions of compounds with opposing charges is the basis for the LBL and self-assembly methods. This technique continues to be the most common and well-known ways for creating multilayer thin films. Micro- and nanoscale superhydrophobic structures with precisely regulated surface morphologies can be easily created using LBL deposition and self-assembly, both of which are low-cost methods. By joining silica particles, Lee et al. [61] created a surface with a dual-size roughness that could be adjusted and contained particles that resembled raspberries. By incorporating silver nanoparticles into a monolayer array of polystyrene microspheres, it was possible to create bionic superhydrophobic coatings [62]. Using amine groups, Ming et al. [63] also created particles that resembled raspberries. Through the interaction between the epoxy and amine groups, silica particles with amine- and epoxy functionalization of sizes 70 and 700 nm were covalently grafted together in this procedure. Once the surface had been altered with PDMS, it showed superhydrophobicity. These techniques produced a WCA of roughly 165°.

6.4.2   ROUGHENING THE SURFACE OF LOW-SURFACE-ENERGY MATERIAL

This section primarily focuses on various techniques used in the past few years to roughen surfaces of low-surface-energy materials to produce superhydrophobic membranes. Materials used for the following purpose are discussed briefly.

6.4.2.1   Fluorocarbons

The extremely low surface energies of fluorinated polymers are the focus of a great deal of current study. These polymers can be roughened to create superhydrophobic surfaces. Shiu et al. [64] used an oxygen plasma treatment on Teflon to make it rough. Attaining a WCA of 168° was possible using this method. Due to their poor solubility, many fluorinated materials are used indirectly, mixed with other abrasive substances to make superhydrophobic surfaces. Superhydrophobicity was achieved by Zhang et al. [65] by stretching a Teflon (polytetrafluoroethylene) film. The superhydrophobic quality of the material is due to the presence of fibrous crystals with large amounts of surface area but no atoms.

6.4.2.2   Organic Materials

Recent studies have demonstrated that hydrophobicity may also be produced utilizing polyamide, alkyl ketene, polycarbonate, and paraffinic hydrocarbons. By electrostatic spinning and spraying PS solution in dimethylformamide (DMF), Jiang et al. [66] demonstrated how to create the superhydrophobic coating. Then, the surface obtained was composed of porous microparticles and nanofibers.

6.4.2.3   Inorganic Materials

A small number of inorganic materials have also shown to be superhydrophobic. Recent research on ZnO and TiO_2 has led to the development of films with reversible wettability. Feng et al. [67] used a two-step solution technique to produce ZnO nanorods. X-ray diffraction analysis showed that the ZnO nanorod films were extremely hydrophobic because the (001) plane of the nanorods on the film's surface had low surface energy. A hydroxyl group was deposited on the ZnO surface because of UV exposure creating electron/hole pairs. This results in a transformation of the film from being superhydrophobic to superhydrophilic.

6.4.2.4   Silicones

Polydimethylsiloxane (PDMS) is a member of the class of organosilicon compounds known as silicones. PDMS is an excellent material for creating superhydrophobic surfaces due to its inherent deformability and hydrophobic characteristics. Various methods are practiced to fabricate the superhydrophobic surfaces using PDMS. A CO_2 pulsed laser was used by Khorasani et al. [68] to modify the surface of PDMS by adding peroxide groups as an excitation source. The graft polymerization of 2-hydroxyethyl methacrylate (HEMA) onto the PDMS can be started by these peroxides. The treated PDMS measured WCA 175°. The porosity and chain ordering on the surface of PDMS were the cause of such an increase in WCA.

6.5   CHARACTERIZATION TECHNIQUES

The phobic and philic properties of a surface toward water and oil primarily control its ability to self-clean. Researchers use a variety of characterization approaches to evaluate surface roughness, surface energy, and liquid droplet contact angle to quantify this phenomenon and correlate it with the self-cleaning phenomenon. A crucial aspect of describing hydrophobic coatings is the use of contact angle measurements. In the subject of self-cleaning coatings, contact angle estimate is extremely important because it is the only measurement that directly evaluates a surface's phobicity or philicity. The simple drop contact angle test, the sessile drop-let test, and the advancing/receding contact angle hysteresis are frequently used in the estimating procedure [69,70]. Atomic force microscopy, also known as topog-raphy, is frequently utilized in the process of evaluating self-cleaning coatings with regard to their surface roughness. Teshima et al. made use of this procedure in order to evaluate the degree of roughness present on the superhydrophobic surface that was produced via PET etching (polyethylene terephthalate). In addition, envi-ronmental ellipsometric porosimetry, grazing incidence X-ray analyses at low and wide angles (GI-SAXS and GI-WAXS), electronic and near-field microscopy, field-emission scanning electric microscopy (FE-SEM), transmission electron micros-copy (TEM), and Fourier transform infrared spectroscopy (FTIR) are frequently used in character analysis.

6.6   APPLICATIONS OF SUPERHYDROPHOBIC COATINGS FOR PIPELINES

Several distinguishing characteristics of superhydrophobic surfaces such as water repellency, self-cleansing, and anti-sticking behavior increase their utilization in various sectors within the industry. Superhydrophobic coatings find a wide range of applications in almost every realm of industry. The self-cleaning superhydrophobic coatings can be of significant use in materials ranging from walls, window glass, mini boats, solar cell panels, cotton shirts, paper, wood, sponges, fabric shoes, mar-ble, and plastic. Researchers are trying their best to develop surfaces that turn out to be highly durable, long-lasting as well as possess great mechanical strength. Apart from this, their resistance to water makes these surfaces even more valuable since they inhibit microorganism growth over them. Hence, they have exceptional anti-fouling and antibacterial characteristics. Superhydrophobic coatings are increasingly being utilized to protect the surfaces of marine vessels and submarines since they are in constant contact with the alkaline environment and may also get covered by several microorganisms. They are of importance as coatings to building materials, for instance, marbles and sandstone as they can help protect these surfaces from environmental damage and pollution as well as acid rain. They can even be applied over currency notes that suffer deterioration due to dust and sweat from people, thus improving their life of use.

6.6.1 Self-Cleaning

The ability of a surface to eliminate foreign substances on its own, such as dirt and pollutants, without the assistance of an external source is referred to as self-cleaning. This ability is most commonly observed in superhydrophobic surfaces. Creation of a coating with a surface that exhibits hydrophobic qualities [71] and integration of that surface roughness with low-surface-energy molecules increases the superhydrophobicity of surfaces for applications involving self-cleaning. This will allow surfaces to have a higher capacity for repelling water. Electrodeposition was the method that Bai and Zhang [72] used to create a self-cleaning, superhydrophobic reduced graphene oxide/nickel coating for a 304 stainless steel plate. In the process of electrodeposition, stainless steel plates were first polished with silicon carbide sandpaper before being used as the cathode. The nickel plate that was used as the anode was first carefully cleaned by ultrasonic before being used in the process. The self-cleaning capabilities of the coating as it was manufactured were evaluated by introducing graphite granules, which represent impurities, to the surface of the coating. As a result of the coating's poor surface adhesion, the contaminants were thoroughly removed by water droplets without the surface being tilted in any manner.

6.6.2 Anti-Biofouling

The build-up of bacteria, germs, algae, or seaweed on damp surfaces is known as biofouling [73]. When a surface (natural or artificial) is submerged in seawater, biofilm adhesion has reportedly been observed to occur. The oil and gas industries, the maritime industry, and the aquatic industry are all impacted by marine biofouling, which has negative effects on desalination facilities, marine vessels, and underwater structures [74,75]. Superhydrophobic surfaces have recently come to light as a possible solution to the challenges posed by biofouling [76,77]. Due to their engineered micro-nanoscale stratified structures and low wettability, the quantity of microbes clinging to these surfaces is drastically reduced. Silver nanoparticles (AgNPs), which can be utilized to generate surfaces superhydrophobic, have been found to be extremely resistant to microbial growth [78]. Using silane and the thiol-ene click chemistry method, Pang et al. [79] created grafted ionic liquids (ILs) on the surface of stainless steel. This helped to reduce the amount of bacterial fouling that occurred on the surface. It was discovered that the IL-grafted SS surface had good antibacterial resistance to Gram-negative Escherichia coli when its anti-biofouling performance was assessed. It was also discovered that compared to comparable ILs with hexafluorophosphate and tetrafluoroborate anions, the ILs employed in the study with bromide anions demonstrated superior antibacterial resistance against *E. coli*.

6.6.3 Anti-Corrosion

Construction of electric cables, pipelines, buildings, ships, and other industrial structures all requires the utilization of metals and alloys. Metals are of abundant use in the industry ranging from automobiles to coins. Corrosion of metals could be retarded by developing a superhydrophobic coating and hence preventing metal/water contact. Also, fabric could be prevented from getting stained by colored or muddy/dirty water through the deposition of superhydrophobic coatings since normal washing of clothes requires a lot of detergent, energy, and water, and they often tend to lose shine. The deterioration of metals by corrosion is one of the most serious problems confronting the industry today. This issue can be solved by using superhydrophobic coatings. Metals often tend to deteriorate through oxidation and electrochemical processes in an aqueous media, drastically reducing their life span. For instance, one of the main problems related to the wear and tear of automobiles is the regular corrosion of ships and docks in seawater. The superhydrophobic coating on metal can prevent corrosion by limiting water molecule diffusion. One of the most widely used materials in industries, stainless steel, can be coated with hybrid silica-sol–gel film in a single-step electrodeposition technique. By intricate setting of deposition potential and time, it was observed that stainless steel had better corrosion resistance in NaCl solution. Through electrodeposition, to prevent Q235 carbon steel from corroding, Ye et al. developed a superhydrophobic oligoaniline-containing electroactive silica coating [80]. It was hypothesized that the coating's corrosion resistance came from a mix of the superhydrophobic properties of the E-M-SiO$_2$ pre-process coating and the unusual catalytic potential of the aniline trimer unit. Passive films are formed on the surface of the Q235 steel thanks to this coating, which also serves to block corrosion medium. Fe$_2$O$_3$ and Fe$_3$O$_4$ make up these films. A superhydrophobic coating with excellent corrosion resistance [81] was developed by Xiang et al. via electrodeposition. The mild steel substrate had a micro-nano structured nickel (Ni) coating applied to it and then released on it to increase its resistance to corrosion and enhance its mechanical properties. Less than 2.25×108 A/cm^2 was the desired corrosion current density. After that, a chromium (Cr) (III) layer was electrodeposited over the nickel. After 20 seconds of electrodeposition, the structure changed from a "fish-scale" shape into a "cone-shaped" shape. Observations of the Ni/Cr$_2$O coating revealed a WCA of 167.9° with a standard deviation of 2.4°.

6.6.4 Anti-Icing

A common problem in colder climates is oil leaks from pipelines due to the formation of ice. Pipelines are vulnerable to icing because it can cause blockages and freeze vital components like valves [82]. Ice formation is an inevitable consequence of low temperatures [83]. Some oil and gas operations may be impacted by the formation and deposition of ice on the exterior of pipelines transporting crude oil and other petroleum products in colder climes [83,84]. Many people, both in academia and the business world, are interested in anti-icing coatings such as superhydrophobic coatings [84]. Pan et al.'s anti-icing superhydrophobic coating [85] required nothing more than spraying a solution of hydrophobic silica nanoparticles (SNPs) and poly (methyl methacrylate) (PMMA) onto a steel substrate. During manufacturing, this coating was put through a condensing test and a cold water dripping test to determine how well it prevented ice. Only a small number of areas on the superhydrophobic coating froze during the cold water dripping test, which was conducted in an anti-icing room with an interior temperature of –20°C and the droplet temperature maintained at 0°C. The samples were dried out in a humidity chamber at a temperature of 40°F and a humidity level of 80%. Water vapor condensed into a liquid on the samples' exteriors when the temperature was lowered to –20°C. The superhydrophobic covering supposedly showed only a trace amount of moisture on its surface. This led researchers to conclude that the coatings, as they currently stand, effectively prevent ice.

6.6.5 OTHER SIGNIFICANT APPLICATIONS

6.6.5.1 Superhydrophobic Coatings in Glasses

Glass finds immense application as a material of fabrication for solar cell panels, windshields of automobiles, electronic appliances, windshields of automobiles, etc. Cleaning normal glass takes a lot of resources, time, and energy. This even causes a lot of accidents. To get rid of all these, self-cleaning coatings were used on glasses. The transparent and highly durable superhydrophobic coatings on glass are of wide applications in the industrial arena. Self-cleaning glasses can be synthesized via two pathways: the first one is based upon the lotus effect which means developing a surface possessing the tendency to keep itself unaffected by dirt even post-rainfall owing to its high dual degree surface roughness and water-resistant wax coating. Several methodologies such as plasma etching and CVD may be utilized to create a rough surface.

6.6.5.2 Superhydrophobic Coatings in Textiles

The development of textiles that are stain and water-resistant has been the subject of extensive research. Often, this entails coating the target fabric with a superhydrophobic material, ideally one that doesn't alter the fibers' other characteristics. Self-cleaning fabrics can repel dirt far better than traditional fabrics, and owing to catalytic mechanisms, they can also remove a variety of stains. The majority of currently used methods rely on coating the textile with low-surface-energy substances like fluorinated polymers. These fabrics are anticipated to be stronger and survive longer because of the lower cleaning requirements. Because of the catalytic principles underlying their self-cleaning behavior, which may also result in the eradication of germs, odors, etc., they can be used to manufacture clothing for medical staff. Additionally, the survivability of microorganisms on such surfaces is decreased by water repellency. These textiles have the potential for use in the defense industry as well.

6.6.5.3 Superhydrophobic Coatings on Vehicles

Vehicles can become quite dirty due to the various types of pollutants and impurities that get collected over time. Washing them daily involves a lot of time and cost, and they even tend to affect the shine of these vehicles. For this, we can go for the utilization of a transparent and durable self-cleaning coating that is superhydrophobic. These can be fabricated on the surface of the vehicles using the technique of spraying at room temperature followed by overnight drying. When compared to the vehicles that were not coated, it was observed that water was sufficient to wash the oil while the soil left an imprint on the vehicle in the latter non-coated vehicle case.

6.6.5.4 Superhydrophobic Coatings on Building Walls

In the walls of the building, the phenomena of loss of shine are quite frequent because of the dust and carbonaceous materials that get accumulated over time on the walls.

Paints composed of polymer cannot completely prevent the walls from getting affected and can help achieve self-cleaning only to some extent. Thus, again we switch to superhydrophobic coatings for the same that could self-clean the walls and retain their shine as well. This means that a drizzle or water spray is enough to bring back the shine of these superhydrophobic-coated building walls. These coatings can be synthesized by utilizing SNP suspension. These walls repelled dirty and muddy water. However, this was not the case in walls without a coating which tends to get dirty.

6.7   CHALLENGES AND FUTURE OUTLOOK

As far as the issues of corrosion, biofouling, and dirt collection are concerned, superhydrophobicity seems to offer solutions to various issues that we often encounter. Most of the methods currently employed to create these coatings for uses ranging from self-cleaning glasses, fabrics, and anti-fouling coatings are only appropriate for laboratory work. Even though there have been substantial improvements in the synthesis of biomimetic self-cleaning coatings with super hydrophobic and super hydrophilic properties, there are still several obstacles that need to be removed before they can be used in a variety of commercial applications. The techniques are extremely delicate, and they frequently depend on the meticulous management of numerous parameters. Experimentation and result collection have been conducted inside the laboratory, but the applications are for outdoors. However, the raw materials are quite costly. Therefore, before the more widespread usage of these coatings is practicable, it is necessary to obtain less expensive but similarly effective raw ingredients. Whenever we are worried about the technological applications of superhydrophobic coatings, there are a few key concerns to consider, which are listed as follows:

a. One can go for developing these coatings on a huge scale on kinds of materials. Most of the techniques employed are specific to the purpose and the type of substrate under use. The focus must be paid toward generating such coatings without any kind of deformities and irregularities as well as the adherence of coatings toward the specific material. The most widely used techniques for manufacturing coatings for utilization on a large scale are sol–gel techniques, plasma treatment, and some spraying techniques. The methodology followed for the usage of Teflon coatings on non-stick cookware is one such excellent technique.

b. Another important aspect is regarding the long-term stability of such coatings. Such coatings work as per the basis identified on micro- and nanolevel surface roughness that is susceptible to get deteriorated by the application of mechanical stresses. This brings the need to take care of mechanical stability. Though these coatings aren't exposed to a high temperature, but, due to the effect of temperature on surface tension, one must also regulate the thermal stability of these coatings. Because of the surface tension effect, cold-water-resistant surfaces display a higher affinity toward hot water.

Researchers and analysts have been visualizing the future of self-cleaning coatings, so they've been working to gain a deeper understanding of the connection between the structure of the coating substrate and its capacity, to enhance the coatings' self-cleaning ability and durability, and to create surfaces with unique and additional functionalities. Additional study of multifunctional coatings is in store for the future, with potential applications in the glass and solar cell industries, as well as the textile and medical fields. In addition, it is essential to create novel synthesis and surface modification procedures to build the substrates that may be used for the coating's higher adhesion and strength. Beyond this, there is a need for greater research into the toxicity of coatings so that they may be safely employed in water purification membranes, self-repair, self-healing, and self-lubricating coatings, among other uses. Coatings that are both high quality and affordable are urgently needed in the industry.

6.8 CONCLUSION

Wettability is a complicated property that influences several surface properties. It is vital to have a thorough understanding of surface phenomena to optimize their utility in potential societal applications and to boost their ability to clean themselves more effectively. Such applications include, but are not limited to, self-cleaning automobile windshields, anti-reflective solar panels, anti-biofouling paints for boats, anti-dust architectural coatings, stain-resistant clothes, anti-stick coatings for outdoor usage including commercial and residential facades, etc. Even though there has been considerable forward movement in the research and development of self-cleaning coatings, there are still a great many obstacles that need to be conquered before these coatings may be utilized in a wide variety of industrial applications. First and foremost, researchers should investigate the relationship between coating material structure and capacity at a higher level. Next, they should work to improve the longevity of surfaces as well as their capacity for self-cleaning. Finally, they should develop surfaces that have additional capabilities in addition to their capacity for self-cleaning.

REFERENCES

[1] Barthlott, Wilhelm, and Christoph Neinhuis. "Purity of the sacred lotus or escape from contamination in biological surfaces." *Planta* 202, no. 1 (1997): 1–8.
[2] Gao, Xuefeng, and Lei Jiang. "Water-repellent legs of water striders." *Nature* 432, no. 7013 (2004): 36–36.
[3] Byun, Doyoung, Jongin Hong, Jin Hwan Ko, Young Jong Lee, Hoon Cheol Park, Bong-Kyu Byun, and Jennifer R. Lukes. "Wetting characteristics of insect wing surfaces." *Journal of Bionic Engineering* 6, no. 1 (2009): 63–70.
[4] Parkin, Ivan P., and Robert G. Palgrave. "Self-cleaning coatings." *Journal of Materials Chemistry* 15, no. 17 (2005): 1689–1695.
[5] Hashimoto, Kazuhito, Hiroshi Irie, and Akira Fujishima. "TiO2 photocatalysis: A historical overview and future prospects." *Japanese Journal of Applied Physics* 44, no. 12R (2005): 8269.
[6] Popoola, Lekan Taofeek, Alhaji Shehu Grema, Ganiyu Kayode Latinwo, Babagana Gutti, and Adebori Saheed Balogun. "Corrosion problems during oil and gas production and its mitigation." *International Journal of Industrial Chemistry* 4, no. 1 (2013): 1–15.
[7] Beavers, John A., and Neil G. Thompson. "External corrosion of oil and natural gas pipelines." *ASM Handbook* 13, no. 05145 (2006): 1–12.

[8] Zelmati, Djamel, Omar Bouledroua, Oualid Ghelloudj, Abdelaziz Amirat, and Milos B. Djukic. "A probabilistic approach to estimate the remaining life and reliability of corroded pipelines." *Journal of Natural Gas Science and Engineering* 99 (2022): 104387.

[9] Wang, Hairui, F. Liu, Y.P. Zhang, D.Z. Yu, F.P. Preparation and properties of titanium oxide film on NiTi alloy by micro-arc oxidation. *Applied Surface Science* 257, no. 13 (2011): 5576–5580, ISSN 0169-4332.

[10] Yun, Hong, Jing Li, Hong-Bo Chen, and Chang-Jian Lin. "A study on the N-, S-and Cl-modified nano-TiO2 coatings for corrosion protection of stainless steel." *Electrochimica Acta* 52, no. 24 (2007): 6679–6685.

[11] Ferrari, Michele, Alessandro Benedetti, and Francesca Cirisano. "Superhydrophobic coatings from recyclable materials for protection in a real sea environment." *Coatings* 9, no. 5 (2019): 303.

[12] Zhang, Dawei, Luntao Wang, Hongchang Qian, and Xiaogang Li. "Superhydrophobic surfaces for corrosion protection: A review of recent progresses and future directions." *Journal of Coatings Technology and Research* 13, no. 1 (2016): 11–29.

[13] Zhang, Xue-Fen, Yi-Qing Chen, and Ji-Ming Hu. "Robust superhydrophobic SiO2/ polydimethylsiloxane films coated on mild steel for corrosion protection." *Corrosion Science* 166 (2020): 108452.

[14] Drelich, Jaroslaw W. "Contact angles: From past mistakes to new developments through liquid-solid adhesion measurements." *Advances in Colloid and Interface Science* 267 (2019): 1–14.

[15] Nuraje, Nurxat, Waseem S. Khan, Yu Lei, Muhammet Ceylan, and Ramazan Asmatulu. "Superhydrophobic electrospun nanofibers." *Journal of Materials Chemistry A* 1, no. 6 (2013): 1929–1946.

[16] Ijaola, Ahmed Olanrewaju, Peter Kayode Farayibi, and Eylem Asmatulu. "Superhydrophobic coatings for steel pipeline protection in oil and gas industries: A comprehensive review." *Journal of Natural Gas Science and Engineering* 83 (2020): 103544.

[17] Sethi, Sushanta Kumar, and Gaurav Manik. "Recent progress in super hydrophobic/hydrophilic self-cleaning surfaces for various industrial applications: A review." *Polymer-Plastics Technology and Engineering* 57, no. 18 (2018): 1932–1952.

[18] Whyman, Gene, Edward Bormashenko, and Tamir Stein. "The rigorous derivation of Young, Cassie–Baxter and Wenzel equations and the analysis of the contact angle hysteresis phenomenon." *Chemical Physics Letters* 450, no. 4–6 (2008): 355–359.

[19] Gowri, Sorna, Luís Almeida, Teresa Amorim, Noémia Carneiro, António Pedro Souto, and Maria Fátima Esteves. "Polymer nanocomposites for multifunctional finishing of textiles—A review." *Textile Research Journal* 80, no. 13 (2010): 1290–1306.

[20] Sas, Iurii, Russell E. Gorga, Jeff A. Joines, and Kristin A. Thoney. "Literature review on superhydrophobic self-cleaning surfaces produced by electrospinning." *Journal of Polymer Science Part B: Polymer Physics* 50, no. 12 (2012): 824–845.

[21] Jafari, Reza, Siavash Asadollahi, and Masoud Farzaneh. "Applications of plasma technology in development of superhydrophobic surfaces." *Plasma Chemistry and Plasma Processing* 33, no. 1 (2013): 177–200.

[22] Smith, Tennyson, and G. Lindberg. "Effect of acoustic energy on contact angle measurements." *Journal of Colloid and Interface Science* 66, no. 2 (1978): 363–366. ISSN 0021-9797.

[23] Robert, N. Wenzel. "Resistance of solid surfaces to wetting by water." *Industrial & Engineering Chemistry* 28, no. 8 (1936): 988–994.

[24] Callies, Mathilde, and David Quéré. "On water repellency." *Soft Matter* 1, no. 1 (2005): 55–61.

[25] Li, Xue-Mei, David Reinhoudt, and Mercedes Crego-Calama. "What do we need for a superhydrophobic surface? A review on the recent progress in the preparation of superhydrophobic surfaces." *Chemical Society Reviews* 36, no. 8 (2007): 1350–1368.

[26] Cassie, A. B. D., and S. Baxter. "Wettability of porous surfaces." *Transactions of the Faraday Society* 40 (1944): 546–551.

[27] Quéré, David. "Non-sticking drops." *Reports on Progress in Physics* 68, no. 11 (2005): 2495.

[28] Yan, Yu Ying, Nan Gao, and Wilhelm Barthlott. "Mimicking natural superhydrophobic surfaces and grasping the wetting process: A review on recent progress in preparing superhydrophobic surfaces." *Advances in Colloid and Interface Science* 169, no. 2 (2011): 80–105.

[29] Cassie, A. B. D. "Contact angles." *Discussions of the Faraday Society* 3 (1948): 11–16.

[30] Yan, Yu Ying, Nan Gao, and Wilhelm Barthlott. "Mimicking natural superhydrophobic surfaces and grasping the wetting process: A review on recent progress in preparing superhydrophobic surfaces." *Advances in Colloid and Interface Science* 169, no. 2 (2011): 80–105.

[31] Marmur, Abraham. "Wetting on hydrophobic rough surfaces: To be heterogeneous or not to be?." *Langmuir* 19, no. 20 (2003): 8343–8348.

[32] Sun, Taolei, Lin Feng, Xuefeng Gao, and Lei Jiang. "Bioinspired surfaces with special wettability." *Accounts of Chemical Research* 38, no. 8 (2005): 644–652.

[33] Gu, Zhong-Ze, Hiroshi Uetsuka, Kazuyuki Takahashi, Rie Nakajima, Hiroshi Onishi, Akira Fujishima, and Osamu Sato. "Structural color and the lotus effect." *Angewandte Chemie* 115, no. 8 (2003): 922–925.

[34] Feng, Lin, Yanan Zhang, Jinming Xi, Ying Zhu, Nü Wang, Fan Xia, and Lei Jiang. "Petal effect: A superhydrophobic state with high adhesive force." *Langmuir* 24, no. 8 (2008): 4114–4119.

[35] Sun, Manhui, Chunxiong Luo, Luping Xu, Hang Ji, Qi Ouyang, and Dapeng Yu. "Artificial lotus leaf by nanocasting." *Langmuir* 21, no. 19 (2005): 8978–8981.

[36] Guo, Zhiguang, and Weimin Liu. "Biomimic from the superhydrophobic plant leaves in nature: Binary structure and unitary structure." *Plant Science* 172, no. 6 (2007): 1103–1112.

[37] Gao, Jian, Yiliu Liu, Huaping Xu, Zhiqiang Wang, and Xi Zhang. "Biostructure-like surfaces with thermally responsive wettability prepared by temperature-induced phase separation micromolding." *Langmuir* 26, no. 12 (2010): 9673–9676.

[38] Söz, Cagla Kosak, Emel Yilgör, and Iskender Yilgör. "Influence of the coating method on the formation of superhydrophobic silicone–urea surfaces modified with fumed silica nanoparticles." *Progress in Organic Coatings* 84 (2015): 143–152.

[39] Yu, Gan, Xiaolin Chen, and Jie Xu. "Acoustophoresis in variously shaped liquid droplets." *Soft Matter* 7, no. 21 (2011): 10063–10069.

[40] Gao, Xuefeng, and Lei Jiang. "Water-repellent legs of water striders." *Nature* 432, no. 7013 (2004): 36–36.

[41] Li, Yunfeng, Junhu Zhang, and Bai Yang. "Antireflective surfaces based on biomimetic nanopillared arrays." *Nano Today* 5, no. 2 (2010): 117–127.

[42] Parker, Andrew R., and Helen E. Townley. "Biomimetics of photonic nanostructures." *Nature Nanotechnology* 2, no. 6 (2007): 347–353.

[43] Li, Yunfeng, Junhu Zhang, Shoujun Zhu, Heping Dong, Zhanhua Wang, Zhiqiang Sun, Jinrui Guo, and Bai Yang. "Bioinspired silicon hollow-tip arrays for high performance broadband anti-reflective and water-repellent coatings." *Journal of Materials Chemistry* 19, no. 13 (2009): 1806–1810.

[44] Gao, Xuefeng, Xin Yan, Xi Yao, Liang Xu, Kai Zhang, Junhu Zhang, Bai Yang, and Lei Jiang. "The dry-style antifogging properties of mosquito compound eyes and artificial analogues prepared by soft lithography." *Advanced Materials* 19, no. 17 (2007): 2213–2217.

[45] Guo, Zhi-Guang, Wei-Min Liu, and Bao-Lian Su. "A stable lotus-leaf-like water-repellent copper." *Applied Physics Letters* 92, no. 6 (2008): 063104.

[46] Qian, Baitai, and Ziqiu Shen. "Fabrication of superhydrophobic surfaces by dislocation-selective chemical etching on aluminum, copper, and zinc substrates." *Langmuir* 21, no. 20 (2005): 9007–9009.

[47] Qu, Mengnan, Bingwu Zhang, Shiyong Song, Li Chen, Junyan Zhang, and Xiaoping Cao. "Fabrication of superhydrophobic surfaces on engineering materials by a solution-immersion process." *Advanced Functional Materials* 17, no. 4 (2007): 593–596.

[48] Li, Mei, Jianhai Xu, and Qinghua Lu. "Creating superhydrophobic surfaces with flowery structures on nickel substrates through a wet-chemical-process." *Journal of Materials Chemistry* 17, no. 45 (2007): 4772–4776.

[49] Liang, Jun, Zhiguang Guo, Jian Fang, and Jingcheng Hao. "Fabrication of superhydrophobic surface on magnesium alloy." *Chemistry Letters* 36, no. 3 (2007): 416–417.

[50] Hou, Xianming, Feng Zhou, Bo Yu, and Weimin Liu. "Superhydrophobic zinc oxide surface by differential etching and hydrophobic modification." *Materials Science and Engineering: A* 452 (2007): 732–736.

[51] Wang, Shutao, Yanlin Song, and Lei Jiang. "Microscale and nanoscale hierarchical structured mesh films with superhydrophobic and superoleophilic properties induced by long-chain fatty acids." *Nanotechnology* 18, no. 1 (2006): 015103.

[52] Khorsand Shohreh, Raeissi Keyvan, Ashrafizadeh Fakhredin. "Corrosion resistance and long-term durability of super-hydrophobic nickel film prepared by electrodeposition process." *Applied Surface Science* 305 (2014): 498–505.

[53] Li, Ying, Wen-Zhi Jia, Yan-Yan Song, and Xing-Hua Xia. "Superhydrophobicity of 3D porous copper films prepared using the hydrogen bubble dynamic template." *Chemistry of Materials* 19, no. 23 (2007): 5758–5764.

[54] Han, Jinpeng, Mingyong Cai, Yi Lin, Weijian Liu, Xiao Luo, Hongjun Zhang, Kaiyang Wang, and Minlin Zhong. "Comprehensively durable superhydrophobic metallic hierarchical surfaces via tunable micro-cone design to protect functional nanostructures." *RSC Advances* 8, no. 12 (2018): 6733–6744.

[55] Martines, Elena, Kris Seunarine, Hywel Morgan, Nikolaj Gadegaard, Chris DW. Wilkinson, and Mathis O. Riehle. "Superhydrophobicity and superhydrophilicity of regular nanopatterns." *Nano Letters* 5, no. 10 (2005): 2097–2103.

[56] Ma, Minglin, Yu Mao, Malancha Gupta, Karen K. Gleason, and Gregory C. Rutledge. "Superhydrophobic fabrics produced by electrospinning and chemical vapor deposition." *Macromolecules* 38, no. 23 (2005): 9742–9748.

[57] Vourdas, Nikolaos, Angeliki Tserepi, and Evangelos Gogolides. "Nanotextured superhydrophobic transparent poly (methyl methacrylate) surfaces using high-density plasma processing." *Nanotechnology* 18, no. 12 (2007): 125304.

[58] Teshima, Katsuya, Hiroyuki Sugimura, Yasushi Inoue, Osamu Takai, and Atsushi Takano. "Transparent ultra water-repellent poly (ethylene terephthalate) substrates fabricated by oxygen plasma treatment and subsequent hydrophobic coating." *Applied Surface Science* 244, no. 1–4 (2005): 619–622.

[59] Huang, Zhongbing, Ying Zhu, Jihua Zhang, and Guangfu Yin. "Stable biomimetic superhydrophobicity and magnetization film with Cu-ferrite nanorods." *The Journal of Physical Chemistry C* 111, no. 18 (2007): 6821–6825.

[60] Li, Yue, Weiping Cai, and Guotao Duan. "Ordered micro/nanostructured arrays based on the monolayer colloidal crystals." *Chemistry of Materials* 20, no. 3 (2008): 615–624.

[61] Tsai, Hui-Jung, and Yuh-Lang Lee. "Facile method to fabricate raspberry-like particulate films for superhydrophobic surfaces." *Langmuir* 23, no. 25 (2007): 12687–12692.

[62] Li, Yue, Eun Je Lee, and Sung Oh Cho. "Superhydrophobic coatings on curved surfaces featuring remarkable supporting force." *The Journal of Physical Chemistry C* 111, no. 40 (2007): 14813–14817.

[63] Ming, Wu, Di Wu, R. van Benthem, and G. De With. "Superhydrophobic films from raspberry-like particles." *Nano Letters* 5, no. 11 (2005): 2298–2301.

[64] Shiu, Jau-Ye, Chun-Wen Kuo, Peilin Chen, and Chung-Yuan Mou. "Fabrication of tunable superhydrophobic surfaces by nanosphere lithography." *Chemistry of Materials* 16, no. 4 (2004): 561–564.

[65] Zhang, Jilin, Jian Li, and Yanchun Han. "Superhydrophobic PTFE surfaces by extension." *Macromolecular Rapid Communications* 25, no. 11 (2004): 1105–1108.

[66] Jiang, Lei, Yong Zhao, and Jin Zhai. "A lotus-leaf-like superhydrophobic surface: A porous microsphere/nanofiber composite film prepared by electrohydrodynamics." *Angewandte Chemie* 116, no. 33 (2004): 4438–4441.

[67] Feng, Xinjian, Lin Feng, Meihua Jin, Jin Zhai, Lei Jiang, and Daoben Zhu. "Reversible super-hydrophobicity to super-hydrophilicity transition of aligned ZnO nanorod films." *Journal of the American Chemical Society* 126, no. 1 (2004): 62–63.

[68] Khorasani Mohammad Taghi, Mirzadeh Hamid, Kermani, Z. "Wettability of porous polydimethylsiloxane surface: morphology study." *Applied Surface Science* 242, no. 3–4 (2005): 339–345.

[69] Devesh Tripathi, Frank R Jones. "Single fibre fragmentation test for assessing adhesion in fibre reinforced composites." *Journal of Materials Science* 33, no. 1 (1998): 1–16.

[70] Kako, T., A. Nakajima, H. Irie, Z. Kato, K. Uematsu, T. Watanabe, and K. Hashimoto. "Adhesion and sliding of wet snow on a super-hydrophobic surface with hydrophilic channels." *Journal of Materials Science* 39, no. 2 (2004): 547–555.

[71] Nguyen-Tri, Phuong, Hai Nguyen Tran, Claudiane Ouellet Plamondon, Ludovic Tuduri, Dai-Viet N. Vo, Sonil Nanda, Abhilasha Mishra, Huan-Ping Chao, and Bajpai Anil Kumar. "Recent progress in the preparation, properties and applications of superhydrophobic nano-based coatings and surfaces: A review." *Progress in Organic Coatings* 132 (2019): 235–256.

[72] Samal Subhranshu Sekhar, and Manohara Shambonahalli Rajanna. "Nanoscience and nanotechnology in India: A broad perspective." *Materials Today: Proceedings* 10 (2019): 151–158.

[73] Sun, Ke, Huan Yang, Wei Xue, An He, Dehua Zhu, Wenwen Liu, Kenneth Adeyemi, and Yu Cao. "Anti-biofouling superhydrophobic surface fabricated by picosecond laser texturing of stainless steel." *Applied Surface Science* 436 (2018): 263–267.

[74] Banerjee, Indrani, Ravindra C. Pangule, and Ravi S. Kane. "Antifouling coatings: Recent developments in the design of surfaces that prevent fouling by proteins, bacteria, and marine organisms." *Advanced Materials* 23, no. 6 (2011): 690–718.

[75] Callow, J. A., and M. E. Callow. "Trends in the development of environmentally friendly fouling-resistant marine coatings." Nature Communications 2 (2011): 244.

[76] Trdan, Uroš, Matej Hočevar, and Peter Gregorčič. "Transition from superhydrophilic to superhydrophobic state of laser textured stainless steel surface and its effect on corrosion resistance." *Corrosion Science* 123 (2017): 21–26.

[77] Qi, Ruolong, Weijun Liu, Hongyou Bian, and Lun Li. "Five-axis rough machining for impellers." *Frontiers of Mechanical Engineering in China* 4, no. 1 (2009): 71–76.

[78] Jasim, Khlowd Mohammed, and Luma M. Ahmed. "TiO2 nanoparticles sensitized by safranine O dye using UV-A light system." *IOP Conference Series: Materials Science and Engineering* 571, no. 1 (2019): 012064.

[79] Pang, Li Qing, Li Juan Zhong, Hui Fang Zhou, Xue E. Wu, and Xiao Dong Chen. "Grafting of ionic liquids on stainless steel surface for antibacterial application." *Colloids and Surfaces B: Biointerfaces* 126 (2015): 162–168.

[80] Ye, Yuwei, Zhiyong Liu, Wei Liu, Dawei Zhang, Haichao Zhao, Liping Wang, and Xiaogang Li. "Superhydrophobic oligoaniline-containing electroactive silica coating as pre-process coating for corrosion protection of carbon steel." *Chemical Engineering Journal* 348 (2018): 940–951.

[81] Xiang, Tengfei, Depeng Chen, Zhong Lv, Zhiyan Yang, Ling Yang, and Cheng Li. "Robust superhydrophobic coating with superior corrosion resistance." *Journal of Alloys and Compounds* 798 (2019): 320–325.

[82] Xu, Hongfei, Ben Bbosa, Eduardo Pereyra, Michael Volk, and M. Sam Mannan. "Oil transportation in pipelines with the existence of ice." *Journal of Loss Prevention in the Process Industries* 56 (2018): 137–146.

[83] Wu, Xinghua, Vadim V. Silberschmidt, Zhong-Ting Hu, and Zhong Chen. "When super-hydrophobic coatings are icephobic: Role of surface topology." *Surface and Coatings Technology* 358 (2019): 207–214.

[84] Latthe, Sanjay S., Rajaram S. Sutar, Appasaheb K. Bhosale, Saravanan Nagappan, Chang-Sik Ha, Kishor Kumar Sadasivuni, Shanhu Liu, and Ruimin Xing. "Recent developments in air-trapped superhydrophobic and liquid-infused slippery surfaces for anti-icing application." *Progress in Organic Coatings* 137 (2019): 105373.

[85] Pan, Sai, Nan Wang, Dangsheng Xiong, Yaling Deng, and Yan Shi. "Fabrication of superhydrophobic coating via spraying method and its applications in anti-icing and anti-corrosion." *Applied Surface Science* 389 (2016): 547–553.

7 Microstructural and Chemical Characterization Techniques of Coatings
State of the Arts

Yogendra Kumar, Prerna Yogeshwar,
Ganesh Kumar, and Jitendra S. Sangwai
Indian Institute of Technology Madras

Deepak Dwivedi
Rajiv Gandhi Institute of Petroleum Technology

CONTENTS

ABBREVIATIONS

AFM	Atomic force microscopy
CVD	Chemical vapour deposition
DLS	Dynamic light scattering
DSC	Differential scanning calorimetry
EIS	Electrochemical impedance spectroscopy
FTIR	Fourier transform infrared spectroscopy
HRTEM	High-resolution transmission electron microscopy
NMR	Nuclear magnetic resonance
NPs	Nanoparticles
OM	Optical microscopy
PD	Physical deposition
PVD	Physical vapour deposition
RBS	Rutherford backscattering spectroscopy
SEM	Scanning electron microscopy
XPS	X-ray photoelectron spectroscopy
XRD	X-ray diffraction
XRF	X-ray fluorescence

7.1 INTRODUCTION

Surface coating is instrumental to a variety of applications such as membrane preparation process, corrosion protection and creation of omniphobic surfaces. Surface coatings can be applied to nanomaterials, films, membranes, plates and rough surfaces to improve their microstructural and chemical properties. In context with nanomaterials, surface coating improves tunability, functionalization and effectiveness of system and brings down their toxicity for biomedical applications, attributing to the concealment of surface charge and composition by the coatings [1–3]. The coatings on nanomaterials are often degradable, leaving toxic residual materials in their original environment. Surface coatings of nanomaterials have been reported to cause severe inflammatory and immunological responses [4]. In the case of flat surfaces, coating improves surface rigidity, corrosion resistance, ablation resistance and lustrous properties. The development of nanostructured coatings through functionalized nanoparticles for applications ranging from barrier properties, wear, oxidation and abrasion resistance to corrosion resistance has gained increasing interest recently [5]. Coating preference depends on a variety of factors, including service environment, component shape and size, and substrate material compatibility. Coating processes vary widely in thickness, from a few microns to several millimetres, allowing many different types of materials to be deposited. The categorization of coating is possible in several ways, for instance, depending on how the coating material is applied to the substrate surfaces; one common deposition approach is used. Deposition methods include atomic deposition, particulate deposition and bulk coatings [6,7].

Coating is characterized based on their thickness and classified into sheet, thick films, foils and thin-film coating as shown in Figure 7.1. Majority of engineering applications use thin-film coatings as they have improved adhesion characteristics

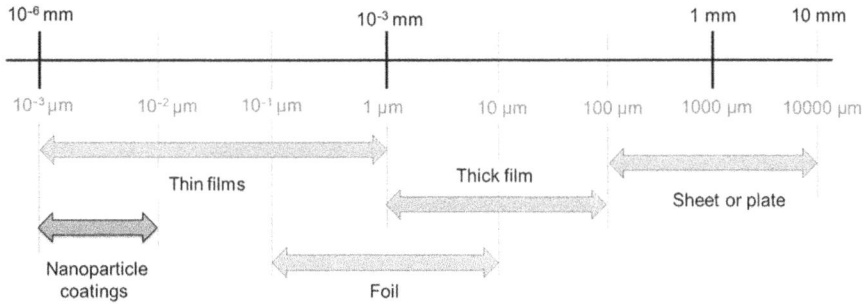

FIGURE 7.1   Scale of coatings based on thickness of film or coating.

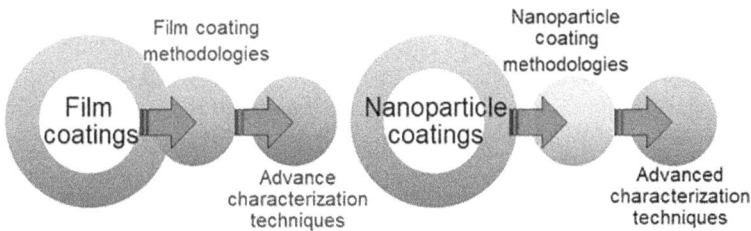

FIGURE 7.2   Flow scheme of current article.

and does not affect size of substrate. Thin films cover substrates with thicknesses ranging from monolayers to hundreds of micrometres. Normally, the coating is considered to serve as a protective layer of the film and wear is the primary function of a hard coating. The thickness can be narrowed to a micrometre, and the materials can be chosen based on their hardness and chemical stability. Depending on their composition, films/coatings may be homogeneous in-depth, layered, or gradient. Thin-film coatings are usually 300–500 µm thick and physically or chemically deposited over substrate via a chemical reaction between the hot substrate and inert gases in a low-pressure chamber. The thin-film coating is done by either (1) chemical deposition or (2) physical deposition. The chemical deposition techniques are further classified into plating, atomic layer deposition, chemical vapour deposition (CVD) and spin coating [6]. Contrarily, in a physical deposition, the precursors (either solid, liquid or gas) are physically moved over the substrate surface. Physical deposition (PD) techniques such as ion plating, thermal evaporation and sputtering were utilized in the recent past. However, the choice of deposition techniques will depend on the source, deposition rate, substrate structure and operating temperature [7]. The patterned deposition of thin films is done via photolithographic or chemical etching techniques, where undesired deposited materials are etched away from the substrate.

Another type of coating that reduces nanoparticle cytotoxicity and associated poisoning is chemical passivation. Nanoparticles coated with porous structure are termed as core-shell systems and widely used in biomedical, sensing and catalysis applications. The scheme of this outlined article is depicted pictorially in Figure 7.2.

The current article discusses several advanced coating and characterization techniques used to prepare a multifunctional surface for a multifarious surface. The article discussed both thin-film and nanoparticle coating and characterization techniques. The future prospects of modern coating and role of characterization techniques in development of modern strategies were also discussed.

7.2   FEATURES AND PROPERTIES OF THIN-FILM COATING

In spite of their extreme thermal stability and reasonable hardness, thin films are fragile [8]. Alternatively, organic materials are tough, but also soft and have reasonable thermal stability [9]. Four material properties play an important role in mechanical device stability: elastic modulus, yield strength, interfacial adhesion and fracture toughness of the film. There is often a difference between the mechanical properties of thin films and those of bulk materials. As a result of the nanostructure and substrate attachment of thin films, this can be partially explained. Typically, thin films can withstand very high residual stresses due to their high yield strengths. Plastic deformation, thin-film fracture or interfacial delamination can relieve this residual stress later in the processing or during actual device operation. In order to characterize thin films, it is important to consider both elastic and plastic properties.

7.3   TYPES OF COATING

The types of coating depend on surface features, properties and characteristics. Depending on surface features and smoothness, the coating is divided into hard and soft coating. Hard coatings are used to reduce wear in mechanical systems, and soft coatings are used to passivate or smoothen substrate structure. Depending on homogeneity/heterogeneity or composition, coatings are divided into single-layer, double-layer, gradient, multilayer and composite coatings. The pictorial depiction of different types of film coated materials is shown in Figure 7.3. The choice of films and materials is depending on application, i.e., wear resistant coatings are usually hard at surface and aesthetic coatings are soft in nature.

   In hard coatings over polymer substrates, the designing strategy of functionalized coating is vital for the durability of the coating. Various stoichiometry forms of SiO_xC_y:H materials were used to design functional protective coatings for polymeric

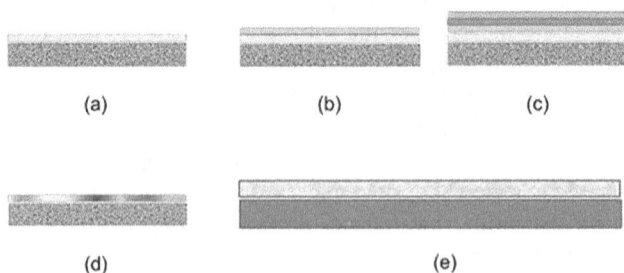

FIGURE 7.3   Types of film coatings: (a) single-layered, (b) double-layered, (c) multi-layered, (d) Gradient, and (e) composite film coating.

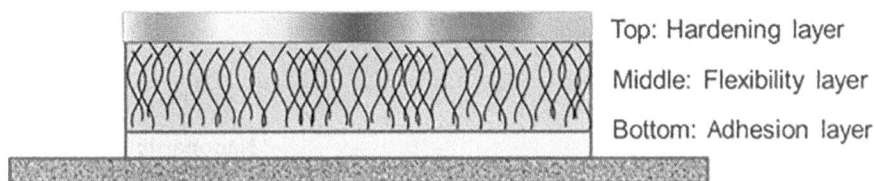

Top: Hardening layer

Middle: Flexibility layer

Bottom: Adhesion layer

FIGURE 7.4 Pictorial depiction of hard coating structure consisting of adhesion, flexibility and hardening layers.

substrates. The coatings had three layers that solve adhesion, flexibility and hardening problems separately. Various ratios of HMDSO are applied over polymer substrates to produce these layers, named for their different functions. The hard coatings are prepared via incorporating three layers that serve three different functions, i.e., adhesion, flexibility and mechanical strength. The pictorial depiction of these three-layer structures is shown in Figure 7.4.

As the most polymeric (richer in carbon and hydrogen) of the three layers, the first layer (adhesion) ensures good bonding to the substrate. For good adhesion to the substrate and good mechanical performance, this layer should be thicker than 600 nm. As the ratio of oxygen to precursor increased from 0.42 to 9.0, the concentration of the reactants was dynamically changed during the deposition process. Through this process, researchers aim to gradually promote the combustion of the organometallic precursor's organic chains, thereby enhancing SiO_2-like structures. A layer of surface hardness is gradually increased by this method, even though some CHx radicals may remain trapped in the film that is being deposited from the precursor. Over half of the overall thickness of the hard coating is made up of the second layer (flexibility), which is about 1 mm thick. In addition to providing flexibility to the coating, the second layer is deposited at the highest rate. To achieve a lower quantity of Si–O–Si bonds, incomplete combustion mechanisms are preferred over Si–CHx mechanisms. The hardness of the coating is provided by a third hardening layer aiming to obtain a layer close to stoichiometric SiO_2 to ensure index matching. The layer should be highly mechanically strong so that it can protect the substrate from external abrasions. The deposition rates of adhesion, flexibility and hardening layers are 0.5–5, 3–5, and 0.5 nm/s, respectively [10]. The brief classification of various coatings is given in Figure 7.5.

7.4 METHODOLOGIES

Dip coating, drop casting, electrospinning/electrospraying, optical deposition and layer-by-layer deposition are among the common coating techniques. The simplest of these techniques is drop casting, while optical deposition is able to monitor the coating process without fully understanding its mechanism. An electrospinning/electrospraying process begins with a large potential difference between the needle of the syringe (which acts as an electrode) and the ground surface. In electrospraying, nanoparticles are formed on the surface depending on the concentration of the solution and the distance between electrodes. In electrospinning, nanofibres are formed, and in electro spraying, nanoparticles are formed [11,12]. An ultra-thin film containing nanostructured materials can be obtained by layer-by-layer deposition, using one or more of

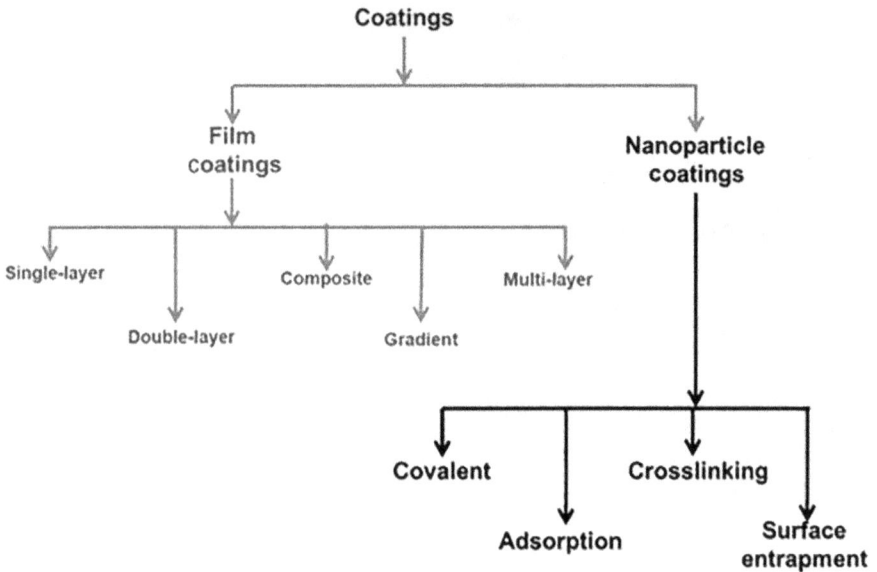

FIGURE 7.5 Classification of film and nanoparticle coating.

the above-mentioned techniques. In layer-by-layer deposition, the oppositely charged materials are deposited by a suitable coating method over a charged substrate.

   In drop casting, small films with low adhesion are formed by directly dropping nanoparticles or polymeric solution over substrate. The solution casting process is followed by subsequent drying for better adhesion over substrate. Contrarily, dip casting is able to deposit monolayered film over substrate. The substrate is directly immersed into solution for material deposition. The immersion process is followed by curing to remove excess material, drying and heating cycles. Firstly, a solution containing coating material must be immersed in the substrate at a constant speed. Upon remaining inside the solution for some time, the substrate is pulled up, at a constant speed, with a thin layer of material over the substrate. Following drainage of excess liquid from the support, the solvent evaporates from the liquid, forming a thin layer. A number of parameters influence the film properties and thickness, such as humidity levels, immersion time, concentration and temperature, withdrawal speed, number of dipping cycles and solution composition [13,14]. Apart from these aforesaid techniques, chemical vapour deposition, physical vapour deposition and sputtering are other coating techniques used extensively in substrate film coating.

7.5 ADVANCED CHARACTERIZATION TECHNIQUES FOR THIN-FILM COATINGS

It might be a good idea to ask the following question: what is the amount/characteristics of property in a thin film? There is a huge difference between properties according to where they are located on a film. Adhesion is primarily a property of the film/substrate interface, while surface tension is a property of the top layer. It is common practice to exclude the interface and top layer from determining film composition. When it comes to properties like wear resistance, the whole structure also includes the substrate. In order to evaluate a characterization technique, we must consider the interaction volume. A large interaction volume will primarily force us to measure the substrate's properties. In general, if it is small, the results apply only to a small fraction of the film (e.g., a single grain) and cannot be extrapolated for the entire film. Auger electron spectroscopy is a typical example. It penetrates a few nanometres into the surface, yielding only the top layer of native oxide. Various microstructural, topographical, mechanical and weathering characterizations are carried out to test effectiveness, microstructural, thermal, intrinsic, interaction and performance properties of nanocoating/thin-film coatings. The pictorial depiction of different characterizing properties and characterization techniques is given in Figure 7.6.

X-ray fluorescence (XRF), X-ray diffraction (XRD), X-ray photoelectron spectroscopy (XPS), atomic force microscopy (AFM), Hall measurements and electrical conductivity measurements were used to characterize thin films. From in situ measurements of electrical conductivity, abnormal electrical transport behaviour was observed. Freestanding films can be tested for tensile properties, microbeam cantilever deflection can be used, but nanoindentation is the easiest way to measure thin-film mechanical properties since no special preparation is necessary and tests can be conducted quickly and affordably. For measuring the mechanical properties of films, nanoindentation is an excellent technique. It is used to measure force and indentation depth simultaneously by forcing a sharp diamond indenter into the test material.

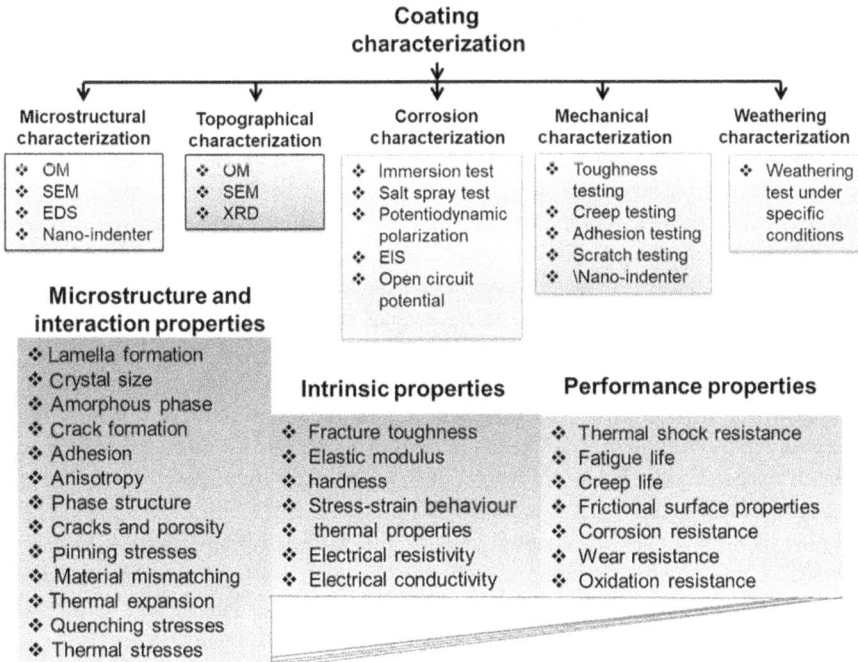

FIGURE 7.6 Pictorial depiction of different characterization techniques for coating with their characterizing properties.

In coating, properties such as toughness, tribological properties, fatigue, adhesion, residual stresses and dimensions are very important since CVD and PVD thin-film materials are very hard and brittle. These parameters have been quantified using experimental-analytical testing procedures. Besides providing information about the actual geometry of the coated tool, these also describe the film's material and functional properties. The combination of these procedures and FEM-supported computations contributes to understanding the failure mechanisms of cutting tool films, thereby reducing the experimental cost of optimizing cutting conditions. Temperature-dependent film strength properties can be estimated by nanoindentation at elevated temperatures. The toughness and brittleness of micro-blasting coatings will be characterized using a nano-impact test. An inclined impact test, a scratch test, and a Rockwell test can be used to determine film adhesion. A nano-scratch test can be used to evaluate the adhesion of thin films (0.5 mm) [15]. Despite the fact that Rockwell and scratch tests are often considered reliable for assessing film adhesion, they are not always accurate. Confocal measurements can be used to investigate cutting-edge roundness changes on coated tools, for example, after micro-blasting. The corresponding tool wedge radii, the average value and the fluctuations of the roundness of the cutting edges can be determined by monitoring successive cross-sections of the cutting edges before and after micro-blasting at various pressures.

X-ray diffraction (XRD) analysis of coating gives insights into the crystallographic structure of coatings. Moreover, XRD scan makes the identification of crystallographic defects and amorphous and crystalline regions possible. The microstructural insights of coatings are usually done using SEM micrographs, and EDS analysis is useful for carrying out elemental mapping of coating. Insights about lamellae, defects and cracks can also be visualized via HRSEM/SEM micrograph. Functional protective coatings on hard polymeric substrates are usually characterized using Fourier transform infrared (FTIR) spectroscopy, X-ray photoelectron spectroscopy (XPS) and Rutherford backscattering spectroscopy (RBS). FTIR helps in detecting functional groups that assist in determining structural stability and flexibility of hardening, flexible and adhesion layers. XPS provides compositional analysis in the atomic percentage of films near the surface (~5 nm). In addition to complementing and further supporting the FTIR and XPS results, RBS provides in-depth compositional analysis [10]. In the case of C and O, RBS analysis is particularly relevant since the XPS signals originate near the surface and may be affected by surface contamination, despite previous sputtering.

FTIR and DSC characterizations play a vital role in characterizing organic coating and curing mechanisms. The presence of crystalline and semicrystalline regions affects organic coatings' flexibility and adhesion characteristics. A complex interplay of intermolecular association and crosslinking is thought to control the mechanical properties of coatings via pendant linear alkyl chains. EIS studies are important for capacitance measurement of coating and differences in barrier properties. Having a lower capacitance means a coating has a higher barrier property and is more corrosion-resistant. Oxidative crosslinking of coating facilitates a higher barrier for the transport of ionic species through the coating and offers the best corrosion resistance [16].

Electron probe microanalysis (EPMA) was used to determine the chemical composition of the samples. In EPMA, an energy-dispersive X-ray (EDX) detector and four wavelength detectors were used. Moreover, oxidation and thermal stability analysis of coatings can be performed by capturing chemical/microstructural transformation under continuous heating through in situ TEM/EELS analysis [17].

According to Masdek and Alfantazi [18], a homogenous passive layer can be grown more effectively on a nanocrystalline structure by controlling alloy composition and grain size and properly refining grain size [18]. Nanocrystalline substances with high grain boundaries promote selective oxidation of protective oxides and can be more readily adhered to surfaces due to their fast diffusion paths. The percentage of passivating metals (such as Al and Cr) for the complete layered film was significantly reduced. There are numerous benefits to incorporating nanoparticles into mechanically, biologically, optically and corrosion-resistant coatings [19]. As a result of inactivating the unprotected metal surfaces, primarily in the faulty areas using the scanning vibrating electrode approach, corrosion inhibition performance was enhanced. Nanocontainers with more layers have a smaller average diameter measured using light scattering experiments. First, PEI and PSS monolayers show an increase of about 8 mm, while the benzotriazole layer shows an increase of about 4 mm. In order to obtain the self-healing property of the ultimate layer, one bilayer is not adequate, while more than three bilayers can lead to an increased collection of nanocontainers during layer deposition and assembly, altering the protective layer matrix.

7.6   COATED NANOPARTICLES

It is common for nanomaterials to have properties that are inherited from isolated particles. Usually, nanoparticles lose their special properties after being combined with a macroscopic workpiece. Consequently, if macroscopic parts are to exhibit the properties of isolated particles, particle interactions must be avoided or reduced. A second layer of ceramic or polymer can be applied to each particle to accomplish this. Nanoparticle coating or passivation is primarily done to reduce poisoning rate and cytotoxicity of nanoparticles. Specialized coatings also facilitate special functionalization for certain applications such as biofuel production, CO_2 capture and biomedical applications. Highly reactive nanoparticles are coated with a porous layer to improve its multifunctional properties, sensing and poisoning resistance. Such systems are termed as core-shell nanoparticles. Similarly, core-corona-canopy functionalized coating improves fluidization characteristics of nanoparticles without dispersing it into fluids. Such types of systems are already utilized in solid electrolytes and CO_2 capture applications. The enzyme immobilization, surface passivation of core via inert shell, dye immobilization and functional group immobilization are quite common coating techniques used these days. Pictorial depictions of some of the coated/immobilized nanoparticles are shown in Figure 7.7.

The coating of nanoparticles is usually done via polymerization or sol–gel methods. Physical absorption, covalent attachment and layer-by-layer nanofabrication are commonly used polymer surface conjugation techniques [20]. In microwave plasma processes, particles leave the plasma zone with electric charges, thus preventing agglomeration, so these materials can only be synthesized using this process.

Moreover, sol–gel methods have more control over surface morphology and texture than polymerization methods [21]. Hydrolysis, condensation and drying are the main steps in the sol–gel method that deliver the final metal oxides. Metal precursors undergo rapid hydrolysis to produce metal hydroxide solutions, followed by immediate condensation, which forms a three-dimensional gel [22]. As a result, the resulting gel is readily converted to a xerogel or aerogel depending on how it is dried. According to the solvent used, sol–gel methods can be classified into aqueous and

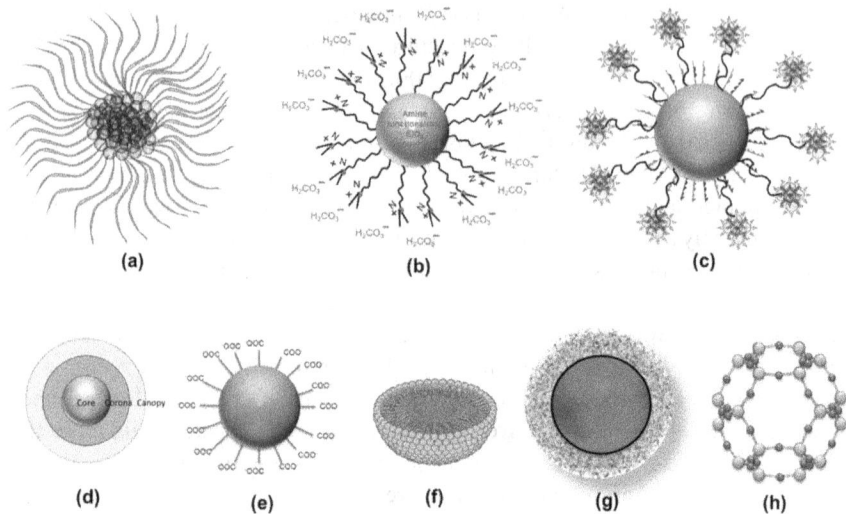

FIGURE 7.7 Pictorial depiction of functionalized/coated nanoparticles: (a) enzyme immobilized nanoparticle, (b) amine group immobilized nanoparticle, (c) dye/biomolecule immobilized nanoparticle, (d) core-corona-canopy nanoparticle, (e) surface stabilized nanoparticle, (f) micelle loaded nanoparticle, (g) core-shell nanoparticle, and (h) nano-MOFs.

non-aqueous. The sol–gel synthesis of polymer encapsulated nanoparticles under probe sonication may be effective in controlling size and mixing of reagents during nucleation. Nanoparticle synthesis in probe sonicated environments may also help in reducing thickness of coating over substrate. The pictorial depiction of probe sonicated sol–gel synthesis of nickel nanoparticles is shown in Figure 7.8a and b.

Many special purpose coated nanoparticles were prepared by many researchers using sol–gel methods. The type of functionalized coated nanoparticles is discussed below.

7.6.1 Functional Groups Immobilized Nanoparticles

Functional group immobilized nanoparticles are vital for biomedical fields where biomolecules are added in conjugation with nanoparticles [23,24]. As nanoparticles are functionalized, molecular interactions between them can be analysed, background signals are reduced, specificity is improved, and the use of nanoparticles for therapy and imaging can be performed simultaneously. In addition to improving targeting efficacy, nanoparticle functionalization reduces toxic effects. Nanoparticles with external moiety improve internalization, biocompatibility, imaging and specificity of targeted drug delivery systems [25]. Surface modification usually begins with the use of homo- or heterobifunctional crosslinkers containing organic functional groups ($R-NH_2$, $R-COOH$, etc.), which are intended to bind biological molecules (e.g., nanodrugs, dendrimers, tumour markers, carbohydrates, dyes, peptides and antibodies) [26–28]. Alternatively, enzyme functionalized nanoparticles have demonstrated their utility in biofuel production stimulation. Biofuels production processes such as hydrolysis, transesterification and methanolysis are stimulated by enzyme immobilized nanoparticles

(a)

$$N_2H_4 + 4\,OH^- \rightarrow 4\,H_2O + N_2 + 4\,e^-$$

N_2H_4

N_2

OH⁻

• e⁻

$$2\,Ni^{2+} + 4\,e^- \rightarrow 2\,Ni$$

Ni^{2+}

PVP encapsulated
transition complex

(b)

NiCl₂.6 H₂O, Water and
Ethanol (1:1) molar

Hydrazine
hydrates

Drop wise hydrazine
casting

PH= 13-14
T =65 °C

Ni 2⁺
Ions

N_2H_4

OH⁻

O

O

O

Nucleation in
Magnetic field

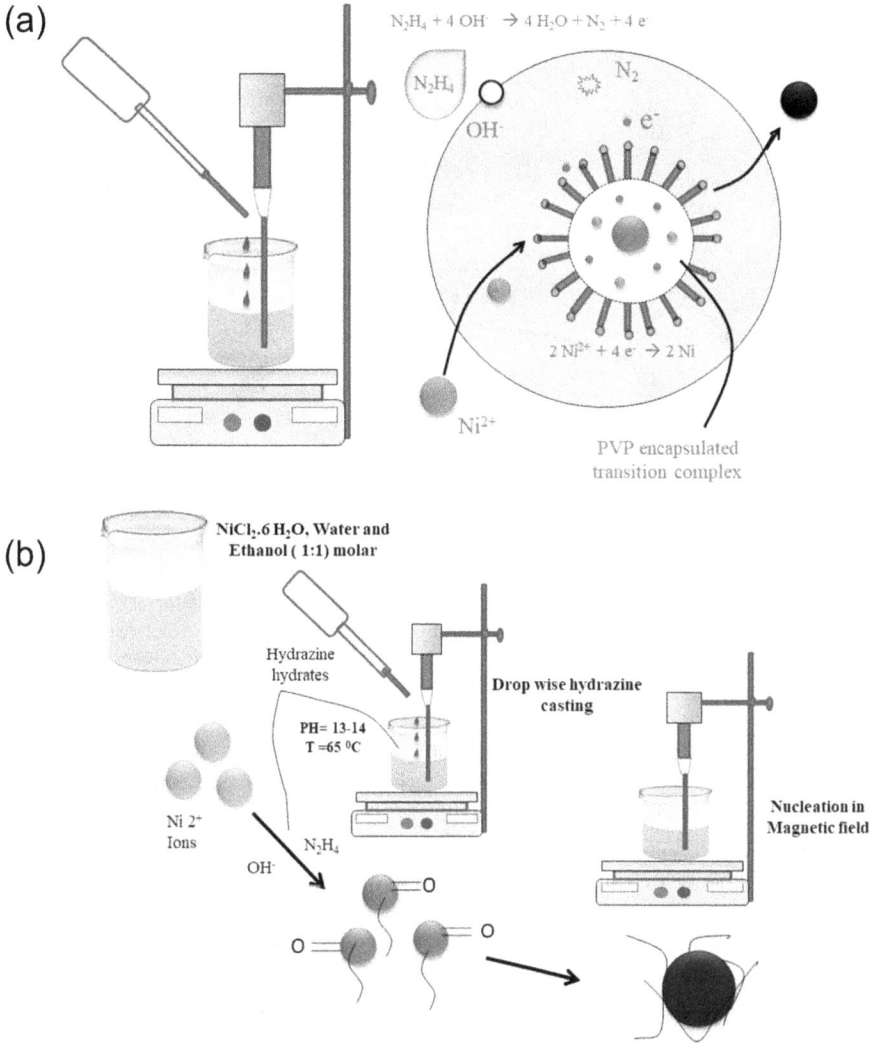

FIGURE 7.8 Pictorial depiction of probe sonication nanoparticle synthesis route (a) probe sonicated synthesis of coated nanoparticles in polymeric solution and (b) probe sonicated synthesis of metallic nickel nanoparticle using hydrazine.

[29]. Immobilization of nanoparticles over the wall of continuous capillary reactors may effectively enhance perturbation and mixing; thus, these systems effectively reduce separability and recovery cost associated with suspended nanoparticle systems [22,30].

Furthermore, immobilization of nanoparticles over substrate material through covalent immobilization chemistry gains much attention to develop functionalized surfaces. A stable complex is formed when enzymes/nanoparticles are covalently immobilized to the support matrix. Among the enzyme/nanoparticles functional groups used in covalent

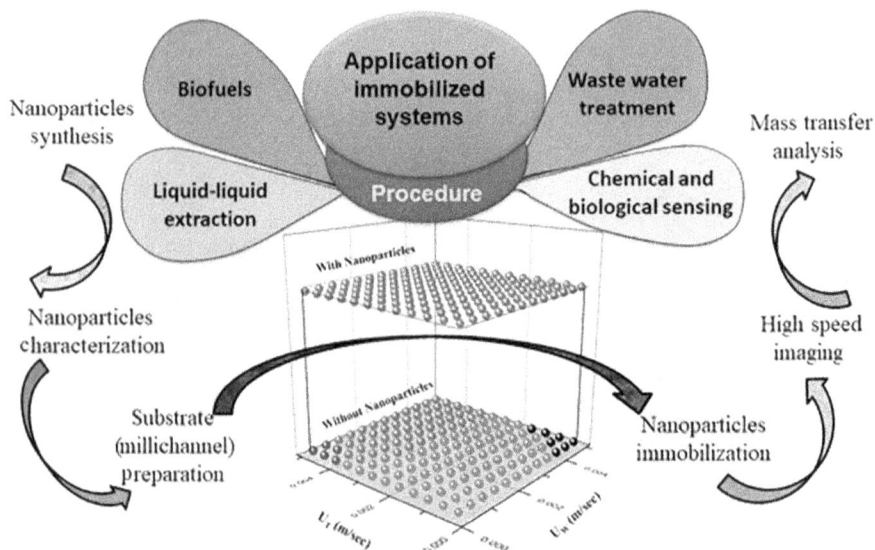

FIGURE 7.9 Nanoparticle coated substrate, process route and application. (Flow pattern diagram is taken from permission Jaiswal et al. [22]. Copyright (2022), American Chemical Society.)

coupling are amino groups, carboxylic groups, sulfhydryl groups, indoles, thiol groups, imidazole groups, phenolic groups and hydroxyl groups [31]. It has generally been found that enzymes are bound to solid supports in two stages. The first step is to activate the solid support using linkers such as glutaraldehyde or carbodiimide, and the second step is to covalently couple the enzyme to the activated support. Through covalent bonding, linker molecules serve as a multifunctional reagent between the enzyme and the surface. Biocatalysts are covalently immobilized when enzymes are attached to NP through covalent bonds. Covalent bonds protect enzyme leakage from surface matrixes when enzymes and support matrixes are strong. They also improve thermal stability. Covalent binding of the enzyme restricts the enzyme's conformation, which deactivates it. The surfaces can be used in catalysis, water treatment, biofunctionalization and separation of chemicals. Pictorial depiction of the process route of nanoparticle immobilized surfaces for multifarious applications is shown in Figure 7.9. Surface immobilization on the wall of capillary microreactors, and membrane systems improve overall efficacy of system. The another benefit of surface immobilized system is reduction in separability cost as compare to nanoparticles dispersed/nanofluidic system. However, nanoparticles leaching associated with surface immobilized system confronting large-scale utilization of immobilized systems.

7.6.2 CORE-SHELL NANOPARTICLES

Core-shell nanoparticles (CSNs) are nanostructured materials with interesting properties and a wide range of applications in catalysis, biology, materials chemistry and sensors. They have recently attracted increased attention due to their interesting properties and broad range of applications. A range of core-shell nanoparticles with tailorable properties can be produced by rationally tuning both the core and shell of such materials [32]. These nanoparticles can play critical roles in a number of catalytic processes and provide sustainable solutions to energy shortages today. These methods include the Stöber method, the solvothermal method, the one-pot method using surfactants, etc. for preparing different classes of CSNs [33]. CSNs are used for oxidation, reduction, coupling reactions and other catalytic and electrocatalytic applications. Core-shell nanostructure properties are determined by the substances in the core or shell. A core-shell function can be controlled by altering the materials constituting the core and shell. Coating substances increase stability and control the release of core materials or functionalize them. In biomedicine, core-shell nanostructures are mostly used for bioimaging, controlled drug release, drug delivery and other biomedical applications [20].

There are two types of core-shell nanoparticles: organic and inorganic, depending on the coating substance.

7.6.3 Organic Coated Core-Shell NPs

The functionalization of NP surfaces decreases as a result of ionization and clustering over a long period of time; coating them with organic materials (e.g., xanthate, thiol-derivatives, etc.) plays an important role in shielding against ionization processes by using the layer-by-layer technique [34–36]. It is possible to enhance their optical properties, dispersibility, inertness and cytotoxicity when they are uniformly coated with silica (instead of a coating of glass) as an insulator material [20].

7.6.4 Inorganic Coated Core-Shell NPs

Nanomaterials with non-organic coatings have been found to have different applications in solid-state electronics, televisions and computers [20]. Aside from their potential applications in biomedical science, these substances could also be applied to coatings depending on their specific configurations. In terms of nanomaterials, any structure with three dimensions at the nanoscale is considered a nanomaterial. Health care, biosensing, drug delivery, medical imaging and coatings use nanomaterials in a variety of shapes, sizes and compositions [37,38]. Furthermore, non-organic substances such as metallic, metal oxide, bimetallic and metal-salt nanoparticles are commonly used as material coatings.

7.6.5 Core-Corona-Canopy System

Core-corona-canopy coatings have a nanoparticle core surrounded by a flexible corona and functional group containing canopy. Corona plays a role in flexibility of coating whereas canopy adjoins different groups. These types of coatings are widely used in developing flexible electrolytes, semisolid-absorbents, cathodes and anodes. The nanoparticle core in this structure facilitates recoverability, separability and vital functions (i.e., magnetism, electrical and thermal conductivity) [39–42].

7.7   ADVANCE CHARACTERIZATION TECHNIQUES FOR COATED NANOPARTICLES

Coated nanoparticles' characterization is essential to visualize their surface features, size, shape, morphology, dispersion stability, functional groups, porosity, concentration and composition. Dynamic light scattering and zeta analyser are vital for detecting nanofluids' dispersion stability and hydrodynamic nanoparticle size. By measuring fluctuations in scattering light intensity from suspended particles, DLS can be used to measure nanoparticle size and size distribution. It is possible to determine the hydrodynamic diameter of particles by analysing fluctuations in intensity caused by Brownian motion [43]. It is possible to detect the amount of material present and infer density using bulk analysis techniques, including TGA, DLS and DCS, and external and internal morphology can be identified with the help of imaging techniques such as SEM and STEM. UV-vis spectroscopy techniques are used to detect particle optical absorbance and concentration. Similar to film coatings, SEM micrographs can be used to determine nanoparticle size distribution, morphology and surface features.

PT was combined with XPS, XAS, electrochemical impedance spectroscopy (EIS), and high-resolution transmission electron microscopy (HRTEM) to study the evolution of surface oxidation state, structure and composition of nanoparticles. By using XAS and XPS, respectively, to measure the oxidation state of bulk volumes and top surface layers (5–10 nm) of nanoparticles, scale-bridging method provides detailed information about changes in the surface states of nanoparticles. APT and HRTEM are also used to analyse nanoscale and atomic-scale elements and structures. A small-angle X-ray scattering (SAXS) technique can be used to monitor the nucleation and self-assembly of functionalized nanoparticles, while TEM and SEM are used to observe surface features. PXRD or XAS can be used to study colloidal reactions using synchrotron-based techniques or powdered X-ray absorption spectroscopy (XAS) [44]. Moreover, to track complex changes in chemical and electronic properties during complex reactions, more advanced high energy-resolution fluorescence-detected X-ray absorption near edge structure (HERFD-XANES) is required to capture subtle changes in the near chemical environment of absorbing atoms. In order to identify nanoparticle biofunctionalization, nuclear magnetic resonance (NMR) is a useful method to identify functional groups, chemical nature and molecular interactions [45]. Furthermore, emerging tools such as magnetic resonance imaging (MRI) and nanoparticle biodistribution (NB) characterization enables us to visualize real-time measurement and characterization at the molecular and cellular scale. NB characterization can contribute to the development of strategies for targeted drug delivery by detecting the bio dispersion of nanoparticles at cellular levels [46] (Table 7.1).

7.8   FUTURE PROSPECTS

Modern coating with tunable microstructural properties can be used to create omniphobic surfaces with very high hydrophobicity. The nanostructured coating can also help in facilitating structural colour aims to produce colour-changing coating or weather-control coating for modern aesthetic applications, especially in future cars. Self-cleaning surfaces and a self-healing nature can also be possible with modern coating

TABLE 7.1
Different Characterization Strategies Used Coated/Functional Nanoparticles. (Adopted from Kumar et al. [29]. Copyright (2021), The Royal Society of Chemistry (CC by 3).)

Characterizing Property	Characterization Tools/Methodology
Size	XRD, TEM, NMR AFM, NTA, ICP-MS, DLS, MALDI, SAXS, HRTEM, SEM, FMR, DCS, UV-Vis
Shape	HRTEM, 3D-tomography, EPLS, FMR, TEM, AFM
Elemental-chemical composition	XRD, NMR. ICP-AES, XPS, SEM-EDX, ICP-MS
Crystal structure	HRTEM, XRD, STEM, Electron diffraction, EXAFS
Size distribution	DLS, ICP-MS, DCS, SEM, FMR, SAXS, NTA, DTA, TRPS, superparamagnetic relaxometry
Concentration	UV-Vis, ICP-AES, ICP-MS, PTA, DCS
Agglomeration state	DCS, SEM, DLS, UV-Vis, TEM, Z-potential, Cryo-TEM
Density	RMM-MEMS, DCS
3D visualization	SEM, AFM, 3D-tomography
Structural defects	EBSD, HRTEM

and coating materials advancements. Coating or passivation of nanomaterials with functional groups or inorganic matter will improve bioavailability, drug elude specificity and reduce cytotoxicity in drug delivery applications. Specialized coatings may play a crucial role in the development of human implants and improve their biofunctionalization. Moreover, dye-coated nanoparticle systems will be crucial for tumour detection and imaging and help in the development of tumour drugs and their repurposing [47]. Coatings are essential in developing new strategies for future applications, and simultaneously development of newer characterization techniques withal play a vital role in assessing the quality of coated materials. Newer characterization strategies are developing with the coating material for modern and sophisticated applications.

7.9 CONCLUSION

From the present discussion, it is evident that coatings play an indispensable role in facilitating tunable and multifunctional attributes in modern applications. Both substrate and nanoparticles coated surfaces require sophisticated characterization tools to assess their stability, thickness, mechanical strength, porosity and chemical characteristics of coatings. Types, properties and characterization methodologies used for microstructural and chemical characterization of coating have been outlined in this article. Adding a functionalized topography and surface chemistry that includes drugs or growth factors to polymeric layers can make a biomaterial more effective. The use of hybrid hierarchical coating systems will improve tissue regeneration and wound healing, and more importantly, will facilitate a more efficient and effective health care system that can meet society's current challenges. Furthermore, effective coating methodologies will improve corrosion resistance, sensing and functionalization efficiency. The functionalized coatings remains active even after hundreds of cycle that will help researchers to devise new strategies for functionalized bio/chem system for effective conversion and hydrophobic functional coating to improve corrosion resistance. Researchers aim to develop simple, fast and inexpensive coating methodologies for creating novel nanostructure systems with desired physicochemical properties; however, they must adhere to the environmental and economic concerns associated with modern developments.

REFERENCES

[1] McNeil SE. Nanoparticle therapeutics: A personal perspective. *Wiley Interdiscip Rev Nanomedicine Nanobiotechnology* 2009;1:264–71. https://doi.org/10.1002/wnan.6.

[2] Vollath D, Szabó D V. Coated nanoparticles: A new way to improved nanocomposites. *J Nanoparticle Res* 1999;1:235–42. https://doi.org/10.1023/A:1010060701507.

[3] Cui F, Qian F, Yin C. Preparation and characterization of mucoadhesive polymer-coated nanoparticles. *Int J Pharm* 2006;316:154–61. https://doi.org/10.1016/j.ijpharm.2006.02.031.

[4] Liu J, Liu Z, Pang Y, Zhou H. The interaction between nanoparticles and immune system: Application in the treatment of inflammatory diseases. *J Nanobiotechnology* 2022;20:127. https://doi.org/10.1186/s12951-022-01343-7.

[5] Farooq SA, Raina A, Mohan S, Singh RA, Jayalakshmi S, Haq MIU. Nanostructured coatings: Review on processing techniques, corrosion behaviour and tribological performance. *Nanomaterials* 2022;12(8):1323. https://doi.org/10.3390/nano12081323.

[6] Martin PM. *Handbook of Deposition Technologies for Films and Coatings: Science, Applications and Technology*. Elsevier, United Kingdom, 2009, 1–912.

[7] Gross ME. *Handbook of Deposition Technologies for Films and Coatings*. Elsevier, United Kingdom, 2010, 532–53.

[8] Adhihetty I, Vella J, Volinsky A. Mechanical properties, adhesion and fracture toughness of low-k dielectric thin films for microelectronic applications. Honolulu, 2001, ICF1001077.

[9] Rao MC, Shekhawat MS. A brief survey on basic properties of thin films for device application. *Int J Mod Phys Conf Ser* 2013;22:576–82. https://doi.org/10.1142/s2010194513010696.

[10] Fernández-Hidalgo P, Martín-Palma RJ, Conde A, Gago R, Simancas J, García-Diego I, et al. Structural and chemical characterization of functional SiO[sub x]C[sub y]:H coatings for polymeric lenses. *J Vac Sci Technol B Microelectron Nanom Struct* 2004;22:2402. https://doi.org/10.1116/1.1795834.

[11] Park JH, Braun P V. Coaxial electrospinning of self-healing coatings. *Adv Mater* 2010;22:496–9. https://doi.org/10.1002/adma.200902465.

[12] Tosi D, Sypabekova M, Bekmurzayeva A, Molardi C, Dukenbayev K. Fiber surface modifications for biosensing. In *Optical Fiber Biosensors* 2022:253–282. https://doi.org/10.1016/b978-0-12-819467-6.00010-x.

[13] Faustini M, Ceratti DR, Louis B, Boudot M, Albouy PA, Boissière C, et al. Engineering functionality gradients by dip coating process in acceleration mode. *ACS Appl Mater Interfaces* 2014;6:17102–10. https://doi.org/10.1021/am504770x.

[14] Bindini E, Naudin G, Faustini M, Grosso D, Boissière C. Critical role of the atmosphere in dip-coating process. *J Phys Chem C* 2017;121:14572–80. https://doi.org/10.1021/acs.jpcc.7b02530.

[15] Bouzakis KD, Michailidis N, Skordaris G, Bouzakis E, Biermann D, M'Saoubi R. Cutting with coated tools: Coating technologies, characterization methods and performance optimization. *CIRP Ann - Manuf Technol* 2012;61:703–23. https://doi.org/10.1016/j.cirp.2012.05.006.

[16] Patel CJ, Mannari V. Air-drying bio-based polyurethane dispersion from cardanol: Synthesis and characterization of coatings. *Prog Org Coatings* 2014;77:997–1006. https://doi.org/10.1016/j.porgcoat.2014.02.006.

[17] Rojas TC, El Mrabet S, Domínguez-Meister S, Brizuela M, García-Luis A, Sánchez-López JC. Chemical and microstructural characterization of (Y or Zr)-doped CrAlN coatings. *Surf Coatings Technol* 2012;211:104–10. https://doi.org/10.1016/j.surfcoat.2011.07.071.

[18] Nik Masdek NR, Alfantazi A. Review of studies on corrosion of electrodeposited nanocrystalline metals and alloys. *ECS Meet Abstr* 2010;MA2010–01:909–909. https://doi.org/10.1149/ma2010-01/16/909.

[19] Bhadouria AS, Kumar A, Raj D, Verma A, Singh S, Tripathi P, et al. Corrosion mitigation in oil reservoirs during CO_2 injection using nanomaterials. In book Nanotechnology for CO_2 Utilization in Oilfield Applications 2022:127–146. https://doi.org/10.1016/b978-0-323-90540-4.00014-4.

[20] Fahmy HM, Mosleh AM, Elghany AA, Shams-Eldin E, Abu Serea ES, Ali SA, et al. Coated silver nanoparticles: Synthesis, cytotoxicity, and optical properties. *RSC Adv* 2019;9:20118–36. https://doi.org/10.1039/c9ra02907a.

[21] Mackenzie JD, Bescher EP. Chemical routes in the synthesis of nanomaterials using the sol-gel process. *Acc Chem Res* 2007;40:810–8. https://doi.org/10.1021/ar7000149.

[22] Jaiswal P, Kumar Y, Shukla R, Nigam KDP, Panda D, Guha Biswas K. Covalently immobilized nickel nanoparticles reinforce augmentation of mass transfer in millichannels for two-phase flow systems. *Ind Eng Chem Res* 2022;61:3672–84. https://doi.org/10.1021/acs.iecr.1c04419.

[23] Georgakilas V, Otyepka M, Bourlinos AB, Chandra V, Kim N, Kemp KC, et al. Functionalization of graphene: Covalent and non-covalent approaches, derivatives and applications. *Chem Rev* 2012;112:6156–214. https://doi.org/10.1021/cr3000412.

[24] Lee JW, Choi SR, Heo JH. Simultaneous stabilization and functionalization of gold nanoparticles via biomolecule conjugation: Progress and perspectives. *ACS Appl Mater Interfaces* 2021;13(36):42311–28. https://doi.org/10.1021/acsami.1c10436.

[25] Veerapandian M, Yun K. Functionalization of biomolecules on nanoparticles: Specialized for antibacterial applications. *Appl Microbiol Biotechnol* 2011;90:1655–67. https://doi.org/10.1007/s00253-011-3291-6.

[26] Mout R, Moyano DF, Rana S, Rotello VM. Surface functionalization of nanoparticles for nanomedicine. *Chem Soc Rev* 2012;41:2539–44. https://doi.org/10.1039/c2cs15294k.

[27] Montaseri H, Kruger CA, Abrahamse H. Review: Organic nanoparticle based active targeting for photodynamic therapy treatment of breast cancer cells. *Oncotarget* 2020;11:2120–36. https://doi.org/10.18632/oncotarget.27596.

[28] Singh U, Morya V, Rajwar A, Chandrasekaran AR, Datta B, Ghoroi C, et al. DNA-functionalized nanoparticles for targeted biosensing and biological applications. *ACS Omega* 2020;5:30767–74. https://doi.org/10.1021/acsomega.0c03656.

[29] Kumar Y, Yogeshwar P, Bajpai S, Jaiswal P, Yadav S, Pathak DP, et al. Nanomaterials: Stimulants for biofuels and renewables, yield and energy optimization. *Mater Adv* 2021;2:5318–43. https://doi.org/10.1039/d1ma00538c.

[30] Kumar Y, Jaiswal P, Panda D, Nigam KDP, Biswas KG. A critical review on nanoparticle-assisted mass transfer and kinetic study of biphasic systems in millimeter-sized conduits. *Chem Eng Process - Process Intensif* 2022;170:108675. https://doi.org/10.1016/j.cep.2021.108675.

[31] Vijayalakshmi S, Anand M, Ranjitha J. Microalgae-based biofuel production using low-cost nanobiocatalysts. In book Microalgae Cultivation for Biofuels Production 2019: 251–63. https://doi.org/10.1016/B978-0-12-817536-1.00016-3.

[32] Ghosh Chaudhuri R, Paria S. Core/shell nanoparticles: Classes, properties, synthesis mechanisms, characterization, and applications. *Chem Rev* 2012;112:2373–433. https://doi.org/10.1021/cr100449n.

[33] Gawande MB, Goswami A, Asefa T, Guo H, Biradar A V., Peng DL, et al. Core-shell nanoparticles: Synthesis and applications in catalysis and electrocatalysis. *Chem Soc Rev* 2015;44:7540–90. https://doi.org/10.1039/c5cs00343a.

[34] Asapu R, Claes N, Bals S, Denys S, Detavernier C, Lenaerts S, et al. Silver-polymer core-shell nanoparticles for ultrastable plasmon-enhanced photocatalysis. *Appl Catal B Environ* 2017;200:31–8. https://doi.org/10.1016/j.apcatb.2016.06.062.

[35] Kvi L, Soukupova J, Vec R, Prucek R, Holecova M, Zbor R. Effect of surfactants and polymers on stability and antibacterial activity of silver nanoparticles (NPs). *J Phys Chem C* 2008;112:5825–34.

[36] Tzhayik O, Sawant P, Efrima S, Kovalev E, Klug JT. Xanthate capping of silver, copper, and gold colloids. *Langmuir* 2002;18:3364–9. https://doi.org/10.1021/la015653n.

[37] Cong JC, Ke CY, Yong XG, Fu ZR, Peng LY. A review on the application of inorganic nanoparticles in chemical surface coatings on metallic substrates. *RSC Adv* 2017;7:7531–9. https://doi.org/10.1039/c6ra25841g.

[38] Luchini A, Vitiello G. Understanding the nano-bio interfaces: Lipid-coatings for inorganic nanoparticles as promising strategy for biomedical applications. *Front Chem* 2019;7:343. https://doi.org/10.3389/fchem.2019.00343.

[39] Song J, Wang C, Hinestroza JP. Electrostatic assembly of core-corona silica nanoparticles onto cotton fibers. *Cellulose* 2013;20:1727–36. https://doi.org/10.1007/s10570-013-9922-6.

[40] Wichaita W, Kim YG, Tangboriboonrat P, Thérien-Aubin H. Polymer-functionalized polymer nanoparticles and their behaviour in suspensions. *Polym Chem* 2020;11:2119–28. https://doi.org/10.1039/c9py01558b.

[41] Jespersen ML, Mirau PA, Von Meerwall E, Vaia RA, Rodriguez R, Giannelis EP. Canopy dynamics in nanoscale ionic materials. *ACS Nano* 2010;4:3735–42. https://doi.org/10.1021/nn100112h.

[42] Sheng L, Chen Z, Wang X, Farooq AS. Transforming porous silica nanoparticles into porous liquids with different canopy structures for CO_2 capture. *ACS Omega* 2022;7:5687–97. https://doi.org/10.1021/acsomega.1c05091.

[43] Cant DJH, Minelli C, Sparnacci K, Müller A, Kalbe H, Stöger-Pollach M, et al. Surface-energy control and characterization of nanoparticle coatings. *J Phys Chem C* 2020;124:11200–11. https://doi.org/10.1021/acs.jpcc.0c02161.

[44] Nakagawa F, Saruyama M, Takahata R, Sato R, Matsumoto K, Teranishi T. In situ control of crystallinity of 3D colloidal crystals by tuning the growth kinetics of nanoparticle building blocks. *J Am Chem Soc* 2022;144:5871–7. https://doi.org/10.1021/jacs.1c12456.

[45] Grote L, Zito CA, Frank K, Dippel AC, Reisbeck P, Pitala K, et al. X-ray studies bridge the molecular and macro length scales during the emergence of CoO assemblies. *Nat Commun* 2021;12:4429. https://doi.org/10.1038/s41467-021-24557-z.

[46] Ojea-Jiménez I, Capomaccio R, Osório I, Mehn D, Ceccone G, Hussain R, et al. Rational design of multi-functional gold nanoparticles with controlled biomolecule adsorption: A multi-method approach for in-depth characterization. *Nanoscale* 2018;10:10173–81. https://doi.org/10.1039/c8nr00973b.

[47] Kumar Y, Sinha ASK, Nigam KDP, Dwivedi D, Sangwai J. Functionalized nanoparticles: Tailoring properties through surface energetics and coordination chemistry for advanced biomedical applications. *Nanoscale* 2023;15:6075–104. https://doi.org/10.1039/D2NR07163K.

8 Corrosion under Insulation (CUI) in Oil and Gas Industries

Siddharth Atal, Deepak Dwivedi,
and Naveen Mani Tripathi
Rajiv Gandhi Institute of Petroleum Technology
Rajiv Gandhi Institute of Petroleum
Technology, Amethi (Assam Center)

CONTENTS

ABBREVIATION

CUI   Corrosion under insulation

8.1   INTRODUCTION

Functional materials are used in a wide variety of technical systems, such as memories, insulation, displays, and communications because they have one or more properties that may be extensively modified in a controlled manner by external stimuli (temperature, electric field, and magnetic field) [1].

DOI: 10.1201/9781003242550-8

The corrosion under insulation (CUI) is specifically huge concern in corrosion for petroleum industries. This basically combines two areas of research: corrosion and insulation. The pipeline's corrosion is divided into two parts: internal and external corrosion [2]. CUI is basically an external corrosion in petroleum pipelines [3]. The thermal insulating materials stop or lessen several types of heat transmission from outside to inside or vice versa via conduction, convection, and radiation, regardless of the surrounding air temperature. Different organic and inorganic thermal insulation materials, including expanded polymers, wool, cork, straw, and industrial flax, are used in several thermal insulation systems. Different new grades of insulating materials such as new glass, mineral fibres, and glass bases are being developed and tested. There are different techniques for assessing the properties of insulating materials and systems such as shape, flammability, composition, and structure [4].

Thermal science researchers are working to reduce heat loss as well as both capital and operational costs. To reduce heat loss and the amount of insulation required, researchers have previously utilized various objective functions in the design study of a piping system. A common tactic in these complex systems is to add up all optimization methods with the appropriate weights, and then the emerging composite functions should be reduced [5]. However, it examines details and is frequently unnecessary from a practical standpoint because many types of insulation are only available in certain specific sizes. The analytical solution should only be used if a very specific thickness value is required. The characteristics of the insulating material and the intended use of the equipment determine the appropriate insulation thickness for any given application. For important and critical processes, the most vital element which should be considered is reliability. If energy or heat conservation is considered factor, then the most decisive element is savings of every year (economical) in contrast to the operational cost factor. Therefore, it can be said that energy is precious and thermal insulation material is required to save energy. In this scenario, thermal conductivity becomes a decisive factor. There are numerous methods to determine the material's thermal conductivity and to utilize it. In Table 8.1, the thermal conductivity of some common insulators is summarized from different literature.

8.2　INSULATION MATERIALS

There is organic [17,18] and inorganic [19,20] types of insulation materials. Hydrocarbon polymers, which may be extended to create high void structures, are the basis for organic insulations, for example, expanded polystyrene (thermocol) with polyurethane foam (PUF). Siliceous/aluminous/calcium compounds in fibre, granular, or powder form are the foundation of inorganic insulation. Examples of inorganic insulating material include calcium silicate and mineral wool. In the following sections, the properties and behaviour of a few inorganic insulating materials are described in detail.

8.2.1　Glass Wool

Due to its great durability, lightweight, high tensile strength, and excellent thermal and acoustic qualities, glass wool (see in Figure 8.1) is one of the most popular types

TABLE 8.1
Thermal Conductivity of Common Insulators at Room Temperature (25°C)

S. No.	Insulators	Capacity to Resist Heat Flow (R-Value) (°F/per inch)	Thermal Conductivity (W/m.K)	References
1	Asphalt	0.32	0.69	[6]
2	Cotton	3.85	0.04	[7]
3	Cork (EPS)	2.22	0.05	[8,9]
4	Ceramic fibre	0.08	0.08	[10]
5	Glass fibre	4.00	0.03	[11]
6	Glass wool	4.00	0.04	[8,9,12]
7	Plywood	1.25	0.13	[13]
8	Rubber	0.20	0.35	[14]
9	Vermiculite	2.08	0.06	[15]
10	Wool	3.5–3.8	0.05	[16]

FIGURE 8.1   Glass wool. (Reproduced with permission from [24].)

of insulation used worldwide. Its service temperature ranges up to 250°C [21]. Glass wool is made up of long, thin inorganic strands that are joined by a high-temperature adhesive. These fibres, which have an average diameter of 6–7 μm, are scattered throughout the material to trap millions of tiny pockets of air, making it an effective thermal and acoustic insulator. Glass wool's small weight also has a lot to gain from in terms of transportation and installation. Glass wool also has no impurities like iron shots, sulphur, or chloride and is chemically inert. The product does not assist

TABLE 8.2
Area of Application of Glass Wool in Different Oil and Gas Sectors

Name of the Industrial Units	Application	Recommended Products	Benefits
Refineries, petrochemical units	Vessels and pipes	24–32 kg/m³ Rolls 52–64 kg/m³ Boards	Energy conservation Temperature control for process control Non-corrosive Durable
LNG storage tanks	Suspended deck	12–20 kg/m³ Rolls	Easy to apply Energy conservation Excellent resilience Temperature maintenance

the growth of mould and is non-corrosive to metal. It is produced using sustainable raw resources and is green at every level [22].

Glass wool is moulded into goods with a range of densities and thicknesses. It is available as slabs and rolls of paper with or without aluminium foil. This material has different types of facing like aluminium foil, glass cloth, and black glass tissue.

8.2.1.1 Temperature Range

Glass wool can be used in applications where the temperature ranges from −195°C to +230°C. However, it can go up to 450°C for very specific applications. It is safe to use aluminium foil facing up to 120°C [23].

Glass wool has insulation applications for different components (pipelines, vessels, etc.) in the oil and gas industries. Table 8.2 summarizes these applications.

8.2.2 Rockwool/Mineral Wool

Rockwool (see Figure 8.2) is made with fibres of connected rock and used to insulate flat or curved surfaces from sound and heat. It has applications in different sectors such as industrial insulation, pipe insulation, insulation over faux walls, insulation for conveyance appliances, and walls. A type of insulation known as 'rockwool' is created using real rocks and minerals. Slag wool insulation, stone wool insulation, and mineral wool insulation are further names for it. Rockwool may be used to create a variety of goods thanks to its strong sound and heat insulation qualities. In industrial settings, automotive applications, and the construction of buildings, rockwool insulation is frequently employed [25].

Minerals and other raw materials are heated to about 2910°F (about 1600°C) in a furnace while being pushed through by an air or steam stream to create rockwool. Advanced production methods, which resemble how cotton candy is manufactured in some respects, spin the molten rock on a rotating wheel at high speeds. The final output is a mass of starch-bound, extremely finely twisted fibres. In order to reduce the generation of dust, oil is also added [26].

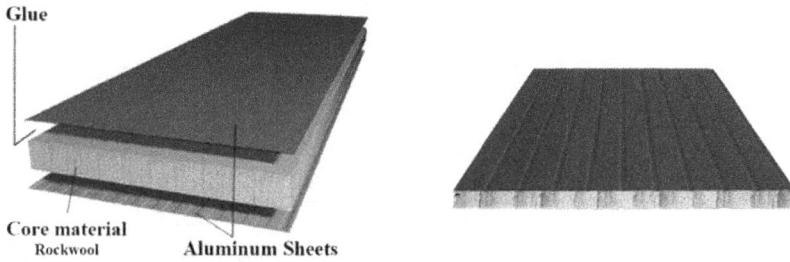

FIGURE 8.2 Rockwool filled between aluminium plates. (Reproduced with permission from [27].)

Although rockwool insulation's individual fibres are great heat conductors on their own, however, the rolls and layers of that kind of insulation are particularly good at blocking the flow of heat. Because of their extremely high melting point, they are frequently employed to stop the spread of fire in buildings. The use of rockwool as insulation significantly lowers the amount of energy consumed in both household and commercial settings.

8.2.2.1 Loose Mineral Wool

It has the potential to protect hot, abrasive objects including machinery, tanks, pipes, ovens, furnaces, buildings, or corrosive surfaces because of its extraordinarily long fibres and strong technical features required to withstand high service temperatures.

They are sound-absorbing, non-settling, non-corrosive, odourless, water-repellent, incombustible, and fire-resistant and the service temperature –50°C to 750°C.

Resin-bonded slabs: Long, delicate fibres produced from molten natural rock are combined with a thermosetting epoxy slab to create a resin-bonded slab, which excels at thermal insulation, fire prevention, and sound absorption. Slabs are typically used in sound-absorbing projects for theatres, auditoriums, and commercial buildings when noise reduction is required. Service temperature of this is –50°C to 800°C.

Lightly resin-bonded (LRB): Mattresses made of rockwool that are lightly resin bound (LRB) are made of fine fibres that are produced from a variety of carefully chosen rocks and melted at a high temperature. They are baked to create mattresses with a specific density and thickness in a machine-laid fibre pattern. Then, mattresses are cut and sewed to the required sizes.

They are not capillary or hygroscopic; therefore, they do not take in moisture from the environment. Mattress stability is not impacted by moisture, and its operating temperature is –50°C to 750°C.

8.3 CORROSION UNDER INSULATION (CUI)

In all industrialized countries, corrosion causes significant economic losses, notably in coastal businesses. Different countries have a significant impact on their economy due to CUI. According to International Zinc Association (2021), only corrosion is

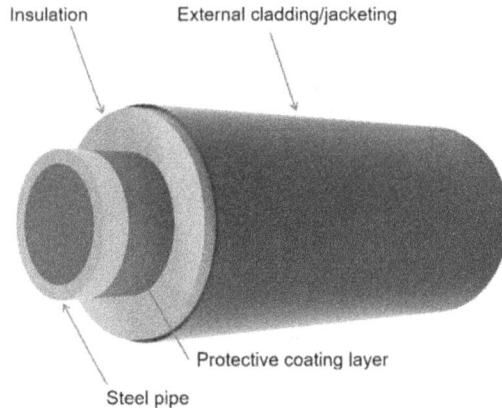

FIGURE 8.3   Insulated pipe. (Reproduced with permission from [33].)

affecting up to 3.4% of USA GDP and 5%–7% of India's GDP [28]. Therefore, especially in India, it is main concern to normalize and reduce the losses due to corrosion. The industry needs to take essential and all possible steps to prevent the formation of corrosion. Corrosion-related failures at processing facilities are a substantial source of danger on- and offshore oil and gas projects. Forty to 60% of pipe maintenance costs are related to CUI only [29,30]. Figure 8.3 depicts the schematic diagram of steel pipe insulation with some insulating material, for example, rockwool and glass wool. Further, an external jacket covers the protective/insulating material.

Conventional insulation materials which permit moisture most often lead the underlying steel surface to deteriorate more rapidly, which can lead insulated structures like pipes and vessels to fail substantially. The potential of a catastrophic collapse poses a serious threat to the environment, the workers' safety, and the facility itself if such equipment is operating under high pressure [31].

The offshore oil and gas industry generally recognizes corrosion as a major safety risk. Even in situations when 'water-resistant' versions are used, open cell insulating materials frequently have high amounts of moisture before being applied. It is conceivable that the insulation picked up humidity from the air and may have even been subject to marine rain. Moisture will enter the space around the pipe simply from the application and sealing of these materials [32].

As discussed before, CUI is the underlying cause of 40%–60% of pipe maintenance expenses. In addition, repairing deterioration from CUI consumes about 10% of the maintenance budget.

CUI and stress corrosion cracking (SCC) can be prevented by some pre-checked industrial checking pathways, which often gets prepared based on day-to-day industrial operation which provides support to researchers/engineers to choose appropriate material for construction of pipelines. For example, if the carbon steel is being used as materials of construction, then the temperature ranges must be considered for deciding whether corrosion will occur or not. Along with this, there are many factors to influence the CUI; in addition, it gets inspected with different methods.

8.3.1 Factors Affecting the CUI

CUI poses a risk to numerous businesses. CUI might be taken as a source of catastrophic failure which causes serious threats to the pipelines and accessories without any prior warning.

There are many factors which affect the corrosion under the insulation [34–36] such as:

- The regularity and length of the exposure to moisture
- Operational temperature
- Protective barrier conditions and type of insulations
- Environment corrosivity
- Moisture content where the machine is installed
- Proximity to machinery that produces humidity, such as cooling towers

Hence, early-stage identification of CUI becomes imperative for oil and gas sectors.

8.4 NON-DESTRUCTIVE TECHNIQUES (NDT)

To characterize CUI, a range of non-destructive techniques (NDT) (with advantages and limitations) as listed in Figure 8.4 could be used as CUI monitoring techniques.

8.4.1 Infrared Thermography

Since there is typically a measurable temperature differential between the dry insulation and the wet insulation, the infrared can be utilized under the correct

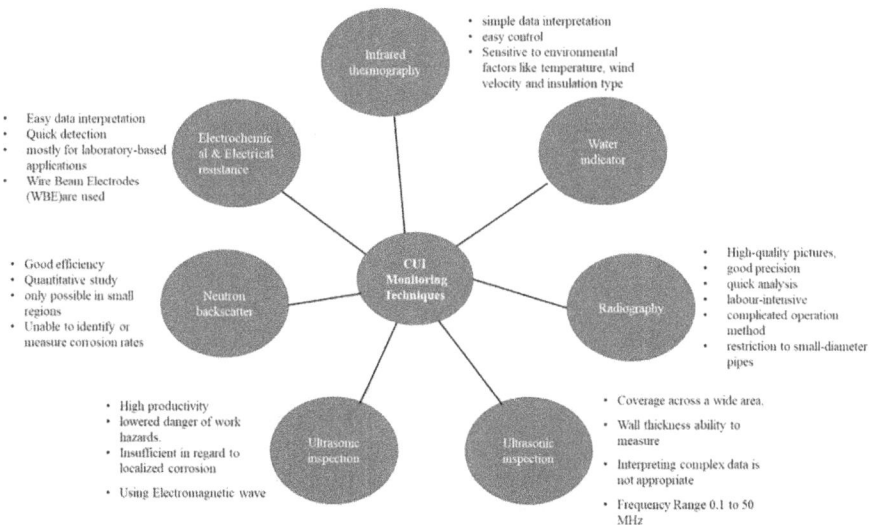

FIGURE 8.4 CUI monitoring techniques.

circumstances to find moist patches in the insulation. The regions underneath the wet insulation are clearly susceptible to corrosion [37].

8.4.2 RADIOGRAPHY

There are four different types of radiography methods that are regularly used: computed radiography (X-rays), real-time radiography (X-rays), digital detector array (X-rays), and profile radiography (either X-rays or gamma rays) [38]. In all four radiography methods, to measure the reduction in wall thickness, the profile radiography method is most effective [39].

8.4.3 OTHER IMPORTANT TECHNIQUES

In addition to above NDT techniques, other useful techniques such as ultrasonic inspection, eddy current, etc., is being used in industries. However, these techniques have identified limitations which need to be considered during industrial operations. It is noteworthy that ultrasonic inspection has a relatively tiny covering area. Insulation is cut into pockets where monitoring probes are placed to take measurements just immediately. Additionally, this approach works well, but it can only be used in a narrow region. Cut pockets provided for measurements undermine the insulation's intended function [40]. On the other hand, pulsed eddy current has a time-varying magnetic field pulse can be reflected and then further captured by the receiver coil by producing an eddy current from an excitation coil in the direction of the designated objects. If there are imperfections or flaws, the magnetic field fluctuates [37].

In the current scenario, visual inspection of the piping conditions is the only way to guarantee the proper screening of CUI without removing external jacketing or cladding. NDT is a cost-effective as well as effective corrosion detecting method that can be used without causing any harm to virgin materials. Each sort of NDT method has advantage and disadvantages of its own. To identify CUI, radiography is frequently being utilized by oil and gas industries. The utilization of an electrochemical probe to identify local corrosion and the application of modelling to foretell CUI are current trends, however, yet to be popularized in different segments of oil and gas sectors. Application of AI-ML can be seeded by data obtained using NDT techniques as delineated earlier.

8.5 THE TEMPERATURE RANGE FOR CORROSION UNDER INSULATION (CUI)

Stainless and austenitic steels are fragile in the 140°F (60°C) to 300°F (149°C) temperature range, while carbon steel and low-alloy steels are more vulnerable in the 32°F (0°C) to 300°F (149°C) temperature range. This information indicates that the ideal temperature range for CUI in most steels is between 200°F (93°C) and 240°F (116°C). In this temperature range, the system gets enough heat energy to boost up chemical reactions without letting moisture escape before it can reach the equipment's surface. Because intermittent boiling and flashing procedures can provide an aggressive CUI environment. CUI prevention focuses on to design the whole system preventing moisture in the insulation system at temperatures below 300°F (149°C) [41,42].

Under the category of CUI, stainless steel corrodes due to the presence of chloride. Once the concentration of chloride reaches to a critical level, it causes the stress corrosion cracking under the insulation. Critical concentration of the chloride depends on the types of the stainless steel alloy and its operating temperature range [43]. It is noteworthy that the maximum operating temperature range for 316 austenitic stainless steel is found between 50°C and 60°C, whereas the same is observed different for duplex stainless steel. This also gets affected by alloy's composition. For an example, stainless steel containing 6% Mo exhibits the operating temperature between 100°C and 120°C, whereas duplex stainless steel consisting of 22% Cr and 25% Cr exhibits the operating temperature range 80–100°C and 90–110°C, respectively [43].

8.6   OUTLOOK AND PROSPECTS

To lower corrosion costs and avoid long-term damage, maintenance is crucial. In the situation of high humidity or spill danger, thermally insulated systems require special care. We can resolve this problem using integrated BIM (Building Information Modelling) and sensor for corrosion prediction and visualisation. A mathematical model for predicting corrosion and a prototype is created to verify the effectiveness of the suggested strategy by Tsai and other researcher [44].

Vacuum jacketed piping is one type of the advanced insulated piping system that offers an alternate method of lowering the danger of CUI. These double-walled piping systems use vacuum barriers rather than insulation to decrease heat transfer. The vacuum is kept intact by the tightly sealed double-walled pipes [45].

The role of liquefied natural gas (LNG) in the world's energy markets is anticipated to grow over time. Unlike any other market for energy resources, the market for LNG is expanding. Cryogenic vacuum pipes with layered insulation have recently been created to improve structural stability and insulation performance. Excellent thermal installation carried out in line with exacting specifications is necessary to ensure the LNG cold temperature of 160°C [46]. By limiting the amount of air heat that can enter the pipe or other system components, cryogenic insulation keeps the liquid cool and enables it to maintain its shape. Cryogenic insulation is a vital route activity in a gas plant, and access is usually constrained and limited in those areas where insulation is done.

8.7   CONCLUSION

CUI continues to be a major industry issue. Despite improvements in materials and monitoring techniques, CUI affects the oil and gas industries severely by impacting the overall maintenance budget of the industries. There are many insulation materials such as glass wool and mineral wool that are being used in and oil and gas sectors, particularly in the downstream sectors. This chapter brings the importance of CUI monitoring techniques with the associated limitations. Radiography and visual inspections are identified as industrially viable techniques for CUI identification. On the other hand, electrochemical techniques have also been emerged as one of the possible alternatives for existing CUI monitoring techniques.

In addition to this, this chapter also highlighted the challenges involved with CUI identification and monitoring techniques and provided valuable information regarding cryogenic piping system.

REFERENCES

[1] Duyar, M., Arellano Trevino, M., and Farrauto, R.J. "Dual function materials for CO_2 capture and conversion using renewable H_2." *Applied Catalysis B: Environmental* 168 (2015): 370–376.

[2] Bhadouria, A.S., Kumar, A., Raj, D., Verma, A., Singh, S., Tripathi, P., Kumar, Y., et al. Corrosion mitigation in oil reservoirs during CO injection. (2022).

[3] Vanaei, H. R., Eslami, A., and Egbewande, A. "A review on pipeline corrosion, in-line inspection (ILI), and corrosion growth rate models." *International Journal of Pressure Vessels and Piping* 149 (2017): 43–54.

[4] Javaherdashti, R. "Corrosion under Insulation (CUI): A review of essential knowledge and practice." *Journal of Materials Science and Surface Engineering* 1.2 (2014): 36–43.

[5] Bahadori, A. *Thermal insulation handbook for the oil, gas, and petrochemical industries.* Amsterdam: Gulf Professional Publishing, 2014.

[6] Wilds, N. "Corrosion under insulation." *Trends in Oil and Gas Corrosion Research and Technologies* (2017): 409–429.

[7] Pfundstein, M. et al. *Insulating materials: Principles, materials, applications.* Walter de Gruyter, 2012.

[8] Schiavoni, S., Bianchi, F., and Asdrubali, F. "Insulation materials for the building sector: A review and comparative analysis." *Renewable and Sustainable Energy Reviews* 62 (2016): 988–1011.

[9] Jelle, B.P. "Traditional, state-of-the-art and future thermal building insulation materials and solutions–Properties, requirements and possibilities." *Energy and Buildings* 43.10 (2011): 2549–2563.

[10] Headley, A.J. et al. "Thermal conductivity measurements and modeling of ceramic fiber insulation materials." *International Journal of Heat and Mass Transfer* 129 (2019): 1287–1294.

[11] Tavman, I. H., and Akinci, H. "Transverse thermal conductivity of fiber reinforced polymer composites." *International Communications in Heat and Mass Transfer* 27.2 (2000): 253–261.

[12] Papadopoulos, A.M. "State of the art in thermal insulation materials and aims for future developments." *Energy and Buildings* 37.1 (2005): 77–86.

[13] Demirkir, C. E. N. K., Colakoglu, G., Colak, S. E. M. R. A., Aydin, I., and Candan, Z. Influence of aging procedure on bonding strength and thermal conductivity of plywood panels. (2016).

[14] Cheheb, Z. et al. "Thermal conductivity of rubber compounds versus the state of cure." *Macromolecular Materials and Engineering* 297.3 (2012): 228–236.

[15] Guan, W.-M. et al. "Preparation of paraffin/expanded vermiculite with enhanced thermal conductivity by implanting network carbon in vermiculite layers." *Chemical Engineering Journal* 277 (2015): 56–63.

[16] Ye, Z. et al. "Thermal conductivity of wool and wool–hemp insulation." *International Journal of Energy Research* 30.1 (2006): 37–49.

[17] Berardi, U., and Naldi, M. "The impact of the temperature dependent thermal conductivity of insulating materials on the effective building envelope performance." *Energy and Buildings* 144 (2017): 262–275.

[18] Omer, S.A., Riffat, S.B., and Qiu, G. "Thermal insulations for hot water cylinders: A review and a conceptual evaluation." *Building Services Engineering Research and Technology* 28.3 (2007): 275–293.

[19] Karamanos, A., Hadiarakou, S., and Papadopoulos, A. M. "The impact of temperature and moisture on the thermal performance of stone wool." *Energy and Buildings* 40.8 (2008): 1402–1411.

[20] Wieland, H. et al. "Perspektiven für Dämmstoffe aus heimischen nachwachsenden Rohstoffen." *Landtechnik* 55.1 (2000): 22–23.

[21] Yuan, J. "Impact of insulation type and thickness on the dynamic thermal characteristics of an external wall structure." *Sustainability* 10.8 (2018): 2835.

[22] Kumar, D. et al. "Comparative analysis of building insulation material properties and performance." *Renewable and Sustainable Energy Reviews* 131 (2020): 110038.

[23] Villasmil, W., Fischer, L.J., and Worlitschek, J. "A review and evaluation of thermal insulation materials and methods for thermal energy storage systems." *Renewable and Sustainable Energy Reviews* 103 (2019): 71–84.

[24] Kapoor, A., Mudgal, S., and Muruganandam, L. "Impact of shock waves on glass wool composition and properties." *Materials Today: Proceedings* 46 (2021): 7056–7060.

[25] Yang, S. J., and Zhang, L. W. "Research on properties of rock-mineral wool as thermal insulation material for construction." *Advanced Materials Research* 450 (2012): 618–622.

[26] Bussell, W.T., and Mckennie, S. "Rockwool in horticulture, and its importance and sustainable use in New Zealand." *New Zealand Journal of Crop and Horticultural Science* 32.1 (2004): 29–37.

[27] Yılmaz, E., Aykanat, B., and Çomak, B. "Environmental life cycle assessment of rockwool filled aluminum sandwich facade panels in Turkey." *Journal of Building Engineering* 50 (2022): 104234.

[28] Koch, G. "Cost of corrosion." *Trends in Oil and Gas Corrosion Research and Technologies* (2017):3–30.

[29] Eltai, E., Al-Khalifa, K., Al-Ryashi, A., Mahdi, E., and Hamouda, A. M. S. "Investigating the corrosion under insulation (CUI) on steel pipe exposed to Arabian Gulf sea water drops." In *Key Engineering Materials*, vol. 689, pp. 148–153. Trans Tech Publications Ltd, 2016.

[30] Winnik, S. "Piping system CUI: old problem; different approaches." *European Federation of Corrosion Conference Corrosion in the Refinery Industry. Budapest Congress Centre, Hungary*, 2003.

[31] De Vogelaere, F. "Corrosion under insulation." *Process Safety Progress* 28.1 (2009): 30–35.

[32] Caines, S. et al. "Experimental design to study corrosion under insulation in harsh marine environments." *Journal of Loss Prevention in the Process Industries* 33 (2015): 39–51.

[33] Cao, Q. et al. "A review of corrosion under insulation: A critical issue in the oil and gas industry." *Metals* 12.4 (2022): 561.

[34] Lazar, P.I.I.I. "Factors affecting corrosion of carbon steel under thermal insulation." *Corrosion of Metals under Thermal Insulation*. West Conshohocken, PA: ASTM International, 1985.

[35] Burhani, N.R.A. et al. "Application of logistic regression in resolving influential risk factors subject to corrosion under insulation." *Proceedings of the 2016 International Conference on Industrial Engineering and Operations Management, Kuala Lumpur, Malaysia*, 2016.

[36] Dunn, P.J., and Norsworthy, R. "Control of corrosion under insulation." *ASHRAE Journal* 45.3 (2003): 32.

[37] Hernandez, J. et al. "Detection of corrosion under insulation on aerospace structures via pulsed eddy current thermography." *Aerospace Science and Technology* 121 (2022): 107317.

[38] Lenka, S. "Corrosion under insulation (CUI)-inspection technique and prevention." *Non-Destructive Evaluation* 6 (2017): 97–103.

[39] Pechacek, R.W. "Advanced NDE methods of inspecting insulated vessels and piping for ID corrosion and corrosion under insulation (CUI)." *CORROSION 2003.* San Diego, CA: OnePetro, 2003.

[40] Jones, R.E. et al. "Use of microwaves for the detection of water as a cause of corrosion under insulation." *Journal of Nondestructive Evaluation* 31.1 (2012): 65–76.

[41] Delahunt, J.F. "Corrosion under thermal insulation and fireproofing-an overview." *CORROSION 2003*, 2003. San Diego, CA: OnePetro, 2003.

[42] Marquez, A., Singh, J., and Maharaj, C. "Corrosion under insulation examination to prevent failures in equipment at a petrochemical plant." *Journal of Failure Analysis and Prevention* 21.3 (2021): 723–732.

[43] Houben, J. et al. "Deployment of CUI prevention strategies and TSA implementation in projects." *CORROSION 2012.* San Diego, CA: OnePetro, 2012.

[44] Tsai, Y.-H. et al. "A BIM-based approach for predicting corrosion under insulation." *Automation in Construction* 107 (2019): 102923.

[45] Bunton, M.A. "Vacuum jacketed tubing: past, present, and future." *SPE Western Regional Meeting.* San Diego, CA: OnePetro, 1999.

[46] Al-Homoud, M.S. "Performance characteristics and practical applications of common building thermal insulation materials." *Building and Environment* 40.3 (2005): 353–366.

9 Application of Functional Ceramics in Oil and Gas Industries
Manufacturing, Properties, and Current Status

*Swati Chaudhary, Saurabh Kumar,
Vivek Kumar, and Deepak Dwivedi*
Rajiv Gandhi Institute of Petroleum Technology

Mainak Ray
HPCL-Mittal Energy Ltd

Hakeem Niyas
Centre of Rajiv Gandhi Institute of Petroleum Technology
[a]Equal contribution.

CONTENTS

ABBREVIATIONS

FCCU   Fluidized catalytic cracking unit
SRU   Sulfur recovery unit

9.1   INTRODUCTION

In modern civilization, the world cannot function properly without gasoline, diesel, polymers, fertilizers, cement, and glass, and the production of these products involves high temperatures at some stage. Refractories must be used if a process is conducted at a temperature higher than 400°C to safeguard the building material [1]. As the temperature of the refractory material increases, internal stress will come into the picture [2]. Refractories are inorganic, nonmetallic materials that can endure high temperatures while remaining in contact with molten slag, metal, and gases without suffering physical or chemical changes. Refractories serve two purposes: to withstand high temperatures in a corrosive atmosphere and to inhibit the flow of energy from the reactor system. However, it is noteworthy that the refractory behavior gets changes with application requirements and operating conditions. The refractory consumption in India is elucidated in Figure 9.1.

9.1.1   PROPERTIES OF THE REFRACTORIES

Refractories are distinguished by their physical and chemical properties. Physical property requirements for shaped and unshaped refractories are found to be different. For shaped refractories, density and porosity are important properties. On the other hand, flow ability at a specific water addition has been considered an important parameter for unshaped refractory. The chemical properties of the refractory material are primarily deciphered by the chemical composition of the material. Some important properties (e.g., specific gravity, bulk density, apparent density, permeability)

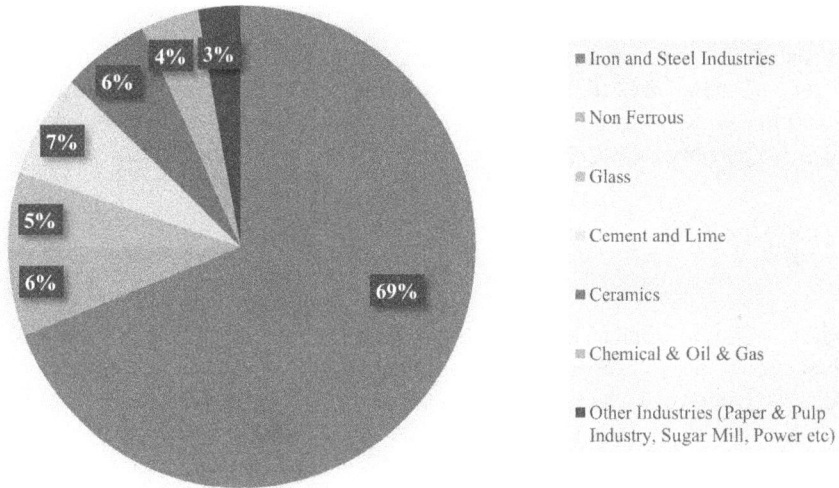

FIGURE 9.1   Refractory consumption in India. (Adopted from industrial data.)

considered important for refractory application in the oil and gas sectors are listed and described in detail.

9.1.1.1   Specific Gravity

Specific gravity is defined as the ratio of the bulk density of the material to the density of the standard fluid (water) or the ratio of the unit weight of the material to the unit weight of water. Refractory materials are identified by their specific gravity. All refractory materials have different densities, and the material is differentiated on the basis of specific gravity.

9.1.1.2   Bulk Density

Bulk density is defined as the ratio of the mass of the material to the volume. In bulk density calculations, the total volume considered in the calculation is different from the true density calculation; here, the total volume is the sum of the volume of material and pores. Due to this, true density is always greater than bulk density. Bulk density is an important property to calculate for lining a furnace, kiln, or reactor [3].

9.1.1.3   Apparent Porosity

The apparent porosity of refractory, often expressed in %, is defined by Eq. (9.1). It is important to note that declining apparent porosity supports the enhancement of bulk density, mechanical strength, thermal conductivity, and corrosion resistance, which is considered imperative for oil and gas sectors' refractory design.

$$\text{Apparent porosity} = \frac{\text{Volume of open pores}}{\text{Total volume}} \times 100 \qquad (9.1)$$

9.1.1.4 Permeability

Permeability is the amount of gas flowing through the pores within the refractory body, and it indicates the extent of pore linkage. The permeability of refractories determines how well they can withstand melting, molten slag, or gas penetration. Specific gas permeability (μ) is defined by Eq. (9.2) with laminar gas flow:

$$\mu = \eta \cdot \frac{V}{t} \cdot \frac{\delta}{A} \cdot \frac{1}{P_1 - P_2} \cdot \frac{2P}{P_1 + P_2} \tag{9.2}$$

where V/t is the volumetric flow of the gas, through the refractory, in m³/s, η is the dynamic viscosity of the gas at the test temperature in Pa·s, δ is the thickness of the refractory in m, A is the cross-sectional area that gas flows through, P is the absolute pressure of the gas, and P_1 and P_2 are the absolute pressure of the gas at the entry and exit points, respectively. The factor $2P/(P_1 + P_2)$ equals 1 for small pressure differences.

9.2 CLASSIFICATION OF REFRACTORIES

Various forms of refractories are available nowadays depending on the need. Classification of refractories is based on several parameters listed below:

1. Basic nature of oxides
2. Physical shape
3. Process of making
4. Based on the application method
5. Based on the chemical composition
6. Based on the insulating value

The physical shape of refractories sets up the ground for grouping them under two different categories; the first is shaped refractories, also known as refractory bricks, whereas the second is known as unshaped refractories or monolithic refractories. Shaped refractories have a fixed geometrical shape, whereas unshaped refractories are found in the form of loose powder, which is mixed with the water or binder to be utilized in the wall areas [4].

9.3 MANUFACTURING OF REFRACTORIES

9.3.1 SHAPED REFRACTORIES

In general, most of the shaped refractories are manufactured using the same routes in which the input raw materials are used in the lumpy form with a size above 50 mm. These lumpy materials are then crushed in a jaw crusher (also known as a primary crusher). After the primary crushing operation, the secondary crushing operation gets performed using a hammer mill. The screening of crushed material is done according to a variety of sizes, such as particles of the sizes 0–1, 1–3, and 3–5 mm. To create a refractory brick, the particles having a range of sizes are combined with a suitable binder (such as molasses mixed with water). In the next step, the mixture is compressed in a mold of desired shape and size using a large pressure. The internal moisture is removed by drying the brick before being fed to a high-temperature (1,150°C–1,750°C) operation, which is also known as firing. Firing imparts strength to the refractory and provides the necessary characteristics to withstand different operational conditions, such as highly corrosive conditions for petroleum refineries, etc.

9.3.1.1 Silica Refractories

Silica refractories contain a minimum of 95% SiO_2. The common impurities are CaO, Fe_2O_3, and Al_2O_3 [2]. The primary raw material is quartzite, which comes in various forms, from coarse, uncemented crystals to fine, cemented ones. The complicated behavior of polymorphism is the most crucial factor in determining if a raw material is suitable for the manufacturing of silica refractories, but other key factors include the rock's fracture behavior, microstructure, chemical composition, refractoriness, and porosity. Alkali should be below 0.3%, and for high-quality products, it should be below 0.1%. Al_2O_3 and TiO_2 should be as low as possible, but they should be below 2%. It has been observed that the properties of the basic bricks depend on the properties of magnesite raw material [5]. The manufacturing process of silica refractories involves crushing raw materials, followed by a screening (based on size). The screen section would undergo the ball milling operations, followed by weighing and binder mixing. The final operation involves drying, firing, and sorting the bounded mixtures to get the finished silica refractory products.

9.3.1.2 Aluminosilicate Refractories

Aluminosilicate refractory comprises mostly alumina (Al_2O_3) and silica (SiO_2). Aluminosilicate refractories are categorized as fireclay refractory when the percentage of Al_2O_3 is less than 50%, and when Al_2O_3 is more than 50%, it is known as high alumina refractory.

9.3.1.3 Fireclay Refractories

Calcined fire clay or chamotte is mainly used as a raw material in manufacturing fireclay and plastic fireclay. The process flow of manufacturing fireclay refractories starts with crushing raw materials, followed by screening (based on size). The screen section is subjected to a ball milling operation followed by weighing and binder mixing. The final operation includes pressing, drying, and heating (1,100°C–1,400°C) to get the final fireclay refractories. In general, the Al_2O_3 content of chamotte is found in the range of 50%–85%. To be blessed with the desired engineering properties to satisfy industrial requirements, high alumina refractories require various raw materials, which end up manufactured refractories with different engineering characteristics. It is worth mentioning that during high-temperature heating, andalusite, sillimanite, and kyanite get easily transformed into mullite. Bricks constructed using these raw materials exhibit excellent resistance to thermal shock and high resistance to creep at a high temperature with good alkali resistance [6]. These properties are highly imperative for refractories manufactured for downstream oil and gas sectors, that is, petroleum refineries.

9.3.1.4 Alumina Refractories

Refractory materials which contain Al_2O_3 more than 90% of are frequently employed in the petrochemical industries. Several raw materials, such as white-fused alumina, tabular alumina, and brown-fused alumina, are available as the primary material for producing corundum refractories [7]. As illustrated earlier in this chapter, other than the manufacturing process, raw materials significantly impact the finished refractory product's qualities. The manufacturing flow process of corundum refractories involves weighing different size fractions of alumina, reactive alumina, and kaolin. These get utilized in preparing the mixture. Subsequently, forming is utilized, followed by drying to get the finished refractory products in the desired shape with desired engineering properties.

9.3.2 UNSHAPED REFRACTORIES

Another way to transport refractories at various industry sites is in the form of powders. Refractories that are not shaped are often produced and supplied as loose powder in bags. Unshaped refractories are easier to manufacture than shaped refractories. While manufacturing unshaped refractories, the necessary components are mixed in a mixer with the desired amount before being bagged. To utilize the product, the bags must first be opened at the installation location, and powders undergo a mechanical mixing operation with the appropriate addition of water/binder. Subsequently, ramming, casting, or spraying get adopted for refractory installation at industrial sites. Unshaped refractories can further be classified into different categories depending on the usage, as illustrated below.:

9.3.2.1 Mortars

Mortars are powders that get used for connecting refractory bricks. In other words, to link two brick surfaces, they can be glued up by placing the powder substance between the surface of the bricks and combining it with water or a binder in liquid form that comes with mortar. Mortar is known to be hardened either at room temperature or at high temperature by heating the bonded brick. The mortar joints found in the brick lining of the furnace facilitate reduced thermo-mechanical stress led by cushioning effect during the furnace operation [8]. This phenomenon is highly relevant for the oil and gas sector, particularly for the downstream (refinery operation).

9.3.2.2 Ramming Masses

The cement industry utilizes ramming masses in very small amounts and for specific applications. Ramming mass is defined as a mixture of fine powders, granular materials, and some additives. It is rammed by hand or with a pneumatic rammer at the desired location after blending with a specified binder such as water. It is typically used to close the gaps between a retainer plate and bricklayers, which otherwise becomes difficult to close.

9.3.2.3 Castables

Another important category of unshaped refractories is castables, which are composed of calcium aluminate cement (also known as high alumina cement) as the binder, aluminosilicate aggregate, and specialized additives. Calcium aluminate cement, such as portland cement, hardens to a solid mass when interacting with water and holds hydraulic properties. Binders in the liquid state are provided separately together with the powder in the absence of cement castables, which must be mixed before real-time application.

9.3.3 Advantages and Industrial Applications

Refractories are widely used for both industrial and experimental purposes due to their superior heat resistance to commercial metals and alloys. Performing heat treatment of shaped refractories leads to improved thermal and structural stability and imparts strength through densification. Hence, such refractories are used in pyro processes such as in refinery furnaces. In addition, heat-treated refractories are known for their faster lining operation inside the furnace than non-heat-treated refractories. It is worth mentioning that the application of heat-treated and shaped refractories in refining furnace's lining makes the refining operation more facile and offers greater safety than non-heat-treated and unshaped refractories.

9.3.4 Limitations of Applications of Shaped Refractories in Industries

Despite the potential engineering properties of shaped refractories, it has certain techno-commercial limitations. Some of these limitations are discussed below:

a. While applying the shaped refractories in geometrically intricated furnaces, large inventory space is often a requirement. This enhances the land and labor cost.
b. It is difficult to automate the installation process of refractories. Due to this limitation, installation demands a huge workforce.
c. Degradation of shaped refractories led by slag and gas penetration through joints becomes one of the major concerns for furnaces or reactor cells.
d. Repairing or patching the refractory lining becomes time-consuming due to slow furnace cooling. Until the furnace is completely cooled down, repairing refractory lining becomes difficult.

9.4 INSTALLATION OF REFRACTORIES

9.4.1 Installation of Shaped Refractories

For any downstream oil and gas sector, building and repairing the refractory lining consider one of the major areas of economic expenditure. However, the same gets compensated by ensuring plant and workforce safety. However, it is noteworthy that the quality of the installation impacts the refractory performance (design, installation, and maintenance of refractories) significantly, irrespective of the refractory materials' quality. For a sound installation, a detailed characterization of refractory materials is required [4].

9.4.2 Monolithic Refractories Installation

Monolithic refractory installation involves using a wide range of engineering materials, for example, mortars, ramming masses, mortars, low cement castables, shotcreting material, gunning, and standard castables. These refractories are also known as partially finished refractories; wherein further finishing is required to be performed at the application site. The monolithic refractories are often provided to the customer as dry powder and consist of a blend of coarse aggregates and fine powders. Before application, the loose powder must be combined on-site with water or another designated liquid binder. Its consolidation, compaction, drying, and preheating all significantly impact its final properties.

9.5   PREHEATING OF REFRACTORIES

During the installation, the refractory lining has significant moisture content due to its inherent physical and chemical characteristics. It is noteworthy that the available moisture found in the refractory lining is either physically or chemically bonded. The porosity of refractories leads to moisture absorption from the air in shaped refractories. However, mortars with 25%–35% moisture content get used to unite bricks. Preheating the reactor or furnace is performed to reduce the moisture-led problems in the refractory lining. Preheating becomes more crucial for a bigger-scale reactor with a thicker refractory lining. It is noteworthy that the refractory lining also experiences mechanical stress during preheating operation, which may become another mode of failure.

9.6   PREHEATING OF CASTABLES

Castables get settled because of the formation of various calcium aluminate hydrated compounds during the hardening process. However, during the heating operation, several hydrates, including $CaO.Al_2O_3.10H_2O$, decompose at different temperatures. Hence, the thermal decomposition of hydrates helps in releasing water. Fast preheating of the castable liner may cause the internal steam to develop, which leads to very high steam pressure development. The castable liner may suddenly steam up and produce a very high steam pressure which the furnace operator must avoid to prevent accidents [9].

9.7   REFRACTORIES IN PETROLEUM REFINERIES

To make crude petroleum useful for different purposes, the crude oil is refined in a refinery where hydrocarbon compounds get separated. Crude petroleum is a mixture of numerous distinct hydrocarbons that can be found in nature. It must be noted that in 2018, 120 million barrels or 19.1 billion liters of daily crude oil were refined. There are two major units of refinery other than process heaters that operate at higher

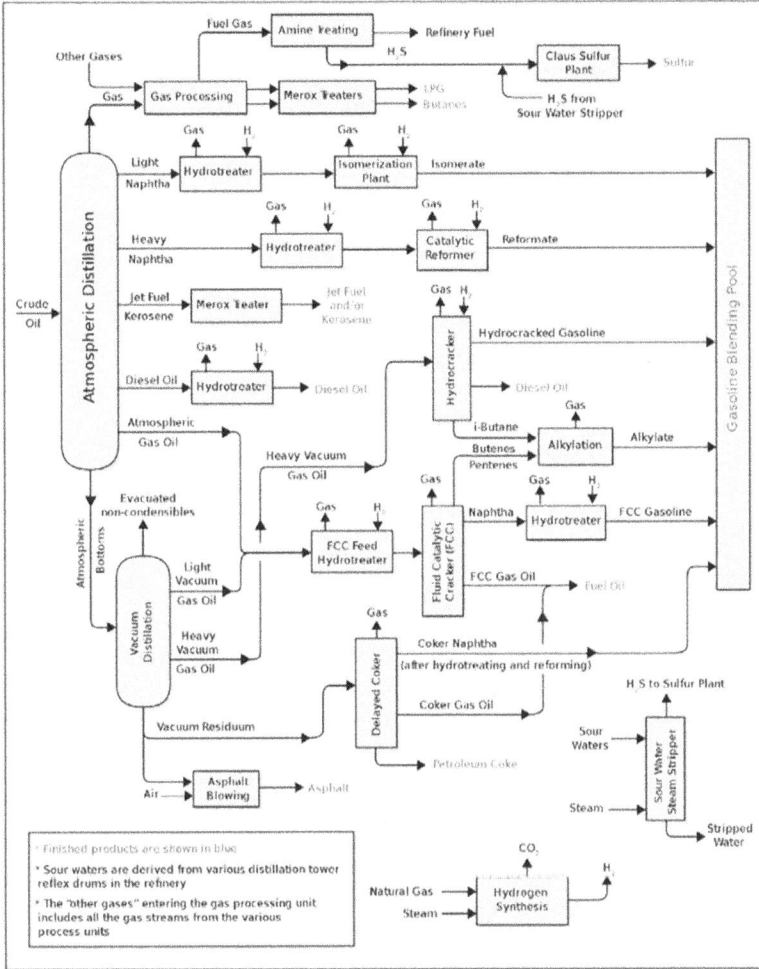

FIGURE 9.2   Flow diagram of refinery. (Adopted from Ref [10], with permission.)

temperatures, for example, the fluidized catalytic cracking unit (FCCU) and the sulfur recovery unit. The refinery process is explained through the flow diagram illustrated in Figure 9.2 [10].

9.7.1   Fluid Catalytic Cracking Unit

Lighter petroleum products such as gasoline and liquefied petroleum gas get converted using a fluid catalytic cracking unit (FCCU) from crude oil feedstock. The FCCU process unit offers various catalytic reactions, utilization of catalysts, separation, and fractionation. Hence, it is known as one of the most complex process units. FCCU performance can be improved by appropriate catalyst selection, optimum adjustment of operating conditions, pretreatment of feedstock, etc. Unit operations can be optimized and controlled using the computer-programmed interface, which will be the most economical alternative. In addition to using efficient catalysts, the selection of refractory lining also holds significant importance. Refractory materials must exhibit sustainability, durability, resistance to moisture absorption, high formability, and high resistance to mechanical shock.

9.7.2 SULFUR RECOVERY UNIT

In the sulfur recovery unit, hydrogen sulfide (H_2S) gets converted to elemental sulfur. Hydrogen sulfide is formed as a byproduct during natural gas processing and high-sulfur crude oil refining. The Claus process is the most often utilized conversion technique. As prescribed in environmental regulations, sulfur recovery units from H_2S gas create sulfur to stop acidic gas emissions. HYSYS's SULSIM program, specifically designed for sulfur, is used to simulate and optimize the plant. Since H_2S gas is corrosive, selecting the refractory lining becomes important to avoid any degradation-induced failure.

9.7.3 SELECTION OF REFRACTORIES

The main purpose of refractories' applications is to provide a longer and more stable economic lifetime to furnaces [1]. There are many reasons why furnaces shut down. However, if a shutdown happens due to refractory, it significantly affects operational costs. It is important to note that cooling the furnace may take a long time and adversely affect production. It will be economical and easy if the refractory life aligns with the planned maintenance schedule of the plant. There is no universally applicable suggestion for refractory materials due to the significant variation in size, raw materials, capacity, and operational procedures. However, it is preferable to adhere to a general rule that it must enhance the performance of refractory and other auxiliary equipment with the least refractory lining failures.

9.8 REFRACTORY PERFORMANCE AND MECHANISM OF DAMAGE

In most cases, instead of one defined cause, a chain of events leads to failure [9]. In other words, many variables caused the failure of the refractory material. The performance of refractories is controlled by the user of the refractories and manufacturer of refractories based on some factors such as operating parameters, quality of refractory, quality of installation, and storage quality, but in general, it is controlled by the user. From a refractory perspective, all these factors have an impact on refractory and can cause failure. Some of these failures can be listed as:

  a. Overheated spots (temperature gets very high)
  b. Excessive cracking due to spalling, thermal shock (a result of temperature fluctuations)
  c. Mechanical shock or compressive forces, linear change, bending, etc.
  d. Due to structural issues

e. Refractory lining Erosion
f. Failure of a steel shell that supports refractory
g. Other mechanical failures

9.9 CONCLUSION

Refractories are important for the oil and gas industries as it helps to regulate the temperature by keeping heat within themselves and preventing furnaces used in different refinery units, such as FCCU, Fired Heaters, Process Heaters, from overheating. A variety of refractories are available as options for the downstream refineries; however, the application depends on certain operating factors such as temperature, pressure, feedstocks, and solution media. Therefore, furnace operators need to have prior knowledge about these factors to avoid any accidental damages due to refractory lining failures. This chapter described the types of refractories that could be used in the refineries and deciphered the associated failure mechanisms.

REFERENCES

[1] P. Sengupta, "Refractory: Characterization," in *Refractories for the Chemical Industries*, P. Sengupta, Ed. Cham: Springer International Publishing, 2020, pp. 1–29. doi: 10.1007/978-3-030-61240-5_1.

[2] N. Akira, *Technology of Monolithic Refractories*. Tokyo: Plibrico Japan Co., Ltd., 1984.

[3] P. Sengupta, "Classification of refractories," in *Refractories for the Chemical Industries*, P. Sengupta, Ed. Cham: Springer International Publishing, 2020, pp. 31–41. doi: 10.1007/978-3-030-61240-5_2.

[4] R. C. Bradt, "Elastic moduli, strength and fracture characteristics of refractories," *Key Engineering Materials*, vol. 88, pp. 165–192, 1993, doi: 10.4028/www.scientific.net/KEM.88.165.

[5] F. Goorman, J. Visser, M. Ruer, C. G. Aneziris, and J. Ulbricht, "The performance of high quality magnesia raw materials in cement applications," in *Proceedings of the Unified International Technical Conference on Refractories (UNITECR 2013)*. Hoboken, NJ: John Wiley & Sons, Ltd, 2014, pp. 189–192. doi: 10.1002/9781118837009.ch33.

[6] P. Sengupta, "Manufacturing and properties of refractories," in *Refractories for the Chemical Industries*, P. Sengupta, Ed. Cham: Springer International Publishing, 2020, pp. 43–84. doi: 10.1007/978-3-030-61240-5_3.

[7] P. Sengupta, "Design, installation, and maintenance of refractories," in *Refractories for the Chemical Industries*, P. Sengupta, Ed. Cham: Springer International Publishing, 2020, pp. 85–134. doi: 10.1007/978-3-030-61240-5_4.

[8] H. Gheisari, S. Ghasemi-kahrizsangi, E. Karamian, and A. Nemati, "Recent advancement in monolithic refractories via application of nanotechnology 'A review Paper,'" *Journal of Nanoanalysis*, vol. 6, no. 1, pp. 1–20, 2019, doi: 10.22034/jna.2019.664385.

[9] T. D. V. Lin, "FCCU advanced control and optimization. [Fluid Catalytic Cracking Unit]," *Hydrocarbon Processing; (United States)*, vol. 72:4, 1993, Accessed: November 27, 2022. [Online]. Available: https://www.osti.gov/biblio/6535017.

[10] A. H. Al-Moubaraki and I. B. Obot, "Corrosion challenges in petroleum refinery operations: Sources, mechanisms, mitigation, and future outlook," *Journal of Saudi Chemical Society*, vol. 25, no. 12, p. 101370, 2021, doi: 10.1016/j.jscs.2021.101370.

10 Recent Progress in Superhydrophobic Macroporous Sorbents for Oil Spill Remediation

Love Dashairya and Partha Saha
National Institute of Technology, Rourkela

CONTENTS

ABBREVIATIONS

AC	Absorption capacity
CA	Contact angle
COVID-19	COronaVIrus Disease of 2019
CVD	Chemical vapor deposition
DA	Dopamine
DTMS	DodecylTriMethoxySilane

DOI: 10.1201/9781003242550-10

GN@PU   Graphene-coated polyurethane sponge
MF   Melamine-formaldehyde
MTCS   Methyltrichlorosilane-functionalized silica nanoparticle-coated superhydrophobic
RGO   reduced graphene oxide
TEOS   tetraethyl orthosilicate

10.1   INTRODUCTION

Over the last few decades, numerous oil spill accidents have been witnessed on the sea and seashores during oil extraction, storage, and transportation throughout the world [1–3]. In this regard, oil spill in seawater has been witnessed during the accidental explosion in Mexican Ixtoc I platform in 1979 (~4.9 million barrels), the Gulf of Mexico in 2010, and 3,000 barrels of oil spillage into Bangladesh-Sundarbans reserve in 2014, invariably affected the existence of marine lives (see 10.1a) [4–7]. In May 2020, during the challenging time of COVID-19 global pandemic and lockdown, Russia witnessed one of the largest Arctic oil spills, which called for a state of emergency (see Figure 10.1b) [8]. The Norilsk oil spillage accident was humankind's most serious industrial disaster and a future lesson for developing nations that must be learned and prevented [9]. About 20,000 tonnes of diesel was spilled into Ambarnaya River, flowing directly into the environmentally sensitive Arctic Ocean [8]. The Mauritius oil spill on the country's southeast coast also occurred during the COVID-19 pandemic, catapulting a massive blow to biodiversity since the coastline is known as wetlands of international importance [8]. The oil spill accident resulted in the death of 39 dolphins on the beach [8, 9].

In recent times, India has witnessed massive oil spillage during oil barrel transportation, including Mumbai oil spill in 2010, which resulted in around 810 tons of oil leakage over 2 days [10]. Oil spill accident in Chennai, India, back in 2017 resulted in approximately ~75 metric tons of crude oil spilling into the Bay of Bengal, affecting a population of over 7 million [11]. Oil spill-related accident affects biodiversity and a substantial economic loss attached to such disasters [11]. It also imposes a severe threat to indigenous people dependent on seafood and living near the seashore, reflecting violation of SDG 2.0 Goal 3 "Good Health and Well-Being" [12]. The oil contamination in and around coastal wetlands also affects power plants for drawing or discharging seawater, the fishing industry, and tourism, referring to Goal 15, "Life on Land." In addition to oil spillage, manufacturing industrial-grade chemicals and leakage of water-insoluble organic solvents such as benzene, toluene, cycloalkane, and dichloromethane to the river beds is a significant challenge that must be addressed with suitable technological innovation and products [2,7,13,14]. The oil spill and remediation is a global problem that has aroused researchers and policymakers around the world to come up with innovative strategies and groundbreaking technologies for oil spill cleanup [2,7,13,14]. The oil spill and hazardous organic solvent-based industrial waste not only impact human life but also affect marine lives (fishes, shoreline, mammals, sea birds, trees, and so on), so there is an imminent need to explore macroporous sorbents that can effortlessly separate oil and organic solvents from fresh or seawater [2,7,13,14]. To date, there are limited solutions to resolve the above problems [3]. Traditional separation technologies such as straw, centrifugation, gravity settling, booms, coagulation, skimming, wool fibers, zeolites, and fly ash have all been

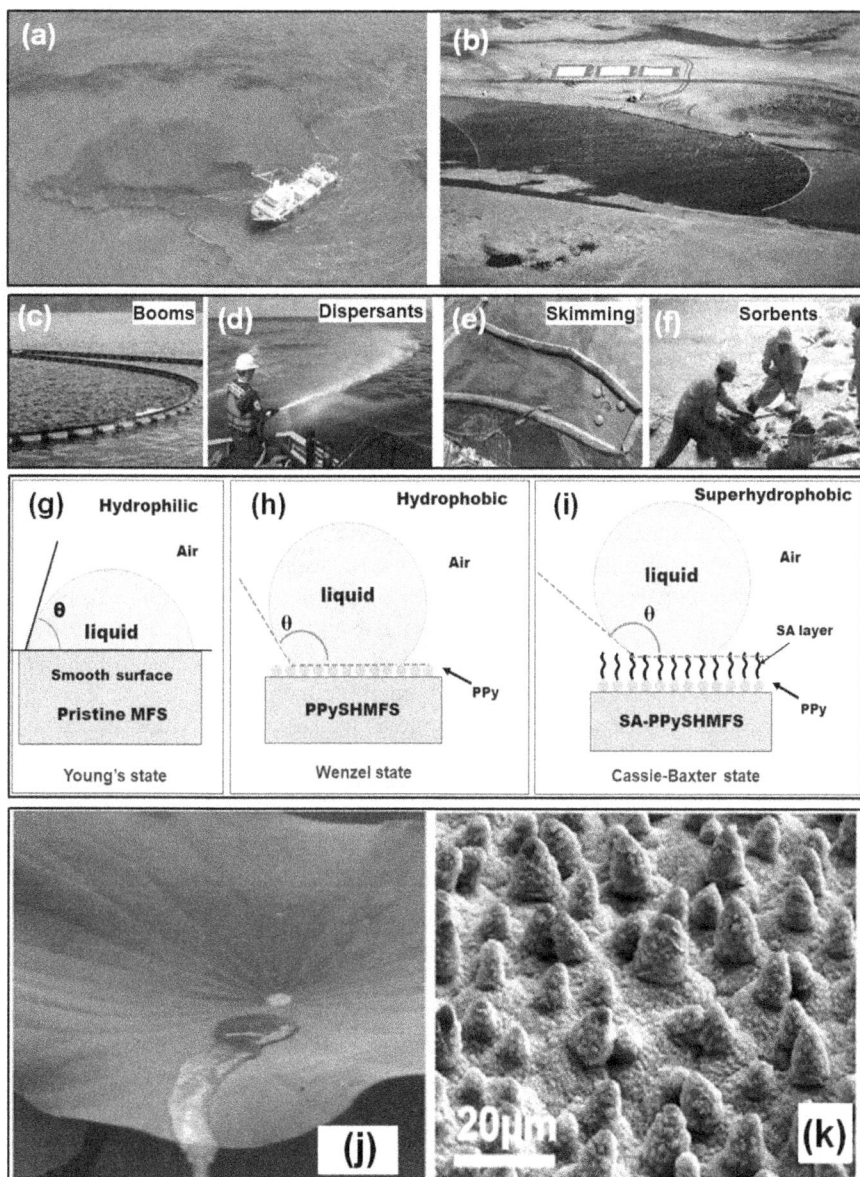

FIGURE 10.1 Digital images of the massive oil spill in (a) Gulf of Mexico; (b) Russia Arctic; (c–f) oil spill cleanup technologies (booms, dispersant, skimming, and sorbents); (g–i) schematic representation for wettability and nonwettability states of macroporous sorbent surfaces with contact angle [21]; and (j, k) digital image of lotus leaf showing self-cleaning superhydrophobic surface properties and scanning electron microscope (SEM) image of lotus leaf rough surface [22]. (Reprinted with permission from Dashairya et al. [21]. Copyright (2020), Elsevier; L.-Y. Meng, S.-J. Park, Multifunctional surface structures of plants: an inspiration for biomimetics, Progress in Materials science, 54 (2009) 137-174. Copyright 2008 Elsevier.)

employed in the past (see Figure 10.1c–e) [3,7,15]. However, the above technologies have their shortcomings, such as incompatibility with the environment, low absorption capacity (AC), and recyclability [3,7,15,16]. As a result, there is a substantial need for "smart sorbents" that can be intentionally designed and manufactured to meet the stringent criteria for selective removal of oil and organic solvents from wastewater as and when needed (see Figure 10.1f) [3,7,15–17].

The sorbents being used are currently restricted into two classes: (1) recoverable and (2) nonrecoverable [3,16,18]. Nonrecoverable processes include chemical dispersants, *in situ* burning, solidifiers, bioremediation, and mechanical technologies (booms, barriers, natural or synthetic sorbents) [3,16,18]. Chemical techniques, such as *in situ* burning and solidifiers, have received much attention due to the ease of removing oil and water mixture, hastening their natural biodegradation [3,16,18]. However, *in situ* burning and solidifiers have drawbacks such as secondary pollutants, limited separation efficiency, and resource loss [7]. Biological remediation, a recently developed technique, transfers or splits hydrocarbon chains into fatty acids, carbon dioxide, and water [7]. This clean technology has all the advantageous parameters necessary for oil spill cleanup. However, sustaining the escalation of microorganisms necessitates complex and stringent circumstances since several variables, such as pollution, accessibility to microorganisms, and the opportunistic usage of bacteria, restrict their employment [7]. As a result, the biological method has several drawbacks that limit its use in oil biodegradation. However, due to environmental protection and economic growth, there is an increasing demand for cost-effective oil spill remediation materials with improved recyclability that can efficiently separate oil-water mixture with greener pasture. In the last several decades, mechanical cleanup has exceeded the performance of all other techniques, with booms, barriers, and natural or synthetic absorbents being the most frequent methods [7]. However, with inherent hydrophilicity and poor AC, such materials cannot recover oil and organic solvents [4,5,7]. In addition, the effectiveness of oil separation utilizing oil booms, barriers, and skimmers is poor, especially in complicated oil-water emulsion systems [4,5].

On the other hand, physical absorption has been a hot topic in the oil-water separation sector for a long time [18,19]. The petroleum industry must develop new oil-water separation processes to expedite oil recovery [18,20]. The absorption of spilled oil or organic solvents and associated methodology must possess a few attributes, *viz.*, robustness, low cost, wider availability, environmental friendliness, and excellent reusability [2,7]. Due to low cost and ease of fabrication, the global market for macroporous sorbents has exponentially increased in recent years [2,7]. Therefore, macroporous sorbents are considered promising future technology for oil spill cleanup compared to known methods owing to the complete removal of oil from an emulsion mixture, including water-insoluble organic species, with hardly any environmental impact [3,13,18].

10.2   WETTABILITY AND NONWETTABILITY: FUNDAMENTAL CONCEPT OF CONTACT ANGLE AND SUPERHYDROPHOBICITY

Wettability and nonwettability are the essential parameters to determine the true nature of any surface and are equally applicable to macroporous sorbents. Hydrophilic and hydrophobic behaviors are commonly used terminologies to define the surface properties of macroporous sorbents. If any material absorbs water and remains wetted, it will be known as hydrophilic. However, if the surface repels water and does not get wetted, the material is called hydrophobic. In particular, the surface roughness and surface energy play crucial roles in determining the interaction between solid with vapor or liquid phase and subsequently the hydrophobic and hydrophilic behavior [23,24].

The well-known Young, Wenzel, and Cassie-Baxter equations are commonly used to define the wettability/nonwettability of any porous or solid surface. If the surface of any substrate does not get wetted owing to lower surface energy and higher surface roughness, the surface is defined as hydrophobic. Hydrophobicity of any surface is characterized by contact angle (CA) measurements, which generally determine the degree of interaction of a liquid or water droplet with a solid surface. The CA and hydrophobicity of solid surfaces are increased by lowering the surface energy given by Young's equation. The surface of any substrate is called hydrophilic if the CA is less than 90°, as described by Young's equation (Eq. 10.1) as given below and shown in Figure 10.1g [21,23–25].

$$\cos\theta = \frac{\gamma_{sv} - \gamma_{sl}}{\gamma_{lv}} \tag{10.1}$$

where γ_{sv}, γ_{sl}, and γ_{lv} are the solid-vapor, solid-liquid, and liquid-vapor interfacial free energy, respectively. Young's equation is applicable only to evaluate the CA of smooth and flat surfaces [24]. However, the practical surface of any materials is not smooth, and thus, Young's equation was further modified by the Wenzel equation illustrated in Eq. (10.2) to determine the CA of liquid on a rough surface, as depicted in Figure 10.1h [21,23–25].

$$\cos\theta_w = r\left(\frac{\gamma_{sv} - \gamma_{sl}}{\gamma_{lv}}\right) = r\cos\theta \tag{10.2}$$

where r represents the surface roughness factor, which is well defined as the ratio of the actual rough surface area to the geometric projected area. θ_w is the CA of a liquid droplet on a rough surface. If the CA is greater than 90°, then the material is hydrophobic [18]. For a material to be superhydrophobic, the CA should be above 150° [18,21,23–25]. But in practical cases, the surfaces are not homogenous but rather heterogeneous. So, in the case of heterogeneous surfaces exhibiting superhydrophobicity, the wettability is explained using a Cassie-Baxter equation, as depicted in Figure 10.1i [21,23–25].

$$\cos\theta_c = f\left(\cos\theta + 1\right) - 1 \tag{10.3}$$

where f denotes the fraction of the solid surface in contact with liquid, θ, and θ_c represents the water CA on smooth and rough surfaces, respectively. For example, the lotus effect ideally defined the Cassie-Baxter state with excellent self-cleaning and anti-wetting capability, as shown in Figure 10.1j [19,22,26]. The water droplets can smoothly roll over on the lotus leaf surface owing to the high surface roughness and low surface energy with a branch-like micropapillary surface structure. Figure 10.1k demonstrates a FESEM image of a lotus leaf surface, showing the presence of a branch-like micropapillary surface and nanoscale tomenta structure providing high surface roughness, leading to the formation of superhydrophobicity [19,22,26]. Notably, the critical factors of the lotus leaf effect can be extended onto macroporous sorbents, which can be helpful in constructing self-cleaning, anti-wetting, and superhydrophobic structures for practical oil spill cleanup [18,19,22]. Therefore, it is paramount to develop superhydrophobic surface on macroporous sorbents enabling wetting/de-wetting characteristics for selective oil-water separation applications [18].

10.3 FABRICATION STRATEGIES OF MACROPOROUS SORBENTS FOR OIL SPILL REMEDIATION

As mentioned before, superhydrophobic sorbents are useful for oil spill remediation based on the absorption and filtration phenomena [27]. Numerous techniques and fabrication strategies have extensively been explored in the past decades to develop superhydrophobic sorbents, including dip coating, wet chemical route, electrospinning, chemical vapor deposition (CVD), hydrothermal, and sol-gel (see Figure 10.2a–d). Macroporous sorbents developed using electrospinning, electrodeposition, CVD, and spray pyrolysis exhibit a more novel microstructure and morphology than other methods. However, the complicated synthesis process, high precision, and the exorbitant cost of instruments restrict their widespread use for large-scale synthesis. Therefore, facile, inexpensive, and convenient approaches are the need of the hour to synthesize superhydrophobic macroporous sorbents with high throughput.

10.3.1 DIP COATING

The dip-coating technique is the most widely used method for fabricating superhydrophobic-superoleophilic sorbents owing to its usage of simple apparatus in a time-saving manner. During dip coating, macroporous sorbents are immersed in a reactive medium for a predetermined duration followed by drying to obtain the desired microstructure. Xiaotan *et al.* [28] synthesized a graphene-coated polyurethane sponge (GN@PU) via a one-step dip coating using aqueous graphene suspension and cellulose nano-whiskers precursor. The prepared GN@PU exhibited a CA of ~152° and an AC of ~31 g/g for lubricating oil separation. The AC of macroporous sorbents is calculated using the given formula [5].

$$AC = \left[\text{Final weight } \left(W_f \right) - \text{initial weight } \left(W_i \right) \right] / \text{initial weight}(W_i)$$

Recently, Saha's group fabricated methyltrichlorosilane-functionalized silica nanoparticle-coated superhydrophobic/superoleophilic cotton fibers (MTCS/SiO$_2$-treated cotton) via a simple dip-coating process for the selective removal of kerosene, hydraulic oil, pump oil, engine oil, diesel, and vegetable oil from water, as shown in

Figure 10.2a [6]. The formation of layer-by-layer (–C–Si–O–) assembly onto the cotton fiber surface enhanced the surface roughness leading to superhydrophobic and selective wettability of MTCS/SiO$_2$-treated cotton. The as-synthesized MTCS/SiO$_2$-treated cotton demonstrated excellent superhydrophobicity with a water CA greater than 173° and an oil-water separation capacity of ~30–42 times its weight after ten reusable cycles at different pH ranges [6]. Jintao's groups [29] fabricated a superhydrophobic stearic acid (SA)-functionalized PU sponge using a one-step dip-dry method for the purification of oil-water mixtures and emulsions. Also, a superhydrophobic SA-coated sponge was able to demonstrate continuous separation of oil-water mixtures using vacuum suction and oil droplets from oil-in-water emulsions. On the other hand, Dashairya *et al.* [25] fabricated SA-functionalized polypyrrole-coated superhydrophobic melamine-formaldehyde (MF) sponge using a facile self-polymerization process, as shown in Figure 10.2b. The synthesized superhydrophobic MF sponge demonstrated a water CA of ~170° and ACs of ~70–90 times its mass for oil-water separation. In recent times, several superhydrophobic macroporous sorbents have been synthesized, including bioinspired dopamine (DA) and dodecyltrimethoxysilane (DTMS) sponge, DA-nickel foam, carbon nanotubes (CNTs)/poly(dimethylsiloxane)-coated PU sponge, polydimethylsiloxane (PDMS) PU sponge,

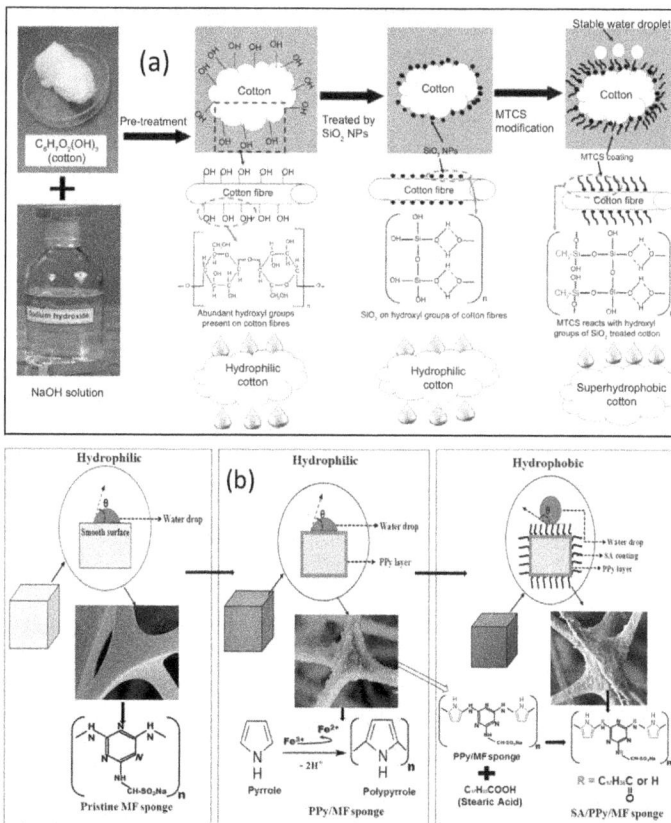

FIGURE 10.2   The proposed synthesis mechanism of superhydrophobic macroporous sorbents (a) MTCS-functionalized SiO$_2$-coated cotton [6]; (b) SA/PPy/MF sponge [25];

(Continued)

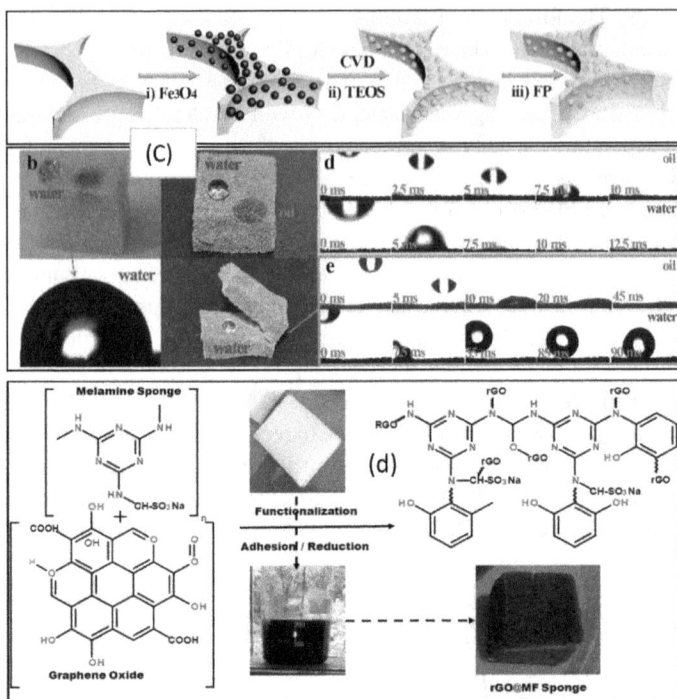

FIGURE 10.2 *(Continued)* (c) PU@Fe$_3$O$_4$@SiO$_2$@FP sponge [35]; (d) self-assembly and hydrothermal synthesis of superhydrophobic rGO@MF sponge [4]. (Reprinted with permission from Dashairya et al. [6]. Copyright (2019), Springer Nature; Reprinted with permission from Dashairya etal. [25]. Copyright (2019), Springer Nature; Reprinted with permission from Wu et al. [35]. Copyright (2015), American Chemical Society; Reprinted with permission from Saha and Dashairya [4]. Copyright (2018), Springer Nature.)

polydopamine (PDA) PU sponge, octadecyltrichlorosilane (OTS) cigarette filters, and PPy-perfluorooctyltriethoxysilane sponge for excellent oil spill remediation [30–33]. Moreover, the dip-coating process was also explored for the synthesis of superhydrophobic nanomaterial-coated macroporous sponges [34]. Recently, polyvinylidene fluoride-hexafluoro-propylene (PVDF-HFP)-functionalized CNTs and SiO$_2$ nanoparticle-coated PU sponges have been fabricated by dip-coating technique [34]. Recently, our group synthesized Zr/Cl dual ion-mediated PPy-functionalized SA-coated MF sponge for efficient oil spill cleanup [21]. Zr/Cl dual ion-polymerized PPy coating enhanced the pristine MF sponge's surface roughness, leading to the formation of a superhydrophobic surface with a water CA and oil AC of ~162.6° and ~100–110 g/g, respectively. Notably, the macroporous superhydrophobic MF sponge retained ~93%–95% oil after the 20th reusable cycle.

10.3.2 Wet Chemical Method

In the wet chemical reaction, pristine macroporous sorbents can act as a reactant and participate in the chemical reactions for superhydrophobic modifications, which involve *in situ* growth and polymerization [16,36]. Chaofan *et al.* [37] applied wet chemical reaction and plasma treatment methods for the fabrication of ZnO_2-coated cotton fabric and filter paper fiber, followed by SA-functionalized superhydrophobic surface modification. The as-synthesized superhydrophobic cotton fabric and filter paper fiber exhibited water CA of ~151° and ~154°, respectively, and greater than 90% oil-water separation efficiency. Furthermore, a wet chemical reaction process was used to synthesize polymer molecular brush-grafted superhydrophobic PU sponge, which demonstrated excellent oil-water separation capacity, reusability, and capacity retention [38]. Besides, Wang *et al.* [39] synthesized thiol-coated metal oxide/metallic nano-crystals sponge and fabrics via *in situ* growth process. Recently, urea-cross-linked modified sponge, oleophilic monomer lauryl methacrylate fluoroalkylsilane-function-alized TiO_2@fabric composites, and poly(ethylene terephthalate) fabric surface were successfully synthesized via a wet chemical method-induced superhydrophobic surface modification [40–42]. These wet chemical reaction-fabricated superhydrophobic-superoleophilic macroporous sorbents were successfully applied for selective oil-water separation, emulsion purification, and underwater oil removal.

10.3.3 Chemical Vapor Deposition (CVD)

CVD methods have been widely used to develop superhydrophobic macroporous sorbents toward selective oil-water separation [17]. Macroporous sorbents synthesized by the CVD method produce a controllable and uniform superhydrophobic coating. Wu and co-workers [35] developed magnetic and superhydrophobic PU sponge via CVD of tetraethoxysilane-derived SiO_2 to tightly coat Fe_3O_4 in fluoropolymer (FP) aqueous solution on PU sponge (PU@Fe_3O_4@SiO_2@FP sponge), as displayed in Figure 10.2c. Pristine PU sponge immediately absorbed both water and oil (see inset Figure 10.2c), whereas oil droplets selectively penetrated PU@Fe_3O_4@SiO_2@FP sponge within 5 ms, and water droplet remained stable. Superhydrophobic PU@Fe_3O_4@SiO_2@FP sponge demonstrated water CA and high AC of ~157° and 13.2−44.5 g/g, respectively (see inset Figure 10.2c). Moreover, the superhydrophobic PU sponges were effectively explored as a membrane and continuous vacuum suction-based oil-water separation. Zhang *et al.* [43] synthesized superoleophilic and superhydrophobic poly-ester materials via CVD method using tetraethyl orthosilicate (TEOS)-derived nano-silicon growth on polyester for oil spill cleanup. Compared to uncoated polyester, CVD-derived superhydrophobic polyester exhibited highly rough surfaces and excel-lent water-repellent properties. Zhou *et al.* [44] fabricated CVD-derived superhydro-phobic cotton fabric using polyaniline and fluorinated alkylsilane, which achieved a static water CA of 156°. CVD-synthesized superhydrophobic cotton fabrics were successfully applied for efficient oil-water separation with a separation efficiency of ~97.8% for 30 reusable cycles. Similarly, Crick's group [45] explored Cu filtration mesh for oil spill cleanup with excellent separation of organic solvents and crude oil from water. The superhydrophobic Cu mesh was fabricated via the CVD method using PDMS-deposited surface modification with a static water CA of ~152°–167°. Besides, Cortese *et al.* [46] modified cotton textiles by growing diamond-shaped car-bon thin film using a CVD method with superior superhydrophobicity for effective oil-water separation. Avargani *et al.* [47] also developed a superhydrophobic poly-ester textile via a CVD method. The surface of polyester textile was modified using CVD-assisted PVC coating on the fabric for separating oil from oil-water mixtures. More recently, various macroporous sorbents and filtration membranes have been reported for facile oil spill remediation using the CVD [47–53]. Superhydrophobic materials developed using CVD produce a uniform and controllable surface coating [17,51]. However, complicated and sophisticated instruments (vacuum conditions and high temperature) are usually required to conduct a CVD-based synthesis [17,51]. Also, CVD-based synthesis processes partially enhance surface roughness; hence, chemical etching and pretreatment of the substrate are essentially required [17,51].

10.3.4 HYDROTHERMAL AND SOL-GEL METHOD

Superhydrophobic macroporous sorbents synthesized by the hydrothermal and sol-gel method demonstrated excellent physicochemical properties such as narrow particle size distribution, uniform microstructure, high specific surface area, excellent porosity, compressibility, flexibility, and mechanical stability [54]. A significant advantage of hydrothermal and sol-gel synthesis is that the method can be hybridized with other processes to enhance the reaction kinetics and increase the ability to prepare a self-assembled novel microstructure. In addition, the hydrothermal method and sol-gel route do not require any seed, catalyst, or harmful and expensive reagents, thus promising for large-scale production catering to the industry needs. For example, Wang et al. [55] constructed superoleophilic ZnO-coated cotton fibers via a facile hydrothermal route followed by silane functionalization for oil-water separation and emulsion treatments. These as-prepared superhydrophobic/superoleophilic cotton fibers demonstrated excellent selective sorption capacities and reusability performance under various corrosive oil-water mixtures, hot water, and harsh water environments. Recently, our group successfully synthesized superhydrophobic-reduced graphene oxide-modified MF sponge (rGO@MF) and cotton fibers (rGO@cotton) via a one-step hydrothermal route for engine oil, pump oil, castor oil, kerosene, acetone, and toluene mixed with saline water mixture separation, as depicted in Figure 10.2d [4,5]. These as-prepared superhydrophobic/superoleophilic rGO@MF and rGO@cotton macroporous sorbents possess excellent porosity, flexibility, and mechanical squeezing properties for oil spill remediation. Moreover, the as-prepared superhydrophobic sorbets exhibited a water CA of ~150°–162° and an AC of ~90–120 times oil by its weight. Furthermore, Wan et al. [56] fabricated MoS_2-coated MF sponge for oil spill remediation and effective dye removal. Hydrothermally synthesized superhydrophobic MoS_2-coated MF sponge exhibits high AC of ~66–157 times its own weight for different organic solvents as well as 98% dye removal efficiency.

Besides, superhydrophobic cellulose membrane was synthesized via a one-step facile sol-gel route by hydrolysis and polycondensation reactions of TEOS and hexadecyltrimethoxysilane (HDTMS) on the membrane surface [57]. Xiaojing et al. [58] also fabricated durable and robust superhydrophobic PDMS-functionalized silica surfaces of polyester textiles via a facile vapor-liquid sol-gel route for oil-water separation. With the attachments of TEOS-derived SiO_2 particles on the surface of polyester textiles, surface roughness was increased, which leads to an enhanced water contact of ~150° and separation efficiency. Similarly, Rajaram groups also synthesized superhydrophobic PMHS-functionalized SiO_2 coated onto PU sponge [59]. The as-prepared sponges exhibited superhydrophobic properties and can separate oil from oil-water and oil-muddy water mixture. Recently, several hydrothermal and sol-gel synthesized macroporous superhydrophobic sorbents, viz., NiOOH@ sponge, PMHS–TEOS-derived xerogel, TEOS/HDTMS sponge, graphene foam, ternary silicone sponge, polyvinyl alcohol sponge, and 3D porous metal foam, have been explored successfully for facile oil spill remediation and emulsion purification, which revealed that hydrothermal and sol-gel methods are the most suitable low cost and time-saving route for the fabrication of macroporous sorbents for oil remediation [60–65].

10.4 OIL SPILL REMEDIATION USING SUPERHYDROPHOBIC-SUPEROLEOPHILIC MACROPOROUS SORBENTS

Over the last few decades, numerous superhydrophobic-superoleophilic macroporous sorbents have been proposed for oil spill remediation, *viz.*, porous carbon-based materials, polyurethane (PU) sponge, MF sponge, cellulose sponge, cotton fibers, etc., as listed in Table 10.1. These macroporous sorbents possess a high surface area and porosity, demonstrating excellent AC and reusability. In addition, these sorbents exhibited excellent mechanical flexibility under load during oil spill remediation [7]. Figures 10.3–10.6 display various macroporous sorbents (carbon monolith, PU sponge, MF sponge, PVA sponge, and cotton fiber) explored for oil spill remediation with enhanced performance [6,25,63,66,67]. These macroporous sorbents possess excellent superhydrophobic and superoleophilic surface properties by virtue of the modifier used during the synthesis procedure. In the past decades, numerous methodologies and modifiers have been explored for the superhydrophobic and superoleophilic surface modification of macroporous sorbents for oil-water separation. Consequently, the modified macroporous sorbents demonstrated excellent superhydrophobic property, selectivity, thermal stability, anticorrosive, and magnetic properties based on the synthesis and modification technique.

10.4.1 Carbon-Based Macroporous Sorbents

Carbon-based materials (graphene, CNTs, porous carbon, activated carbon, etc.) exhibit promising properties, including large surface roughness, specific surface area, excellent flexibility, low density, porosity, and chemical stability for oil spill remediation, organic solvent removal, and water purification application [20,69,70]. Carbon-based macroporous sorbents derived from aerogels, monolith, sponges, and foams have been widely explored as macroporous substrate owing to their excellent superhydrophobicity, AC, reusability, and selectivity towards oil-water separation [20,70].

To date, carbon-based monoliths, aerogels, and sponges have been synthesized by freeze-drying, polymer pyrolysis, and template methods [17]. Zhenzhen *et al.* [66] synthesized superhydrophobic polycarbonate CNTs monolithic porous sponge for selective oil-water separation via thermally activated nonsolvent-induced phase separation method, as shown in Figure 10.3a. The prepared polycarbonate CNTs' monolithic sponge exhibited excellent porosity (~90.1%), superhydrophobicity (CA ~159°), and good AC (~12 g/g) for selective oil-water separation (see Figure 10.3a). Moreover, the as-synthesized superhydrophobic polycarbonate CNTs' monolithic sponge was effective for underwater quick absorption and separation of tetra-chloromethane (dyed with Sudan III) from water as displayed in Figure 10.3a. Similarly, fire-resistant and hydrophobic carbon monolith-based macroporous sponge was synthesized via self-assembly and pyrolysis of resol polymer on the melamine sponge [71]. More importantly, fabricated carbon monolith possesses excellent hydrophobicity, porosity,

TABLE 10.1

Recent Progress in Superhydrophobic Macroporous Sorbents for Oil Spill Remediation

Sorbents	Synthesis Method	Wettability	Contact Angle	Absorption Capacity	Ref.
Superhydrophobic monolith	Solution-immersion processes	Superhydrophobic	150°	20–75 times their weight	[69]
Macro/mesoporous fire-resistant carbon monolith	Self-assembly and copolymerization	Hydrophobic	108°–130°	89%	[71]
Cellulose acetate/MWCNTs monolith	Phase separation method	Superhydrophobic	154°	7.39–19.84 g/g	[129]
Polycarbonate/CNT porous monolith	Phase separation method	Superhydrophobic	159°	12.62 g/g	[66]
Porous rGO/polycarbonate monoliths	Phase separation method	Superhydrophobic	161°	17.5 g/g	[72]
Zinc stearate@PU sponge	Dip-coating	Superhydrophobic/superoleophilic	175°	6–81 times	[83]
Magnetic Fe$_3$O$_4$@PU sponge	Co-precipitation and dip-coating	Superhydrophobic/superoleophilic	150°	20–70 times	[84]
Poly(dimethylsiloxane)-TiO$_2$@PU sponge	Sol–gel and in-situ polymerization	Superhydrophobic/superoleophilic	154°	16.7 g/g	[147]
Stearic acid-Fe$_3$O$_4$@PDA@ PU sponge	Self-polymerization	Superhydrophobic/superoleophilic	150°	42 times	[65]
SiO$_2$@MnO$_2$ cubes@PU sponge	Foaming technology	Hydrophobic	124°	31.6 g/g	[90]
Superoleophilic seashell/epoxy resin/PU sponge	Freeze-dried	Superhydrophobic/superoleophilic	174°	28–42 g/g	[89]
Superhydrophobic PU@Fe$_3$O$_4$@ polystyrene sponge	Dip-coating and self-initiated photografting	Superhydrophobic/superoleophilic	154°	26–95 g/g	[88]
rGO@PU sponge	Solvothermal method	Superhydrophobic/superoleophilic	153°	37 times	[148]
CuO@n-hexadecylamine@PU sponge	Self-polymerization	Superhydrophobic/superoleophilic	153°	45 times	[87]
Polydopamine-coated nanodiamond@PU sponge	Dip-coating and polymerization	Hydrophobic	143°	20–30 times	[86]

(*Continued*)

TABLE 10.1 (*Continued*)
Recent Progress in Superhydrophobic Macroporous Sorbents for Oil Spill Remediation

Sorbents	Synthesis Method	Wettability	Contact Angle	Absorption Capacity	Ref.
Fe_3O_4@oleic acid@graphene oxide@PU sponge	Dip-coating	Superhydrophobic/superoleophilic	158°	80–160 g/g	[85]
ZnO/epoxy resin@PU sponge	Immersion method	Superhydrophobic/superoleophilic	156°	98%	[92]
Reduced graphene oxide nanoribbons@PU sponge	Solution-immersion process	Superhydrophobic/superoleophilic	153°–165°	97%	[91]
Fe_3O_4@SiO_2@Fluoropolymer PU sponges	Chemical vapor deposition	Superhydrophobic/superoleophilic	157°	13–44 g/g	[35]
ZnO-coated cotton fibers	Hydrothermal	Superhydrophobic/superoleophilic	147°–153°	30–55 g/g	[55]
rGO@cotton	Hydrothermal	Superhydrophobic/superoleophilic	162.9°	30–40 g/g	[5]
rGO@MF sponge	Hydrothermal	Superhydrophobic/superoleophilic	162°	80–125 g/g	[4]
Superhydrophobic cellulose membrane	Sol–gel	Superhydrophobic/superoleophilic	161°	98%	[57]
MS@PDA@SiO_2	Sol–gel	Superhydrophobic/superoleophilic	155°	52–107 g/g	[58]
PMHS/SiO_2@PU sponge	Sol–gel	Superhydrophobic/superoleophilic	161°	50–90 g/g	[59]
Dopamine-Induced MF sponge	Polymerization process	Superhydrophobic/superoleophilic	153°	63–130 times	[118]
Spiropyran polymer-MF sponge	Radical copolymerization process	Superhydrophobic/superoleophilic	155°	74–155 times	[111]
Polydimethylsiloxane-functionalized MF sponge	Solution-immersion process	Superhydrophobic/superoleophilic	150°	45–75 times	[119]
Superhydrophobic melamine sponges	Solution-immersion process	Superhydrophobic/superoleophilic	156°	70–120 times	[120]
F/PPy/Ag/F sponge	Vapor-phase polymerization	Superhydrophobic/superoleophilic	156°	63–96 g/g	[121]
MSHO sponge	One-step dipping method	Magnetic, superhydrophilic/oleophobic	Oil CA 144°	Soybean oil ~96% Pump oil ~97.6% Silicone oil ~94%	[122]
Protonated MF Sponge	Solution-immersion process	Superhydrophilic and superoleophobic	Oil CA 158°		[149]

(*Continued*)

TABLE 10.1 (*Continued*)
Recent Progress in Superhydrophobic Macroporous Sorbents for Oil Spill Remediation

Sorbents	Synthesis Method	Wettability	Contact Angle	Absorption Capacity	Ref.
Superhydrophobic MF sponge	Photoinitiated thiolene click chemistry	Superhydrophobic/superoleophilic	152.8°	72–160 times	[123]
Superhydrophobic cellulose sponges	Sol-gel and a dip-coating method	Superhydrophobic	151°	31–34 g/g	[60]
MWCNT cellulose acetate monolith	Phase separation method	Superhydrophobic	155°	7–19 g/g	[129]
MSEC sponge	Freeze-drying method	Superhydrophobic	152.8°	37–51 times	[131]
RGO-coated natural rubber latex foam	Solution-immersion and phase separation method	Superhydrophobic/superoleophilic	131°–141°	–	[127]
PVF-H sponge	Grafting polymerization	Superhydrophobic/superoleophilic	133°	13–56 g/g	[137]
PVF-GA-H sponge	Hydroxyl groups in PVF	Superhydrophobic/superoleophilic	135°	89.3 g/g	[136]
PVF-G_n–H_m sponge	Grafting polymerization	Superhydrophobic/superoleophilic	123°–138°	93%	[135]
PVF-GA sponge	Grafting polymerization	Superhydrophobic/superoleophilic	–	97.3 g/g	[128]

FIGURE 10.3 (a) Digital images of the sessile water droplet on the porous carbon mono-lith cross-section (left), water CA measurement (right) and selective separation processes of tetrachloromethane under water (bottom) [66]; and (b) digital image, FESEM image and absorption capacity of CNT-PU sponge [68]. (Reprinted with permission from Li et al. [66]. Copyright (2018), American Chemical Society; H. Wang, E. Wang, Z. Liu, D. Gao, R. Yuan, L. Sun, Y. Zhu, Recyclable carbon nanotube sponges for oil absorption, Acta Materialia, 59 (2011) 4798-4804. Copyright 2011 Elsevier.)

and intrinsic fire-resistant properties for oil spill remediation in harsh environments. Xin et al. [66] fabricated a carboxylate mCNTs and cellulose acetate-functionalized carbon monolith sponge (CA/OMWNTs) via a facile phase separation method, and the obtained CA/OMWNTs monolithic sponge exhibited good oil-water separation efficiency, capacity retention, and reusability. Furthermore, novel micro-nanoscale binary structures derived from reduced graphene oxide-polycarbonate superhydrophobic carbon monolith sponge were fabricated by thermally activated phase separation method [72]. Reduced graphene oxide-polycarbonate superhydrophobic carbon monoliths sponge showed unique pore structure (~91%), high specific surface area (~137 m^2/g), and excellent AC for oil and organic solvent separation from polluted water. Moreover, the as-synthesized carbon monolithic sponge exhibited strong repellent against corrosive liquids such as acidic and alkaline solutions for oil spill cleanup and chemical leakage [72]. Besides monolithic sponge, Jiangdong et al. [73] synthesized a superhydrophobic magnetic carbon aerogel via carbonization and magnetization for selective oil-water separation. Furthermore, the as-synthesized magnetic carbon aerogel exhibited an excellent AC of ~10 g/g for various oil and organic solvents. Dengsen et al. [74] fabricated TiO_2 nanorods coated on biomass carbon aerogel sponge (SHBCT) to form a superhydrophobic surface via the hydrothermal method followed by thermal annealing. In this work, hierarchical biomass carbon@TiO_2 (HBCT) aerogel was achieved by in situ growth of TiO_2 nanorods on biomass carbon aerogel. Further, SHBCT aerogel was used for oil spill remediation, including under-water carbon tetrachloride separation and continuous oil-water separation [74]. Similarly, Dengsen's group [75] synthesized a superhydrophobic SiO_2 and MnO_2-coated carbon aerogel sponge via sol-gel and carbonization, exhibiting water CA of ~155° and excellent AC of 60–120 g/g. Recently, numerous carbon monolith and aerogel-derived superhydrophobic macroporous sponges have been successfully synthesized for facile oil spill remediation, which includes wrinkled graphene monolith, cellulose acetate monolith, 3D magnetic carbon aerogel, sponge-like aerogel, lignin graphene-carbon aerogel, and porous graphene aerogel for oil spill remediation [73,76–80]. Superhydrophobic carbon-based aerogel and monolith derived from the template technique required two-step preparation, namely, a self-supported porous carbon template and removal of the template. On the contrary, the pyrolysis technique is widely used to form carbon-based aerogel and monolith owing to a simple and more time-saving manner than the template removal method. Furthermore, the freeze-drying strategy was also explored for the fabrication of carbon-based superhydrophobic porous sorbents using graphene oxide, CNTs, and polymer precursors. For example, Long *et al.* [81] developed ultralight, superhydrophobic carbon aerogel by liquid nitrogen-based freeze casting using graphene oxide, cellulose fibers, kaolin, and glucose aqueous solution followed by pyrolysis. Freeze casting-derived superhydrophobic aerogel demonstrated a water CA of ~124°, and excellent absorption efficiency (~75–255 times their own weight) for oil-water separation. Ultralight 3D cellulose/graphene aerogel was synthesized via bidirectional freeze-drying method followed by CVD-assisted superhydrophobic surface modification to produce highly compressible, anisotropic, porous sponge for oil spill cleanup (~80–197 times its weight) [82]. These advantageous attributes render that the superhydrophobic carbon-based macroporous sorbents can be an ideal candidate for potential oil spill remediation.

10.4.2 Polyurethane Sponge-Based Macroporous Sorbents

PU sponge is a macroporous polymer that demonstrates outstanding characteristics, including high AC, flexibility, and wide surface area [65,83–90]. Ideally, the PU sponges are hydrophilic in nature and cannot be utilized to separate oil and water mixture [65,85–90]. Therefore, the surface of PU sponges is always modified and altered to change its wettability from hydrophilic to hydrophobic and oleophilic for selective oil and organic solvent removal from an oil-water mixture [35,84,91,92]. Qiu *et al.* [93] reported the one-pot hydrothermal route to synthesize hollow Al_2O_3 spheres followed by surface functionalization of hollow Al_2O_3 spheres using methacryloxypropyltrimethoxysilane onto PU sponge. The water CA and AC of Al_2O_3/PU sponge were measured ~140° and 37 g/g (10 cycles), respectively. The result indicated that the as-prepared PU sponge has tremendous potential for oil-water separation. Zhang *et al.* [94] modified an as-prepared PU sponge with the addition of Al_2O_3 nanoparticles, produced exceptional oleophilic and superhydrophobic modified PU sponges in both air and oil, when combined with the self-assembly of a lengthy carbon chain of palmitic acid. Using a vacuum pump, the modified sponge may absorb floating oil on the water surface and heavy oil beneath the water. It can also be used to continuously separate volumes of oil pollutants from the water surface. Significantly, the resulting sponges are durable in organic solvents and may be reused for oil-water separation over 500 times without losing superhydrophobicity, demonstrating exceptional durability and reusability in organic solvents. In the past few decades, the distinct physical and chemical characteristics of multiwalled carbon nanotubes (MWCNTs) have lately been revealed to a wide variety of potential applications. Among these, hydroxylated multiwalled carbon nanotubes (MWCNTs-OH) have a large surface area and low surface roughness. Chen's group [95] fabricated a superhydrophobic MWCNTs@PU sponge using hydrophobic surface functionalization of long-chain hexadecyl siloxane groups. The as-synthesized, superhydrophobic MWCNTs@PU sponge exhibited remarkable superhydrophobic properties with a water CA of ~157.4°. Similarly, a superhydrophobic PU sponge was fabricated using a combination of interfacial polymerization (IP) and molecular self-assembly to purge and recycle oil and organic solvent from a water surface. IP is a form of step-growth polymerization in which polymerization occurs at the interface between an aqueous and an organic solution containing two monomers [96]. Qiu et al. [97] used a one-step ultrasonic dip-coating method to develop a superhydrophobic sepiolite (SEP) layer onto the skeleton surface of a three-dimensional porous PU sponge. SEP-modified superhydrophobic PU sponge can absorb numerous oils and nonpolar solvents ~29 times their weight while entirely rejecting water. Furthermore, the as-prepared composite may be reused more than ten times for oil-water separation with a high separation efficiency of ~99.45% in corrosive liquids and hot water. Pourjavadi et al. [98] reported a simple technique for the synthesis of a magnetic and superhydrophobic PU sponge with a water CA of ~159° and an oil AC of ~64–176 g/g that can be used as a sorbent to clean up marine oil spill contamination. Pourjavadi's group [98] synthesized superhydrophobic sponge via coating of carbon black (CB), hexagonal boron nitride (h-BN)@Fe_3O_4, and acrylic resin onto PU sponge surface. The modified sponge was stable under corrosive conditions (pH ~1–14), salt solution, and variable temperatures (–12°C to 105°C) due to the chemical and thermal stability of h-BN and CB on the PU sponge. The as-synthesized superhydrophobic sponge was used for oil spill cleaning more than 20 times without losing much of its absorption ability. In addition, PU/CB/h-BN@Fe_3O_4-PU sponge was applied for oil-in-water emulsion purification with a purity of water of ~99%. Also, the absorption effect of the sponge was tested in an actual situation using genuine spilled oils comprising monoaromatics and polyaromatics in the Persian Gulf's water surface, which showed greater efficacy [98]. Gui et al. [68] synthesized a superhydrophobic CNT-coated PU sponge via a CVD method using dichlorobenzene and ferrocene precursors for oil spill cleanup, as shown in Figure 10.3b. The as-synthesized CNT-PU sponge demonstrated an oil-water separation capacity greater than ~100 g/g and maintained AC of ~20–40 g/g after 10 cycles with 98% oil recovery (see Figure 10.3b). Wang and co-workers developed reinforced superhydrophobic and superoleophilic CNT-modified PU sponge (PU-CNT-PDA-ODA sponge) via a facile oxidative self-polymerization technique using DA, PDA, and octadecylamine for selective oil-water separation [99]. Pristine PU sponge and PDA-functionalized CNT sponge immediately absorbed water and lubricating oil droplet with CA of 0° owing to low surface roughness and hydrophilic/lyophilic surface properties [99]. On the other hand, water droplets remained stable on to superhydrophobic PU-CNT-PDA-ODA sponge with a static water CA of ~158°. In comparison, lubricating oil immediately penetrated with a CA of 0° in the oleophilic macroporous surface of octadecylamine-functionalized PU-CNT-PDA sponge [99]. Superhydrophobic PU-CNT-PDA-ODA sponge exhibited excellent water-repellent properties with oil AC ~34.9 times its mass for 150 reusable oil-water separation cycles [99]. Lin et al. [31] synthesized a superhydrophobic poly(dimethylsiloxane)-functionalized CNT-coated PU (CNT/PDMS-coated PU) sponge via a facile dip-coating method for oil spill cleanup and continuous oil absorption from water surfaces using vacuum suction method, as illustrated in Figure 10.4a–i. Figure 10.4a displayed that the as-synthesized CNT/PDMS-coated PU sponge demonstrated excellent superhydrophobicity with a stable water CA of ~162° owing to the presence of siloxane-functionalized

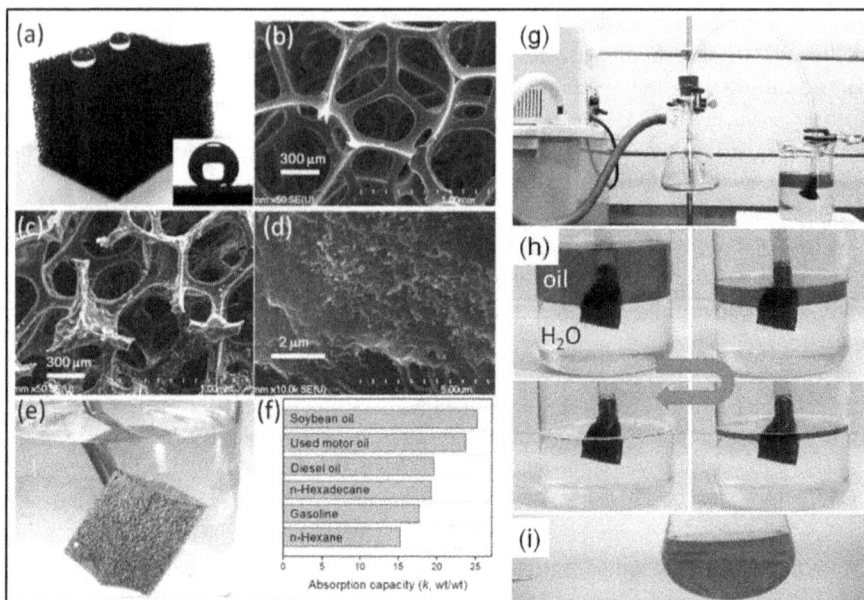

FIGURE 10.4 (a) Digital image showing the hydrophobicity of CNT/PDMS-coated PU sponge using sessile water droplet, FESEM images of (b) pristine PU sponge, (c, d) superhydrophobic CNT/PDMS-coated PU sponge, (e) digital image of the CNT/PDMS-coated PU sponge immersed in a water bath under an external force, (f) absorption capacity of the CNT/PDMS-coated PU sponge for various oil/organic solvent separation, and (g–i) digital image showing the vacuum suction-based continuous oil-water separation using superhydrophobic CNT/PDMS-coated PU sponge [31]. (Reprinted with permission from Wang and Lin [31]. Copyright (2013), American Chemical Society.)

CNT-assisted high surface roughness onto PU sponge. Figure 10.4b–d shows the FESEM images of pristine PU sponge and CNT/PDMS-coated PU sponge, which reveals that the presence of crater-like nanostructures enhanced the surface roughness of CNT/PDMS-coated PU sponge than pristine PU sponge, leads to the formation of superhydrophobic characteristic (Cassie-Baxter state) [31]. Superhydrophobic CNT/PDMS-coated PU sponge exhibited oil AC of ~15–25 times its mass (Figure 10.4e and f). Furthermore, the as-prepared CNT/PDMS-coated PU sponge separated oils up to ~35,000 times their mass using vacuum suction-based continuous oil absorption from the water surface, as shown in Figure 10.4g–i [31].

Developing absorbents with high mechanical robustness for oil-water separation is critical, as maritime oil spill monitoring round the clock and oily industrial effluent have sparked widespread concern. Cao *et al.* [100] developed lightweight CNTs/thermoplastic polyurethane (TPU)/epoxy (EP) composite sponge with a unique three-dimensional herringbone-like structure for oil absorption and oil-water separation using a simple and straightforward ice-templating technique with a density and porosity of 0.11 g cm^{-3} and 63%, respectively. Due to their high porosity and wettability, the super-hydrophobic CNT@PU sponge has significant oil AC and selectivity. Besides, CNTs were investigated for oil and water separation due to their hydrophobic nature and elongated form [5,100,101]. Therefore, CNT-modified superhydrophobic PU sponges were explored for selective oil absorption from pollutant water with two significant benefits. First, CNTs developed superhydrophobic surface on a PU sponge, and second, the number of pores increased, which might be responsible for the capillary action of selective oil absorption [5,100,101]. In recent times, several macroporous CNT-based superhydrophobic PU sponges have been synthesized, *viz.*, MWCNTs/PU sponge, CNTs/thermoplastic PU/EP foam, and so on, for efficient oil-spill cleanup and emulsion purification with excellent superhydrophobicity (CA greater than 150°) [95,99,102–105]. On the other hand, surface modification of PU sponge with a colloidal solution of straw soot and magnetic nanoparticles provided a simple and low-cost method for fabricating an everlasting superhydrophobic magnetic PU sponge [106]. Carbon soot is the most cost-effective material of several carbon-based products. Carbon soot is a by-product of pyrolysis processes involving carbon-containing fuels and incomplete combustion of organic compounds made up of impure carbon particles [106]. Carbon soot is a low-cost, nontoxic, and simple-to-obtain substance utilized in contemporary chemical research and may be used in several scientific breakthroughs [107]. The majority of carbon soot is nonpolar and hence insoluble in polar solvents like water [107]. These characteristics of carbon soot can be exploited for the fabrication of water-resistant or superhydrophobic surfaces [107]. Farshad *et al.* [106] developed a superhydrophobic magnetic PU sponge via the solution immersion method using a colloidal suspension of magnetic nanoparticles and straw soot. Different stabilizers, such as CTAB, SDS, and PEG 6000, were tested to see how they affected magnetic nanoparticles' morphology and geometrical structure. The sponge was further modified with PDMS to increase the hydrophobic properties on the surface. Results indicated that the as-synthesized superhydrophobic PU sponge demonstrated AC ~30 times of own weight with excellent reusability for 30 cycles and a water CA of ~154°.

10.4.3 MELAMINE SPONGE-BASED MACROPOROUS SORBENTS

Among all kinds of porous materials, commercially available MF sponge (also known as Mr. Clean, Magic Eraser) has witnessed significant attention owing to its low cost, flexibility, and macroporous structure for oil-water separation application [108–110]. Akin to PU sponge, surface-modified superhydrophobic MF sponge has exhibited excellent AC and reusability for oil spill remediation (Figure 10.5a) [25]. Furthermore, oil/organic solvents absorbed by superhydrophobic/superoleophilic MF sponge can be recovered by mechanical squeezing for further reuse. Notably, the hydrophilic surface of the MF sponge can be easily modified into a superhydrophobic surface by various methods, including dip coating, CVD, hydrothermal, and sol-gel route [17]. For example, Haiguang *et al.* [111] synthesized a superhydrophobic MF sponge via a copolymerization technique for oil recovery. The as-prepared, superhydrophobic MF sponge exhibited a water CA of ~155.5°, with excellent selectivity and high AC of ~70–154 times its mass. Similarly, a superhydrophobic MF sponge was synthesized via a one-step carbonization at low temperatures of around 500°C–600°C, possessing a high porosity of~99.5%, a water CA of ~140°, and excellent oil removal capacity (90–200 times their own weight) [112].

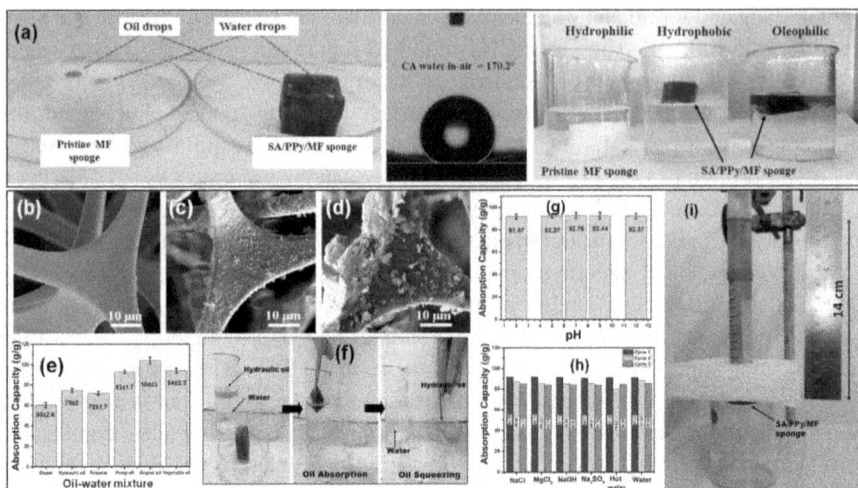

FIGURE 10.5 (a) Digital images of a pristine MF sponge, SA/PPy/MF sponge (left), CA measurement of SA/PPy/MF sponge (right) and hydrophilicity, hydrophobicity, and oleophilicity of pristine and modified sponge (bottom); FESEM images of (b) pristine MF sponge, (c) PPy/MF sponge, (d) SA/PPy/MF sponge, (e) absorption capacity of SA/PPy/MF sponge for different oils and organic solvents, (f) digital images of the step-by-step process for oil-water separation using SA/PPy/MF sponge, (g-h) AC of SA/PPy/MF sponge with a variation of pH and different salts/alkali mixture solution, and (i) measurement of intrusion pressure demonstration of SA/PPy/MF sponge [25]. (Reprinted with permission from Dashairya et al. [25]. Copyright (2019), Springer Nature.)

Also, Fe_3O_4 nanoparticle-coated fluoroalkyl silane-functionalized superhydrophobic MF sponge was fabricated using a facile dip-coating method, in which fluoroalkyl silane was used to develop a binding force between the MF sponge and Fe_3O_4 nanoparticles [113]. The as-synthesized superhydrophobic MF sponge was effectively applied for the removal of various oils (silicone oil, peanut oil, pump oil, kerosene, and gasoline) from pollutant water and displayed AC of ~59–77 g/g. Zeng's groups explored biowaxes surface-modified superhydrophobic MF sponges for underwater oil removal and water-in-oil emulsion purification [114]. Recently, numerous superhydrophobic and superoleophilic surface-modified MF sponges have been explored for oil spill remediation [110,115]. For example, Ruan and co-workers [116] synthesized a superhydrophobic and superoleophilic MF sponge by solution immersion coating with enhanced AC, recyclability, and selectivity for efficient oil-water separation. Pham's and Oribayo's group [36,117] developed silanization and N-acylation-functionalized superhydrophobic MF sponge, respectively, with outstanding recyclability and AC for oil spill cleanup. SA-functionalized and polypyrrole-coated MF (SA/PPy/MF) sponge was synthesized via dip-coating and self-polymerization process for oil-water separation in variable pH and harsh environment (alkali, acidic, and hot water condition) as well as gravity-driven oil spill cleanup, as illustrated in Figure 10.5a–i [25]. Pristine MF sponge and PPy-coated MF sponge showed smooth and relatively low surface, which is directly related to Young's equation and hydrophilic nature

(Figure 10.5b and c). On the contrary, microstructural images of SA/PPy/MF sponge demonstrated that MF sponge surface successfully encapsulated SA-functionalized PPy, leading to enhanced surface roughness and formation of the Cassie-Baxter state (Figure 10.5d). The synthesized superhydrophobic SA/PPy/MF sponge demonstrated an AC of ~104.1 g/g, ~94.13 g/g, ~92.9 g/g, ~71.7 g/g, ~74.6 g/g, and ~60.3 g/g for engine oil, vegetable oil, pump oil, toluene, hydraulic oil, and diesel, respectively (Figure 10.5e). Importantly, absorbed oil can be easily recovered with the mechanical squeezing process (Figure 10.5f). Furthermore, SA/PPy/MF sponge was applied effectively for oil absorption in harsh environments (variable pH, acidic and alkaline condition), and demonstrated excellent absorption capacities with deviation of ~9%, ~7.7%, ~7.7%, ~8.5%, ~6.7%, and ~7.7% for $MgCl_2$, NaCl, Na_2SO_4, NaOH, hot water, and cold water solution condition between the first and third cycles (Figure 10.5g and h). Besides, the intrusion pressure of SA/PPy/MF sponge was calculated ~1.38 kPa during the gravity-driven water flow test (Figure 10.5i).

Recently, Dashairya et al. also developed various superhydrophobic MF sponges, including rGO@MF and Zr/Cl ion-induced SA-PPySHMF sponges via hydrothermal, sol-gel, and polymerization processes for oil spill remediation [4,21,25]. Xu and co-workers [118] have elucidated PDA-functionalized superhydrophobic MF sponge with the lotus leaf-like structure for oil-water separation, contributing towards lower surface energy, high surface roughness, and strong adhesion. The oxygen-induced polymerization process is adopted to coat PDA on the MF sponge surface, which exhibits oil and organic solvents' AC of ~60–130 times its weight. Zhu et al. [111] synthesized superhydrophobic MF sponge via a radical copolymerization process of spiropyran methacrylate monomers on the surface of vinyl-modified MF sponges, which shows the light-controllable oil-water separation properties of the as-synthesized MF sponge exhibiting a water CA of ~155°, with excellent selective oil-water separation, and oil AC of ~70–154 times its mass. Furthermore, Chen groups [119] explored PDMS-functionalized superhydrophobic MF sponge for continuous oil-water separation using a vacuum suction system, which demonstrates continuous removal of oils (toluene, hexane, silicone oil, octadecene, and motor oil) from the surface of the immiscible oil-water mixture. Most importantly, superhydrophobic PDMS-MF sponge possesses porosity, water CA of ~150°, oil ACs of ~45–75 times its weight with excellent reusability through multiple absorption-squeezing cycles. Similarly, polymethylsilsesquioxane-modified superhydrophobic MF sponge was fabricated via an immersion method using potassium methyl silicate aqueous solution [120]. In addition, the as-synthesized MF sponge retained intrinsic fire retardancy and excellent superhydrophobicity in strong acidic/alkali solution, hot water-based oil-contaminated harsh environments. Importantly, a superhydrophobic MF sponge was explored for crude oil spill cleanup as well as surfactant-free water-in-oil emulsion purification [120]. Chen et al. [121] fabricated $AgNO_3$/$FeCl_3$-coated polypyrrole and dodecafluoroheptyl-propyl-trimethoxysilane (G502)-functionalized superhydrophobic MF sponge (MF/PPy/Ag/F sponge) with a two-tier hierarchical structure via vapor-phase polymerization for continuous oil and organic solvents separation from water and under water oil removal. On the other hand, MF sponge was explored for magnetic, superhydrophilic, and oleophobic surface modification using sodium perfluorononanoate, chitosan, acetic acid, and Fe_3O_4 nanoparticles precursor via a facile dip-dry method [122]. The as-synthesized superhydrophilic/

oleophobic melamine (MSHO) sponge exhibits water CA of 0°; however, the oil CA increased to ~144° [122]. Recently, numerous superhydrophobic sponges, including superhydrophobic thiolene MF sponge, lignin-based MF resin sponge, polybenzoxazine MF sponge, Fe_3O_4/Ag-decorated MF sponge, fluorine-free silanized SiO_2 MF sponge, covalent organic framework-coated MF sponge, tetradecylamine-MXene-functionalized MF sponge, PU-PDA-Fe_3O_4-Ag MF sponge, CNT-MF sponge, polystyrene/CNT sponge, PDMS@SiO_2@WS_2 sponge, and MS/PDA/DT sponge, have been explored for effective oil-water separation and emulsion purification [123–126]. The collective results of this study suggest that an economic and efficient superhydrophobic melamine-based sponge exhibiting flexibility, porosity, excellent oil AC, reusability, and mechanical stability would be paradigm shift for oil recovery from oil-water mixture. Moreover, the cheaper raw materials and facile preparation endow the superhydrophobic MF sponge beneficial for practical industrial applications. These findings offer a new responsive absorbent and new approaches for oil recovery.

10.4.4 OTHER MACROPOROUS SORBENTS

Superhydrophobic macroporous sponges, *viz.*, cellulose, polyethylene foam, rubber latex, and polyvinyl-alcohol formaldehyde sponge, have also been explored for oil spill remediation [63,67,127,128]. Recently, Ahuja *et al.* [60] fabricated a cross-linked cellulose sponge from discarded jute bags using the sol-gel method followed by freeze-drying, which was further modified with tetraethyl-orthosilicate/hexadecyl trimethoxysilane (TEOS/HDTMS). The modified cellulose sponge demonstrated a static water CA of ~151° and AC of ~31–35 g/g with separation efficiency reaching ~97%–98.5%. Zhang *et al.* [129] prepared MWCNT-induced superhydrophobic cellulose acetate monolith using a facile-phase separation method with a static water CA of ~155° and oil and organic water mixture separation capacity of ~7–19 g/g. Moreover, (3-aminopropyl)triethoxysilane, hexylamine, dodecylamine, and octadecylamine surface-functionalized hydrophobic cellulose nanofibers and cellulose acetoacetate sponges were synthesized for oil spill remediation with excellent reusability and retention capacity up to ten cycles [130]. Yuan *et al.* [131] fabricated Fe_3O_4 nanoparticle-coated HDTMS-functionalized superhydrophobic/magnetic cellulose (MSEC) sponges for oil spill remediation and emulsion purification, as depicted in Figure 10.6a–h. The presence of Fe_3O_4 nanoparticles and HDTMS on the surface of the cellulose sponge increased the surface roughness and provided superhydrophobic properties, respectively (Figure 10.6b–d) [131]. The as-synthesized MSEC sponge demonstrated a water CA of ~152.8° (Figure 10.6a) and sustained superhydrophobicity in saline, alkali, acidic solution-based corrosive solution with an oil-water separation capacity of ~37–51 times of their mass owing to an enhanced surface roughness of MESC sponge than ethyl cellulose (EC) and magnetic ethyl cellulose (MEC) sponges [131]. Interestingly, superhydrophobic MESC sponge was successfully applied for surfactant-stabilized n-hexane-in-water emulsion purification using UV-vis absorbance spectra with separation efficiency reaching ~99.67%, as illustrated in Figure 10.6e–h [131].

Peng's group [132] synthesized superhydrophobic and MSEC sponge using Fe_3O_4 and HDTMS exhibiting a water CA of ~156° and oil and organic solvents separation efficiency of ~95%. Qin *et al.* [133] synthesized a mechanically robust hydrophobic cellulose sponge by surface modification using reduced graphene oxide and

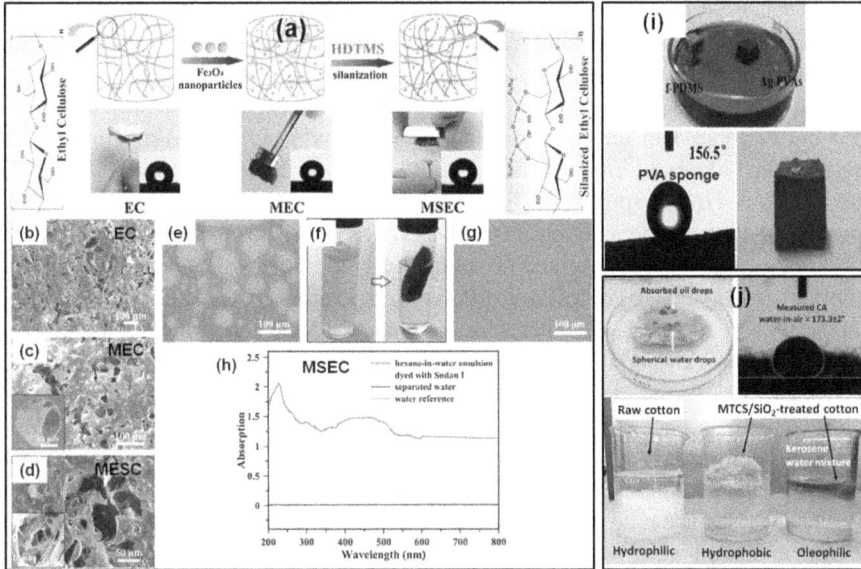

FIGURE 10.6 (a) Synthesis mechanism of ethyl cellulose (EC), magnetic ethyl cellulose (MEC), and magnetic silanized EC (MSEC) sponges along with their wettability using a sessile water droplet and magnetic properties; (b–d) FESEM images of the EC, MEC, and MSEC sponges; (e–g) optical micrographs and digital images of the n-hexane-in-water emulsion purification using superhydrophobic MSEC sponges before and after separation; and (h) UV-vis absorption spectra of polluted n-hexane-in-water emulsion and purified water [131]; (i) digital image showing superhydrophobicity (top), and water CA measurement of PDMS-Ag-PVA sponge [67]; (j) digital images of a superhydrophobic MTCS/SiO$_2$-treated cotton (left), CA measurement of MTCS/SiO$_2$-treated cotton (right) and hydrophilicity, hydrophobicity, and oleophilicity of pristine and modified sponge (bottom) [6]. (Reprinted with permission from Lu et al. [131]. Copyright (2017), Elsevier; J. Chen, J. Xiang, X. Yue, H. Li, X. Yu, Superhydrophobic, compressible, and reusable polyvinyl alcohol-wrapped silver nanowire composite sponge for continuous oil-water separation, Colloids and Surfaces A: Physicochemical & Engineering Aspects, 583 (2019) 124028. Copyright 2019, Elsevier; Reprinted with permission from Dashairya et al. [6]. Copyright (2019), Springer Nature.)

polybenzoxazine to improve surface roughness and water-repellent properties. The modified cellulose sponge exhibited decent oil and organic solvents' AC and excellent recyclability up to 30 cycles. Sun *et al.* [127] fabricated superhydrophobic/superoleophilic rGO@polyethylene aerogel-coated natural rubber latex foam with a static water CA of ~150° and excellent AC for oil-water separation. Junyong and co-workers developed environment-friendly superhydrophobic and superoleophilic PVA sponge via a facile solution immersion route using various solvent precursors (methyltriethoxysilane, hydrochloric acid, and ammonia solution) for continuous oil-water separation [63]. Pristine PVA sponge immediately absorbed water and gasoline droplet with CA of 0° owing to low surface roughness and hydrophilic/lyophilic surface properties. On the other hand, water droplets remain stable on to superhydrophobic PVA sponge with a static water CA of ~152°, while gasoline immediately penetrates the oleophilic mac-roporous surface of the MTCS-PVA sponge. Superhydrophobic PVA sponge exhib-ited excellent water-repellent properties with ~99.6% oil-water separation efficiency and could remove oil up to ~6,200–14,000 times of its mass from contaminated water [63]. Furthermore, Pan and coworkers fabricated several macroporous superhydropho-bic polyvinyl-alcohol formaldehyde sponges, including PVF-H sponge, PVF-GA-H sponge, PVF–G_n–H_m sponge, PVF–GA sponge, and PVF-g-GAM for effective oil spill remediation and organic solvents absorption, as mentioned in Table 10.1 [128,134–137]. For example, a hydrophilic PVF sponge was modified by the reaction of stearoyl chloride with hydroxyl groups at a different temperature to fabricate a superhydro-phobic PVF-H sponge [132]. Then, the as-prepared PVF-H sponge can immediately absorb oil from a contaminated oil-water mixture and attain the AC of ~13.7–56.6 g/g with water CA of ~131°–141°. Besides above, Xi *et al.* [67] also fabricated durable and robust superhydrophobic PDMS-functionalized silver nanowire-coated PVA (PDMS-Ag-PVA) sponge via the hydrothermal and freeze-drying method for oil-water separa-tion. With the attachments of PDMS-functionalized Ag nanowire on the surface of PVA sponge, surface roughness was increased, which led to an enhanced water contact of ~156.5° and oil-water separation capacity of ~1,027%–3,900% of its mass, as shown in Figure 10.6i. Our group recently synthesized MTCS-functionalized SiO_2-coated superhydrophobic cotton fibers for oil spill cleanup, as illustrated in Figure 10.6j [6]. These promising results demonstrate that cellulose sponges, rubber latex foam, PVA formaldehyde sponges, and cotton fiber-based superhydrophobic macroporous sorbents are potentially relevant for large-scale oil spill cleanup and chemical leaks.

10.5 FILTRATION MEMBRANE TECHNOLOGIES FOR SELECTIVE SEPARATION

Besides the macroporous sorbents for oil spill remediation as discussed above, superhydrophobic/superoleophilic filtration membranes have also gained remarkable attention for selective oil-water separation [1,2,13,54]. The devel-opment of filtration membrane technologies involves two critical factors, namely, the formation of a rough surface and superhydrophilic and under-water superoleophobic properties [1,2,54]. In the recent past, there are three types of superhydrophobic and superoleophilic filtration membranes that have been widely explored: meshes (stainless steel [SS] mesh and copper mesh), porous membranes (fabrics and textiles), and thin films [1,54,138,139]. Several meth-odologies have been explored to develop superhydrophobic filtration mem-brane surface, *viz.*, dip coating, wet chemical, electrospinning, solution immersion, polymerization, hydrothermal, spray coating, and sol-gel [1,2,54]. Recently, metallic meshes have been extensively explored for selective oil spill cleanup with excellent separation efficiency. The metallic meshes are a good choice for oil-water separation owing to their multiple advantages [1,13]. Feng *et al.* [140] first proposed an idea for the development of superoleophilic SS mem-branes with a water CA of ~156.2° for water-oil mixture separation. The surface of the pretreated SS mesh (50–200 µm) was modified with PTFE particles, SDBS surfactant, PVAC adhesive, and PVA dispersant in aqueous solution by spray coating followed by calcination [140]. More recently, Bayram group [50] deposited a thin film of poly(1H, 1H, 2H, 2H-perfluorodecyl acrylate) on SS mesh (PPFDA film-coated SS mesh) using CVD technique to synthesize superhydrophobic filtration membrane for oil-water separation. SEM images of modified superhydrophobic SS mesh showed high surface roughness compared to pristine mesh, leading to the formation of the Cassie-Baxter state on the surface of PPFDA film-coated SS mesh with a water CA of ~166.9° [50]. Hence, superhydrophobic PPFDA film-coated SS mesh demonstrated a high oil-water separation efficiency of ~98.5%. Wang's group [141] introduced superhydrophobic properties on SS mesh via dispersion polymerization and emulsion polymerization using TEOS-derived silica particle coating and HDTMS surface functionalization. The superhydrophobic SiO_2-coated SS mesh exhibited excellent water repellency with a water CA of ~161° and selective oil-water separation efficiency [141]. Newton *et al.* [142] fabricated spray-coated PDMS on SS superhydrophobic mesh that exhibits a water CA greater than ~160° and an oil-water separation efficiency of ~96% in corrosive environments with a porous morphology. Furthermore, Zulfiqar and coworkers [143] applied sawdust and polychloroprene adhesive coating on SS mesh, followed by silicone polymer deposition using dip coating, as illustrated in Figure 10.7a–f. After that, a thin layer of candle flame-induced carbon soot was applied on the as-prepared silicone-coated SS mesh, which exhibited a highly porous rough surface and water CA greater than ~151° with self-cleaning properties for a facile oil spill cleanup (Figure 10.7a and b). The robust and mechanically stable superhydrophobic mesh was applied for readily oil-water mixture separation, which instantly passes the oils and blocks the flow of water through its rough surface (Figure 10.7c and d). Moreover, superhydrophobic mesh demonstrated a high organic solvent-water separation (chloroform, toluene, n-hexane, and dichloromethane) efficiency of greater than ~90% (Figure 10.7e and f). Bayram *et al.* [50] reported superhydrophobic-superoleophilic PPFDA thin film-coated SS mesh via a one-step CVD method, which exhibited a water CA of ~166.9°, excellent water-repellent properties, and a high oil-water separation efficiency value of ~98.5%. Superhydrophobic Cu mesh was also prepared by the CVD technique of silicone elastomer on a smooth surface of copper mesh [45].

Zhu *et al.* [144] have synthesized metallic hydroxide nano-needles on the Cu mesh via the solution immersion technique followed by SA-assisted superhydrophobic surface modification. The superhydrophobic SA-functionalized Cu mesh exhibited a water CA of ~157° and a sliding angle of less than ~5°. Metallic meshes demonstrated excellent separation efficiency and promising outcomes for selective oil-water separation. However, metallic mesh-based materials' high material cost, heavy weight, and poor corrosion resistance impede their practical applications [2]. On the contrary, cellulose, polymer, fabrics, and textile-based filtration membrane demonstrated merits over metallic mesh-based materials, owing to lightweight, flexibility, low cost, and high corrosion resistance for efficient oil-water separation [2]. To obtain the superhydrophobic filtration porous membrane surface, Xie *et al.* [57] developed a superhydrophobic

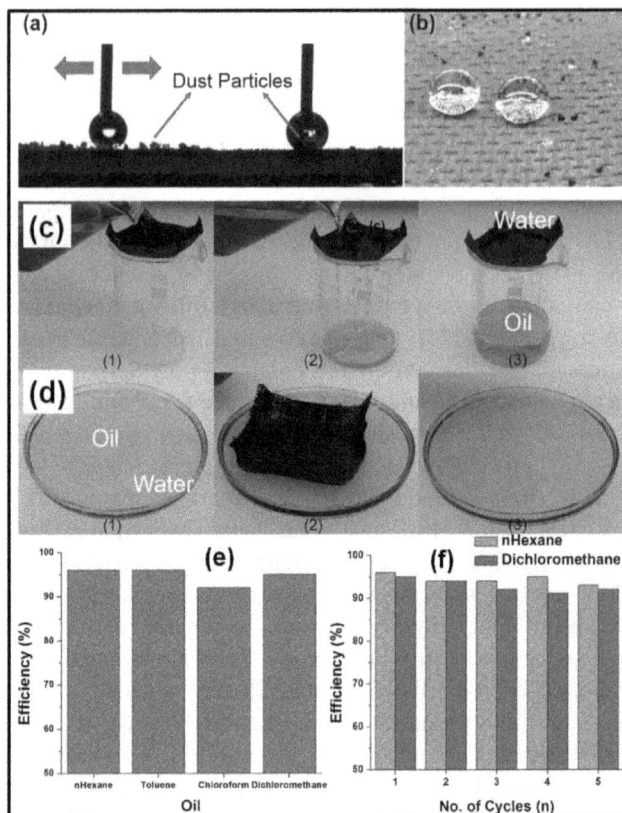

FIGURE 10.7 (a, b) FESEM images of pristine SS and superhydrophobic PPFDA film-coated SS mesh, (c, d) oil and water CA measurement of superhydrophobic PPFDA film-coated SS mesh, (f) digital images showing diesel-water mixture separation performance of pristine SS (left) and superhydrophobic PPFDA film-coated SS mesh (right), and (g, h) continuous oil-water mixture separation on inclined superhydrophobic PPFDA film-coated SS mesh using gravity-driven operation [143]. (Reprinted with permission from Zulfiqar et al. [143]. Copyright (2017), Elsevier.)

and superoleophilic cellulose membrane (SOCM) using a low-cost sol-gel route for efficient oil spill cleanup. The as-synthesized SOCM demonstrated ~98% separation efficiencies with a water CA of ~164° for various oil-water mixtures under harsh environments such as acid, alkali, and high temperatures. Zhang *et al.* [43] reported a facile one-step fabrication of superhydrophobic and superoleophilic polyester textile filtration membrane using MTCS-derived silicone nanofilament growth via CVD method for oil-water separation. More recently, Wang's group [145] synthesized flexible superhydrophobic tree-grape-like poly(tetrafluoroethylene) fibrous membrane (UTPFM) using electro-centrifugal spinning followed by calcination. The as-synthesized UTPFM provided a highly rough surface, porous microstructure, and ultrahigh permeability for oil-water separation. Moreover, UTPFM exhibited a water CA of ~154.6° and ultrahigh permeability of up to 3,200 $Lm^{-2}h$ with a separation efficiency of ~99% [145]. In addition, numerous superhydrophobic/superoleophilic macroporous filtration membranes are reported for the facile selective oil-water separation [138,146]. The advantages of superhydrophobic/superoleophilic macroporous filtration membranes are manifold, including facile fabrication, eco-friendliness, low cost, high separation efficiency, and mechanical integrity, which render filtration membrane technologies a promising strategy for large-scale oil spill remediation.

10.6 CONCLUSIONS AND PROSPECTS

This chapter provides a comprehensive review and overview of various (super) hydrophobic, (super)oleophilic porous substrates, *viz.*, metal meshes, textiles, foams, sponges, and polymeric membranes developed so far for continuous oil/water separation and demulsification, which plays critical roles toward environmental protection. To date, the published literature is inundated with voluminous work on the removal of oil and organic solvents based on either static oil/water mixture or continuous oil/water mixture using a peristaltic pump, which is entirely different from the oil spill accident in the real environment of an offshore oil pollution site. In such a case, the fluidity, viscosity, density, and enormity of the spilled oil would compromise the efficiency, AC, cyclability, and reusability of the porous sorbents. Nonetheless, super(hydrophobic)/ (super)oleophilic materials have shown encouraging results for selective oil/water separation. However, there are still a few bottlenecks that must be overcome and duly addressed. The underlying mechanisms and the interaction of oil droplets and the surface of sorption materials are not clearly understood. Also, the viscosity and surface tension of various oil and organic solvents play a critical role in determining the AC, and subsequent reusability for many cycles must be carefully analyzed and understood before any porous sorbents see the light for commercial applications. From the practical standpoint, the daunting task related to the fabrication of (super)hydrophobic and (super)oleophilic sorbents lies with the increasing use of toxic fluoro compounds and silanes that impede their large-scale production due to exorbitant cost and environmental concerns. Therefore, identifying cost-effective novel compounds, *viz.*, fatty acids, carbonaceous compounds replenishing silanes/fluoro compounds to impart surface hydrophobicity on the membrane surface or porous sorbents, is paramount for the future. In an actual oil spill accident site, it may be likely that various organic solvent mixtures must be removed, separated, and recovered; therefore, how the porous sorbent behaves and performs will be a critical determinant that must be analyzed, mimicking the real-world situation. Membranes and porous sorbents hitherto developed exhibit excellent performance for separating oil/water mixtures and emulsions in a laboratory environment, but their performance should be recorded from actual oil pollutant sites mixed with seawater mainly containing petroleum by-products. Also, oil and organic solvents whose surface tension is close to or below water must be documented to understand the true nature of the as-prepared hydrophobic/hydrophilic sorbents presented here. Furthermore, materials developed to date are either considered fragile, or their hydrophobic coating gets worn out at times, compromising their structural integrity and associated oil/water separation performance. Therefore, the materials could lose their efficacy in an actual situation where high tides and viscous oils are a matter of concern. Based on the challenges associated with (super)hydrophobic/(super)oleophilic materials, future research must fundamentally analyze the surface with the oil/water mixture, especially with oil/water emulsion, for a long duration. Second, green chemistry approaches must devise an industrially viable synthesis method using earth-abundant materials to minimize production costs and bearing on the environment. Although the selective oil/water separation with materials exhibiting wettability/nonwettability is a growing research field, challenges remain. Therefore, the onus is on the scientists and engineers to build mechanically robust, flexible (super)hydrophobic/(super)oleophilic materials for oil-water separation with improved absorption capacity and reusability for thousands of cycles.

REFERENCES

[1] S. Rasouli, N. Rezaei, H. Hamedi, S. Zendehboudi, X. Duan, Superhydrophobic and superoleophilic membranes for oil-water separation application: A comprehensive review, *Materials & Design,* 204 (2021) 109599.

[2] R.K. Gupta, G.J. Dunderdale, M.W. England, A. Hozumi, Oil/water separation techniques: A review of recent progresses and future directions, *Journal of Materials Chemistry A*, 5 (2017) 16025–16058.

[3] N. Zhang, Y. Qi, Y. Zhang, J. Luo, P. Cui, W. Jiang, A review on oil/water mixture separation material, *Industrial & Engineering Chemistry Research*, 59 (2020) 14546–14568.

[4] P. Saha, L. Dashairya, Reduced graphene oxide modified melamine formaldehyde (rGO@ MF) superhydrophobic sponge for efficient oil–water separation, *Journal of Porous Materials*, 25 (2018) 1–14.

[5] L. Dashairya, M. Rout, P. Saha, Reduced graphene oxide-coated cotton as an efficient absorbent in oil-water separation, *Advanced Composites and Hybrid Materials*, 1 (2018) 135–148.

[6] L. Dashairya, D.D. Barik, P. Saha, Methyltrichlorosilane functionalized silica nanoparticles-treated superhydrophobic cotton for oil–water separation, *Journal of Coatings Technology and Research*, 16 (2019) 1–12.

[7] S. Gupta, N.-H. Tai, Carbon materials as oil sorbents: A review on the synthesis and performance, *Journal of Materials Chemistry A*, 4 (2016) 1550–1565.

[8] S. Rajendran, F.N. Sadooni, H.A.-S. Al-Kuwari, A. Oleg, H. Govil, S. Nasir, P. Vethamony, Monitoring oil spill in Norilsk, Russia using satellite data, *Scientific Reports*, 11 (2021) 3817.

[9] Q.P. Le, R.O. Olekhnovich, M.V. Uspenskaya, T.H.N. Vu, Study on polyvinyl chloride nanofibers ability for oil spill elimination, *Iranian Polymer Journal*, 30 (2021) 473–483.

[10] V. Deshmukh, V. Singh, A. Chellappan, G. Dash, Hazardous oil spill in Mumbai Port and adjacent fishing areas, CMFRI Newsletter No. 126, July–September 2010, 126 (2010) 11–11.

[11] Y. Han, I.M. Nambi, T.P. Clement, Environmental impacts of the Chennai oil spill accident–A case study, *Science of the Total Environment*, 626 (2018) 795–806.

[12] M. Dyakova, Investment for health and well-being: A review of the social return on investment from public health policies to support implementing the sustainable development goals by building on health 2020 (2017). Copenhagen: WHO Regional Office for Europe (Health Evidence Network (HEN) synthesis report 51).

[13] N. Baig, Recent progress on the development of superhydrophobic and superoleophilic meshes for oil and water separation: A review, *Contaminants in Our Water: Identification and Remediation Methods* (2020) 175–196. Nadeem Baig.

[14] J.W. Mercer, R.M. Cohen, A review of immiscible fluids in the subsurface: Properties, models, characterization and remediation, *Journal of Contaminant Hydrology*, 6 (1990) 107–163.

[15] M.O. Adebajo, R.L. Frost, J.T. Kloprogge, O. Carmody, S. Kokot, Porous materials for oil spill cleanup: A review of synthesis and absorbing properties, *Journal of Porous Materials*, 10 (2003) 159–170.

[16] M. Peng, Y. Zhu, H. Li, K. He, G. Zeng, A. Chen, Z. Huang, T. Huang, L. Yuan, G. Chen, Synthesis and application of modified commercial sponges for oil-water separation, *Chemical Engineering Journal*, 373 (2019) 213–226.

[17] M. Ge, C. Cao, J. Huang, X. Zhang, Y. Tang, X. Zhou, K. Zhang, Z. Chen, Y. Lai, Rational design of materials interface at nanoscale towards intelligent oil–water separation, *Nanoscale Horizons*, 3 (2018) 235–260.

[18] E.M. Hadji, B. Fu, A. Abebe, H.M. Bilal, J. Wang, Sponge-based materials for oil spill cleanups: A review, *Frontiers of Chemical Science and Engineering*, 14 (2020) 749–762.

[19] L.-Y. Meng, S.-J. Park, Superhydrophobic carbon-based materials: A review of synthesis, structure, and applications, *Carbon Letters*, 15 (2014) 89–104.

[20] E. Piperopoulos, L. Calabrese, E. Mastronardo, C. Milone, E. Proverbio, Carbon-based sponges for oil spill recovery, in: *Carbon Nanomaterials for Agri-Food and Environmental Applications*, Amsterdam: Elsevier, 2020, pp. 155–175.

[21] L. Dashairya, M. Gopinath, P. Saha, Synergistic effect of Zr/Cl dual-ions mediated pyrrole polymerization and development of superhydrophobic melamine sponges for oil/water separation, *Colloids and Surfaces A: Physicochemical and Engineering Aspects*, 599 (2020) 124877.

[22] K. Koch, B. Bhushan, W. Barthlott, Multifunctional surface structures of plants: An inspiration for biomimetics, *Progress in Materials Science*, 54 (2009) 137–178.

[23] B. Arkles, *Hydrophobicity, Hydrophilicity and Silane Surface Modification*, Gelest Inc, Morrisville, 2011.

[24] A. Nakajima, K. Hashimoto, T. Watanabe, Recent studies on super-hydrophobic films, *Monatshefte für Chemie/Chemical Monthly*, 132 (2001) 31–41.

[25] L. Dashairya, A. Sahu, P. Saha, Stearic acid treated polypyrrole-encapsulated melamine formaldehyde superhydrophobic sponge for oil recovery, *Advanced Composites and Hybrid Materials*, 2 (2019) 70–82.

[26] C. Mao, C. Liang, W. Luo, J. Bao, J. Shen, X. Hou, W. Zhao, Preparation of lotus-leaf-like polystyrene micro-and nanostructure films and its blood compatibility, *Journal of Materials Chemistry*, 19 (2009) 9025–9029.

[27] R. Dalapati, S. Nandi, C. Gogoi, A. Shome, S. Biswas, Metal–organic framework (MOF) derived recyclable, superhydrophobic composite of cotton fabrics for the facile removal of oil spills, *ACS Applied Materials & Interfaces*, 13 (2021) 8563–8573.

[28] X. Zhang, D. Liu, Y. Ma, J. Nie, G. Sui, Super-hydrophobic graphene coated polyurethane (GN@ PU) sponge with great oil-water separation performance, *Applied Surface Science*, 422 (2017) 116–124.

[29] J. Wang, Y. Zheng, Oil/water mixtures and emulsions separation of stearic acid-functionalized sponge fabricated via a facile one-step coating method, *Separation and Purification Technology*, 181 (2017) 183–191.

[30] E. Wang, H. Wang, Z. Liu, R. Yuan, Y. Zhu, One-step fabrication of a nickel foam-based superhydrophobic and superoleophilic box for continuous oil–water separation, *Journal of Materials Science*, 50 (2015) 4707–4716.

[31] C.-F. Wang, S.-J. Lin, Robust superhydrophobic/superoleophilic sponge for effective continuous absorption and expulsion of oil pollutants from water, *ACS Applied Materials & Interfaces*, 5 (2013) 8861–8864.

[32] S. Huang, Mussel-inspired one-step copolymerization to engineer hierarchically structured surface with superhydrophobic properties for removing oil from water, *ACS Applied Materials & Interfaces*, 6 (2014) 17144–17150.

[33] X. Zhou, Z. Zhang, X. Xu, X. Men, X. Zhu, Facile fabrication of superhydropho- bic sponge with selective absorption and collection of oil from water, *Industrial & Engineering Chemistry Research*, 52 (2013) 9411–9416.

[34] B. Ge, Z. Zhang, X. Zhu, X. Men, X. Zhou, A superhydrophobic/superoleophilic sponge for the selective absorption oil pollutants from water, *Colloids and Surfaces A: Physicochemical and Engineering Aspects*, 457 (2014) 397–401.

[35] L. Wu, L. Li, B. Li, J. Zhang, A. Wang, Magnetic, durable, and superhydrophobic polyurethane@ Fe3O4@ SiO2@ fluoropolymer sponges for selective oil absorption and oil/water separation, *ACS Applied Materials & Interfaces*, 7 (2015) 4936–4946.

[36] O. Oribayo, Q. Pan, X. Feng, G.L. Rempel, Hydrophobic surface modification of FMSS and its application as effective sorbents for oil spill clean-ups and recovery, *AIChE Journal*, 63 (2017) 4090–4102.

[37] C. Shi, H. Ma, Z. Wo, X. Zhang, Superhydrophobic modification of the surface of cel- lulosic materials based on honeycomb-like zinc oxide structures and their application in oil-water separation, *Applied Surface Science*, 563 (2021) 150291.

[38] G. Wang, Z. Zeng, X. Wu, T. Ren, J. Han, Q. Xue, Three-dimensional structured sponge with high oil wettability for the clean-up of oil contaminations and separation of oil–water mixtures, *Polymer Chemistry*, 5 (2014) 5942–5948.

[39] B. Wang, J. Li, G. Wang, W. Liang, Y. Zhang, L. Shi, Z. Guo, W. Liu, Methodology for robust superhydrophobic fabrics and sponges from in situ growth of transition metal/ metal oxide nanocrystals with thiol modification and their applications in oil/water separation, *ACS Applied Materials & Interfaces*, 5 (2013) 1827–1839.

[40] C.-H. Chung, W.-C. Liu, J.-L. Hong, Superhydrophobic melamine sponge modified by cross-linked urea network as recyclable oil absorbent materials, *Industrial & Engineering Chemistry Research*, 57 (2018) 8449–8459.

[41] J. Huang, S. Li, M. Ge, L. Wang, T. Xing, G. Chen, X. Liu, S.S. Al-Deyab, K. Zhang, T. Chen, Robust superhydrophobic TiO 2@ fabrics for UV shielding, self-cleaning and oil–water separation, *Journal of Materials Chemistry A*, 3 (2015) 2825–2832.

[42] C.-H. Xue, X.-J. Guo, J.-Z. Ma, S.-T. Jia, Fabrication of robust and antifouling super- hydrophobic surfaces via surface-initiated atom transfer radical polymerization, *ACS Applied Materials & Interfaces*, 7 (2015) 8251–8259.

[43] J. Zhang, S. Seeger, Polyester materials with superwetting silicone nanofilaments for oil/water separation and selective oil absorption, *Advanced Functional Materials*, 21 (2011) 4699–4704.

[44] X. Zhou, Z. Zhang, X. Xu, F. Guo, X. Zhu, X. Men, B. Ge, Robust and durable superhy- drophobic cotton fabrics for oil/water separation, *ACS Applied Materials & Interfaces*, 5 (2013) 7208–7214.

[45] C.R. Crick, J.A. Gibbins, I.P. Parkin, Superhydrophobic polymer-coated copper-mesh; membranes for highly efficient oil–water separation, *Journal of Materials Chemistry A*, 1 (2013) 5943–5948.

[46] B. Cortese, D. Caschera, F. Federici, G.M. Ingo, G. Gigli, Superhydrophobic fabrics for oil–water separation through a diamond like carbon (DLC) coating, *Journal of Materials Chemistry A*, 2 (2014) 6781–6789.

[47] F.Z. Pour, H. Karimi, V.M. Avargani, Preparation of a superhydrophobic and supero- leophilic polyester textile by chemical vapor deposition of dichlorodimethylsilane for water–oil separation, *Polyhedron*, 159 (2019) 54–63.

[48] Y. Yi, P. Liu, N. Zhang, M.E. Gibril, F. Kong, S. Wang, A high lignin-content, ultra- light, and hydrophobic aerogel for oil-water separation: Preparation and characteriza- tion, *Journal of Porous Materials,* 28 (2021) 1–14.

[49] Z. Tong, B. Zhang, H. Yu, X. Yan, H. Xu, X. Li, H. Ji, Si_3N_4 Nanofibrous aerogel with in situ growth of SiO_x coating and nanowires for oil/water separation and ther- mal insulation, *ACS Applied Materials & Interfaces,* 13(19) (2021) 22765–22773. doi: 10.1021/acsami.1c05575

[50] F. Bayram, E.S. Mercan, M. Karaman, One-step fabrication of superhydrophobic-superoleophilic membrane by initiated chemical vapor deposition method for oil–water separation, *Colloid and Polymer Science*, 299 (2021) 1–9.

[51] T. Zhu, S. Li, J. Huang, M. Mihailiasa, Y. Lai, Rational design of multi-layered super-hydrophobic coating on cotton fabrics for UV shielding, self-cleaning and oil-water separation, *Materials & Design*, 134 (2017) 342–351.

[52] C.-H. Xue, Q.-Q. Fan, X.-J. Guo, Q.-F. An, S.-T. Jia, Fabrication of superhydrophobic cotton fabrics by grafting of POSS-based polymers on fibers, *Applied Surface Science*, 465 (2019) 241–248.

[53] M.Z. Khan, V. Baheti, J. Militky, A. Ali, M. Vikova, Superhydrophobicity, UV protection and oil/water separation properties of fly ash/Trimethoxy (octadecyl) silane coated cotton fabrics, *Carbohydrate Polymers*, 202 (2018) 571–580.

[54] S.S. Latthe, R.S. Sutar, A. Bhosale, K.K. Sadasivuni, S. Liu, Superhydrophobic surfaces for oil-water separation, in: *Superhydrophobic Polymer Coatings*. Amsterdam: Elsevier, 2019, pp. 339–356.

[55] J. Wang, F. Han, B. Liang, G. Geng, Hydrothermal fabrication of robustly superhydrophobic cotton fibers for efficient separation of oil/water mixtures and oil-in-water emulsions, *Journal of Industrial and Engineering Chemistry*, 54 (2017) 174–183.

[56] Z. Wan, Y. Jiao, X. Ouyang, L. Chang, X. Wang, Bifunctional MoS2 coated melamine-formaldehyde sponges for efficient oil–water separation and water-soluble dye removal, *Applied Materials Today*, 9 (2017) 551–559.

[57] A. Xie, J. Cui, Y. Chen, J. Lang, C. Li, Y. Yan, J. Dai, One-step facile fabrication of sustainable cellulose membrane with superhydrophobicity via a sol-gel strategy for efficient oil/water separation, *Surface and Coatings Technology*, 361 (2019) 19–26.

[58] X. Su, H. Li, X. Lai, L. Zhang, J. Wang, X. Liao, X. Zeng, Vapor–liquid sol–gel approach to fabricating highly durable and robust superhydrophobic polydimethyl-siloxane@ silica surface on polyester textile for oil–water separation, *ACS Applied Materials & Interfaces*, 9 (2017) 28089–28099.

[59] R.S. Sutar, R.C. Salunkhe, S.S. Latthe, V.S. Kodag, P.M. Shewale, S.R. Shinde, M. Sajjan, M. Karennavar, K.K. Sadasivuni, S.V. Mohite, Superhydrophobic PU sponge modified by hydrophobic silica NPs—Polystyrene nanocomposite for oil–water separation, in: *Macromolecular Symposia*. Hoboken, NJ: Wiley Online Library, 2020, p. 2000035.

[60] D. Ahuja, S. Dhiman, G. Rattan, S. Monga, S. Singhal, A. Kaushik, Superhydrophobic modification of cellulose sponge fabricated from discarded jute bags for oil water separation, *Journal of Environmental Chemical Engineering*, 9 (2021) 105063.

[61] S. Yang, L. Chen, C. Wang, M. Rana, P.-C. Ma, Surface roughness induced superhydrophobicity of graphene foam for oil-water separation, *Journal of Colloid and Interface Science*, 508 (2017) 254–262.

[62] M. Li, X. Xu, L. Zhang, Fabricated superhydrophobic three-dimensional rambutan-like-β-NiOOH@ sponge skeletons for multitasking oil–water separation, *Journal of Industrial and Engineering Chemistry*, 84 (2020) 340–348.

[63] J. Chen, J. Xiang, X. Yue, H. Li, X. Yu, Synthesis of a superhydrophobic polyvinyl alcohol sponge using water as the only solvent for continuous oil-water separation, *Journal of Chemistry*, 2019 (2019) 7153109, 8 pages. https://doi.org/10.1155/2019/7153109

[64] S. Sriram, A. Kumar, Separation of oil-water via porous PMMA/SiO2 nanoparticles superhydrophobic surface, *Colloids and Surfaces A: Physicochemical and Engineering Aspects*, 563 (2019) 271–279.

[65] J. Lu, X. Liu, T.C. Zhang, H. He, S. Yuan, Magnetic superhydrophobic polyurethane sponge modified with bioinspired stearic acid@ Fe3O4@ PDA nanocomposites for oil/water separation, *Colloids and Surfaces A: Physicochemical and Engineering Aspects*, 624 (2021) 126794.

[66] Z. Li, B. Wang, X. Qin, Y. Wang, C. Liu, Q. Shao, N. Wang, J. Zhang, Z. Wang, C. Shen, Superhydrophobic/superoleophilic polycarbonate/carbon nanotubes porous monolith for selective oil adsorption from water, *ACS Sustainable Chemistry & Engineering*, 6 (2018) 13747–13755.

[67] G.-Q. Xi, T. Liu, C. Ma, Q. Yuan, W. Xin, J.-J. Lu, M.-G. Ma, Superhydrophobic, compressible, and reusable polyvinyl alcohol-wrapped silver nanowire composite sponge for continuous oil-water separation, *Colloids and Surfaces A: Physicochemical and Engineering Aspects*, 583 (2019) 124028.

[68] X. Gui, H. Li, K. Wang, J. Wei, Y. Jia, Z. Li, L. Fan, A. Cao, H. Zhu, D. Wu, Recyclable carbon nanotube sponges for oil absorption, *Acta Materialia*, 59 (2011) 4798–4804.

[69] B. Ge, X. Men, X. Zhu, Z. Zhang, A superhydrophobic monolithic material with tunable wettability for oil and water separation, *Journal of Materials Science*, 50 (2015) 2365–2369.

[70] D. Kukkar, A. Rani, V. Kumar, S.A. Younis, M. Zhang, S.-S. Lee, D.C. Tsang, K.-H. Kim, Recent advances in carbon nanotube sponge–based sorption technologies for mitigation of marine oil spills, *Journal of Colloid and Interface Science*, 570 (2020) 411–422.

[71] S. Qiu, B. Jiang, X. Zheng, J. Zheng, C. Zhu, M. Wu, Hydrophobic and fire-resistant carbon monolith from melamine sponge: A recyclable sorbent for oil–water separation, *Carbon*, 84 (2015) 551–559.

[72] Y. Wang, B. Wang, J. Wang, Y. Ren, C. Xuan, C. Liu, C. Shen, Superhydrophobic and superoleophilic porous reduced graphene oxide/polycarbonate monoliths for high-efficiency oil/water separation, *Journal of Hazardous Materials*, 344 (2018) 849–856.

[73] J. Dai, R. Zhang, W. Ge, A. Xie, Z. Chang, S. Tian, Z. Zhou, Y. Yan, 3D macroscopic superhydrophobic magnetic porous carbon aerogel converted from biorenewable popcorn for selective oil-water separation, *Materials & Design*, 139 (2018) 122–131.

[74] D. Yuan, T. Zhang, Q. Guo, F. Qiu, D. Yang, Z. Ou, Superhydrophobic hierarchical biomass carbon aerogel assembled with TiO2 nanorods for selective immiscible oil/water mixture and emulsion separation, *Industrial & Engineering Chemistry Research*, 57 (2018) 14758–14766.

[75] D. Yuan, T. Zhang, Q. Guo, F. Qiu, D. Yang, Z. Ou, Recyclable biomass carbon@ SiO2@ MnO2 aerogel with hierarchical structures for fast and selective oil-water separation, *Chemical Engineering Journal*, 351 (2018) 622–630.

[76] L.B. Lv, T.L. Cui, B. Zhang, H.H. Wang, X.H. Li, J.S. Chen, Wrinkled graphene monoliths as superabsorbing building blocks for superhydrophobic and superhydrophilic surfaces, *Angewandte Chemie International Edition*, 54 (2015) 15165–15169.

[77] X. Zhang, B. Wang, X. Qin, S. Ye, Y. Shi, Y. Feng, W. Han, C. Liu, C. Shen, Cellulose acetate monolith with hierarchical micro/nano-porous structure showing superior hydrophobicity for oil/water separation, *Carbohydrate Polymers*, 241 (2020) 116361.

[78] Y. Yu, X. Wu, J. Fang, Superhydrophobic and superoleophilic "sponge-like" aerogels for oil/water separation, *Journal of Materials Science*, 50 (2015) 5115–5124.

[79] Y. Meng, T. Liu, S. Yu, S. Cheng, J. Lu, H. Wang, A lignin-based carbon aerogel enhanced by graphene oxide and application in oil/water separation, *Fuel*, 278 (2020) 118376.

[80] Y. Luo, S. Jiang, Q. Xiao, C. Chen, B. Li, Highly reusable and superhydrophobic spongy graphene aerogels for efficient oil/water separation, *Scientific Reports*, 7 (2017) 1–10.

[81] S. Long, Y. Feng, F. He, S. He, H. Hong, X. Yang, L. Zheng, J. Liu, L. Gan, M. Long, An ultralight, supercompressible, superhydrophobic and multifunctional carbon aerogel with a specially designed structure, *Carbon*, 158 (2020) 137–145.

[82] H.-Y. Mi, X. Jing, A.L. Politowicz, E. Chen, H.-X. Huang, L.-S. Turng, Highly compressible ultra-light anisotropic cellulose/graphene aerogel fabricated by bidirectional freeze drying for selective oil absorption, *Carbon*, 132 (2018) 199–209.

[83] A. Parsaie, Y. Tamsilian, M.R. Pordanjani, A.K. Abadshapoori, G. McKay, Novel approach for rapid oil/water separation through superhydrophobic/superoleophilic zinc stearate coated polyurethane sponges, *Colloids and Surfaces A: Physicochemical and Engineering Aspects*, 618 (2021) 126395.

[84] B. Ge, X. Zhu, Y. Li, X. Men, P. Li, Z. Zhang, Versatile fabrication of magnetic super-hydrophobic foams and application for oil–water separation, *Colloids and Surfaces A: Physicochemical and Engineering Aspects*, 482 (2015) 687–692.

[85] S. Javadian, Magnetic superhydrophobic polyurethane sponge loaded with Fe3O4@ oleic acid@ graphene oxide as high performance adsorbent oil from water, *Chemical Engineering Journal*, 408 (2021) 127369.

[86] N. Cao, B. Yang, A. Barras, S. Szunerits, R. Boukherroub, Polyurethane sponge func-tionalized with superhydrophobic nanodiamond particles for efficient oil/water sepa-ration, *Chemical Engineering Journal*, 307 (2017) 319–325.

[87] H. Li, S. Lin, X. Feng, Q. Pan, Preparation of superhydrophobic and superoleophilic polyurethane foam for oil spill cleanup, *Journal of Macromolecular Science, Part A*, 58 (2021) 1–11.

[88] Y. Zhou, N. Zhang, X. Zhou, Y. Hu, G. Hao, X. Li, W. Jiang, Design of recyclable superhydrophobic PU@ Fe3O4@ PS sponge for removing oily contaminants from water, *Industrial & Engineering Chemistry Research*, 58 (2019) 3249–3257.

[89] A. Jamsaz, E.K. Goharshadi, An environmentally friendly superhydrophobic modified polyurethane sponge by seashell for the efficient oil/water separation, *Process Safety and Environmental Protection*, 139 (2020) 297–304.

[90] D. Yuan, T. Zhang, Q. Guo, F. Qiu, D. Yang, Z. Ou, A novel hierarchical hollow SiO2@ MnO2 cubes reinforced elastic polyurethane foam for the highly efficient removal of oil from water, *Chemical Engineering Journal*, 327 (2017) 539–547.

[91] F. Qiang, L.-L. Hu, L.-X. Gong, L. Zhao, S.-N. Li, L.-C. Tang, Facile synthesis of super-hydrophobic, electrically conductive and mechanically flexible functionalized graphene nanoribbon/polyurethane sponge for efficient oil/water separation at static and dynamic states, *Chemical Engineering Journal*, 334 (2018) 2154–2166.

[92] R. Sun, N. Yu, J. Zhao, J. Mo, Y. Pan, D. Luo, Chemically stable superhydrophobic poly-urethane sponge coated with ZnO/epoxy resin coating for effective oil/water separation, *Colloids and Surfaces A: Physicochemical and Engineering Aspects*, 611 (2021) 125850.

[93] L. Kong, Y. Li, F. Qiu, T. Zhang, Q. Guo, X. Zhang, D. Yang, J. Xu, M. Xue, Fabrication of hydrophobic and oleophilic polyurethane foam sponge modified with hydrophobic Al2O3 for oil/water separation, *Journal of Industrial and Engineering Chemistry*, 58 (2018) 369–375.

[94] L. Zhang, L. Xu, Y. Sun, N. Yang, Robust and durable superhydrophobic polyurethane sponge for oil/water separation, *Industrial & Engineering Chemistry Research*, 55 (2016) 11260–11268.

[95] X.-Q. Chen, B. Zhang, L. Xie, F. Wang, MWCNTs polyurethane sponges with enhanced super-hydrophobicity for selective oil–water separation, *Surface Engineering*, 36 (2020) 651–659.

[96] J. Nan Shen, C. Chao Yu, H. Min Ruan, C. Jie Gao, B. Van der Bruggen, Preparation and characterization of thin-film nanocomposite membranes embedded with poly (methyl methacrylate) hydrophobic modified multiwalled carbon nanotubes by inter-facial polymerization, *Journal of Membrane Science*, 442 (2013) 18–26.

[97] S. Qiu, Y. Li, G. Li, Z. Zhang, Y. Li, T. Wu, Robust superhydrophobic sepiolite-coated polyurethane sponge for highly efficient and recyclable oil absorption, *ACS Sustainable Chemistry & Engineering*, 7 (2019) 5560–5567.

[98] N. Habibi, A. Pourjavadi, Magnetic, thermally stable, and superhydrophobic polyure-thane sponge: A high efficient adsorbent for separation of the marine oil spill pollu-tion, *Chemosphere*, 487 (2021) 132254.

[99] H. Wang, E. Wang, Z. Liu, D. Gao, R. Yuan, L. Sun, Y. Zhu, A novel carbon nanotubes reinforced superhydrophobic and superoleophilic polyurethane sponge for selective oil–water separation through a chemical fabrication, *Journal of Materials Chemistry A*, 3 (2015) 266–273.

[100] N. Baig, F.I. Alghunaimi, H.S. Dossary, T.A. Saleh, Superhydrophobic and superoleophilic carbon nanofiber grafted polyurethane for oil-water separation, *Process Safety and Environmental Protection*, 123 (2019) 327–334.

[101] X. Gui, J. Wei, K. Wang, A. Cao, H. Zhu, Y. Jia, Q. Shu, D. Wu, Carbon nanotube sponges, *Advanced Materials*, 22 (2010) 617–621.

[102] X. Cao, Y. Zhou, X. Wei, W. Zhai, G. Zheng, K. Dai, C. Liu, C. Shen, Lightweight, mechanical robust foam with a herringbone-like porous structure for oil/water separation and filtering, *Polymer Testing*, 72 (2018) 86–93.

[103] T. Zhang, B. Gu, F. Qiu, X. Peng, X. Yue, D. Yang, Preparation of carbon nanotubes/polyurethane hybrids as a synergistic absorbent for efficient oil/water separation, *Fibers and Polymers*, 19 (2018) 2195–2202.

[104] J. Zhao, Y. Huang, G. Wang, Y. Qiao, Z. Chen, A. Zhang, C.B. Park, Fabrication of outstanding thermal-insulating, mechanical robust and superhydrophobic PP/CNT/ sorbitol derivative nanocomposite foams for efficient oil/water separation, *Journal of Hazardous Materials*, 418 (2021) 126295.

[105] B. Lin, J. Chen, Z.-T. Li, F.-A. He, D.-H. Li, Superhydrophobic modification of polyurethane sponge for the oil-water separation, *Surface and Coatings Technology*, 359 (2019) 216–226.

[106] F. Beshkar, H. Khojasteh, M. Salavati-Niasari, Recyclable magnetic superhydrophobic straw soot sponge for highly efficient oil/water separation, *Journal of Colloid and Interface Science*, 497 (2017) 57–65.

[107] H. Omidvarborna, A. Kumar, D.-S. Kim, Recent studies on soot modeling for diesel combustion, *Renewable and Sustainable Energy Reviews*, 48 (2015) 635–647.

[108] Y. Ding, W. Xu, W. Yu, H. Hou, Z. Zhu, One-step preparation of highly hydrophobic and oleophilic melamine sponges via metal-ion-induced wettability transition, *ACS Applied Materials & Interfaces*, 10 (2018) 6652–6660.

[109] O. Oribayo, X. Feng, G.L. Rempel, Q. Pan, Modification of formaldehyde-melamine-sodium bisulfite copolymer foam and its application as effective sorbents for clean up of oil spills, *Chemical Engineering Science*, 160 (2017) 384–395.

[110] Y. Wang, Y. Feng, J. Yao, Construction of hydrophobic alginate-based foams induced by zirconium ions for oil and organic solvent cleanup, *Journal of Colloid and Interface Science*, 533 (2019) 182–189.

[111] H. Zhu, S. Yang, D. Chen, N. Li, Q. Xu, H. Li, J. He, J. Lu, A robust absorbent material based on light-responsive superhydrophobic melamine sponge for oil recovery, *Advanced Materials Interfaces*, 3 (2016) 1500683.

[112] A. Stolz, S. Le Floch, L. Reinert, S.M. Ramos, J. Tuaillon-Combes, Y. Soneda, P. Chaudet, D. Baillis, N. Blanchard, L. Duclaux, Melamine-derived carbon sponges for oil-water separation, *Carbon*, 107 (2016) 198–208.

[113] Z.-T. Li, F.-A. He, B. Lin, Preparation of magnetic superhydrophobic melamine sponge for oil-water separation, *Powder Technology*, 345 (2019) 571–579.

[114] Z.-W.S. Zeng, S.E. Taylor, Facile preparation of superhydrophobic melamine sponge for efficient underwater oil-water separation, *Separation and Purification Technology*, 247 (2020) 116996.

[115] B. Shang, Y. Wang, B. Peng, Z. Deng, Bioinspired polydopamine particles-assisted construction of superhydrophobic surfaces for oil/water separation, *Journal of Colloid and Interface Science*, 482 (2016) 240–251.

[116] C. Ruan, K. Ai, X. Li, L. Lu, A superhydrophobic sponge with excellent absorbency and flame retardancy, *Angewandte Chemie International Edition*, 53 (2014) 5556–5560.

[117] V.H. Pham, J.H. Dickerson, Superhydrophobic silanized melamine sponges as high efficiency oil absorbent materials, *ACS Applied Materials & Interfaces*, 6 (2014) 14181–14188.

[118] Z. Xu, K. Miyazaki, T. Hori, Dopamine-induced superhydrophobic melamine foam for oil/water separation, *Advanced Materials Interfaces*, 2 (2015) 1500255.

[119] X. Chen, J.A. Weibel, S.V. Garimella, Continuous oil–water separation using polydimethylsiloxane-functionalized melamine sponge, *Industrial & Engineering Chemistry Research*, 55 (2016) 3596–3602.

[120] W. Zhang, X. Zhai, T. Xiang, M. Zhou, D. Zang, Z. Gao, C. Wang, Superhydrophobic melamine sponge with excellent surface selectivity and fire retardancy for oil absorption, *Journal of Materials Science*, 52 (2017) 73–85.

[121] J. Chen, H. You, L. Xu, T. Li, X. Jiang, C.M. Li, Facile synthesis of a two-tier hierarchical structured superhydrophobic-superoleophilic melamine sponge for rapid and efficient oil/water separation, *Journal of Colloid and Interface Science*, 506 (2017) 659–668.

[122] C. Su, H. Yang, S. Song, B. Lu, R. Chen, A magnetic superhydrophilic/oleophobic sponge for continuous oil-water separation, *Chemical Engineering Journal*, 309 (2017) 366–373.

[123] K. Hou, Y. Jin, J. Chen, X. Wen, S. Xu, J. Cheng, P. Pi, Fabrication of superhydrophobic melamine sponges by thiol-ene click chemistry for oil removal, *Materials Letters*, 202 (2017) 99–102.

[124] T. Chen, S. Zhou, Z. Hu, X. Fu, Z. Liu, B. Su, H. Wan, X. Du, Z. Gao, A multifunctional superhydrophobic melamine sponge decorated with Fe3O4/Ag nanocomposites for high efficient oil-water separation and antibacterial application, *Colloids and Surfaces A: Physicochemical and Engineering Aspects*, 473 (2021) 127041.

[125] J. Xue, L. Zhu, X. Zhu, H. Li, C. Ma, S. Yu, D. Sun, F. Xia, Q. Xue, Tetradecylamine-MXene functionalized melamine sponge for effective oil/water separation and selective oil adsorption, *Separation and Purification Technology*, 259 (2021) 118106.

[126] J. Yang, Y. Jia, B. Li, J. Jiao, Facile and simple fabrication of superhydrophobic and superoleophilic MS/PDA/DT sponge for efficient oil/water separation, *Environmental Technology*, 43 (2021) 1–17.

[127] Y. Sun, L. Ma, Y. Song, A.D. Phule, L. Li, Z.X. Zhang, Efficient natural rubber latex foam coated by rGO modified high density polyethylene for oil-water separation and electromagnetic shielding performance, *European Polymer Journal*, 147 (2021) 110288.

[128] Y. Pan, K. Shi, Z. Liu, W. Wang, C. Peng, X. Ji, Synthesis of a new kind of macroporous polyvinyl-alcohol formaldehyde based sponge and its water superabsorption performance, *RSC Advances*, 5 (2015) 78780–78789.

[129] X. Zhang, B. Wang, B. Wang, Y. Feng, W. Han, C. Liu, C. Shen, Superhydrophobic cellulose acetate/multiwalled carbon nanotube monolith with fiber cluster network for selective oil/water separation, *Carbohydrate Polymers*, 259 (2021) 117750.

[130] L. Li, L. Rong, Z. Xu, B. Wang, X. Feng, Z. Mao, H. Xu, J. Yuan, S. Liu, X. Sui, Cellulosic sponges with pH responsive wettability for efficient oil-water separation, *Carbohydrate Polymers*, 237 (2020) 116133.

[131] Y. Lu, Y. Wang, L. Liu, W. Yuan, Environmental-friendly and magnetic/silanized ethyl cellulose sponges as effective and recyclable oil-absorption materials, *Carbohydrate Polymers*, 173 (2017) 422–430.

[132] H. Peng, H. Wang, J. Wu, G. Meng, Y. Wang, Y. Shi, Z. Liu, X. Guo, Preparation of superhydrophobic magnetic cellulose sponge for removing oil from water, *Industrial & Engineering Chemistry Research*, 55 (2016) 832–838.

[133] Y. Qin, S. Li, Y. Li, F. Pan, L. Han, Z. Chen, X. Yin, L. Wang, H. Wang, Mechanically robust polybenzoxazine/reduced graphene oxide wrapped-cellulose sponge towards highly efficient oil/water separation, and solar-driven for cleaning up crude oil, *Composites Science and Technology*, 197 (2020) 108254.

[134] Y. Pan, Z. Liu, W. Wang, C. Peng, K. Shi, X. Ji, Highly efficient macroporous adsorbents for toxic metal ions in water systems based on polyvinyl alcohol–formaldehyde sponges, *Journal of Materials Chemistry A*, 4 (2016) 2537–2549.

[135] Y. Pan, C. Peng, W. Wang, K. Shi, Z. Liu, X. Ji, Preparation and absorption behavior to organic pollutants of macroporous hydrophobic polyvinyl alcohol–formaldehyde sponges, *RSC Advances*, 4 (2014) 35620–35628.

[136] Y. Pan, W. Wang, C. Peng, K. Shi, Y. Luo, X. Ji, Novel hydrophobic polyvinyl alcohol–formaldehyde foams for organic solvents absorption and effective separation, *RSC Advances*, 4 (2014) 660–669.

[137] Y. Pan, K. Shi, C. Peng, W. Wang, Z. Liu, X. Ji, Evaluation of hydrophobic polyvinylalcohol formaldehyde sponges as absorbents for oil spill, *ACS Applied Materials & Interfaces*, 6 (2014) 8651–8659.

[138] J. Yang, Y. Tang, J. Xu, B. Chen, H. Tang, C. Li, Durable superhydrophobic/superoleophilic epoxy/attapulgite nanocomposite coatings for oil/water separation, *Surface and Coatings Technology*, 272 (2015) 285–290.

[139] Z. Du, P. Ding, X. Tai, Z. Pan, H. Yang, Facile preparation of Ag-coated superhydrophobic/superoleophilic mesh for efficient oil/water separation with excellent corrosion resistance, *Langmuir*, 34 (2018) 6922–6929.

[140] L. Feng, Z. Zhang, Z. Mai, Y. Ma, B. Liu, L. Jiang, D. Zhu, A super-hydrophobic and super-oleophilic coating mesh film for the separation of oil and water, *Angewandte Chemie*, 116 (2004) 2046–2048.

[141] Q. Wang, M. Yu, G. Chen, Q. Chen, J. Tian, Robust fabrication of fluorine-free superhydrophobic steel mesh for efficient oil/water separation, *Journal of Materials Science*, 52 (2017) 2549–2559.

[142] N.R. Geraldi, L.E. Dodd, B.B. Xu, D. Wood, G.G. Wells, G. McHale, M.I. Newton, Bioinspired nanoparticle spray-coating for superhydrophobic flexible materials with oil/water separation capabilities, *Bioinspiration & Biomimetics*, 13 (2018) 024001.

[143] U. Zulfiqar, S.Z. Hussain, T. Subhani, I. Hussain, Mechanically robust superhydrophobic coating from sawdust particles and carbon soot for oil/water separation, *Colloids and Surfaces A: Physicochemical and Engineering Aspects*, 539 (2018) 391–398.

[144] G. Ren, Y. Song, X. Li, Y. Zhou, Z. Zhang, X. Zhu, A superhydrophobic copper mesh as an advanced platform for oil-water separation, *Applied Surface Science*, 428 (2018) 520–525.

[145] A. Wang, X. Li, T. Hou, Y. Lu, J. Zhou, X. Zhang, B. Yang, A tree-grapes-like PTFE fibrous membrane with super-hydrophobic and durable performance for oil/water separation, *Separation and Purification Technology*, 265 (2021) 119165.

[146] W.-T. Cao, Y.-J. Liu, M.-G. Ma, J.-F. Zhu, Facile preparation of robust and superhydrophobic materials for self-cleaning and oil/water separation, *Colloids and Surfaces A: Physicochemical and Engineering Aspects*, 529 (2017) 18–25.

[147] Q. Shuai, X. Yang, Y. Luo, H. Tang, X. Luo, Y. Tan, M. Ma, A superhydrophobic poly (dimethylsiloxane)-TiO2 coated polyurethane sponge for selective absorption of oil from water, *Materials Chemistry and Physics*, 162 (2015) 94–99.

[148] C. Xia, Y. Li, T. Fei, W. Gong, Facile one-pot synthesis of superhydrophobic reduced graphene oxide-coated polyurethane sponge at the presence of ethanol for oil-water separation, *Chemical Engineering Journal*, 345 (2018) 648–658.

[149] C.-F. Wang, H.-C. Huang, L.-T. Chen, Protonated melamine sponge for effective oil/water separation, *Scientific Reports*, 5 (2015) 1–8.

11 Materials for Renewable Energy Resources for Oil and Gas Industries

Enna Chakervarty, Abhishek Sharma,
and Lalita Ledwani
Manipal University Jaipur

CONTENTS

DOI: 10.1201/9781003242550-11

ABBREVIATIONS

CHP   Combined heat and power
CSP   Concentrated solar power
GHG   Greenhouse gases
NER   Net energy ratio
VPP   Virtual power plant

11.1 INTRODUCTION

Fuels are vital to our economy. In India, fuels directly contributed to a large percentage of the energy needed for the country's transportation system, with 66% of that energy going toward producing electricity, 68% going toward industry, and 27% going toward buildings in 2013 [1]. The oil and gas industry is being pushed to justify the implications of energy shifts on its business and economic models, as well as the contributions it can make to reducing greenhouse gas (GHG) emissions and reaching the Paris Agreement's targets [2]. Various oil and gas companies are witnessing rising social and environmental challenges, pertaining to the role of fuels in a changing energy financial system, as well as their place in the society in which they function. However, in the face of escalating GHG emissions, the crucial issue remains: should today's oil and gas sectors be viewed only as a contributor to the problem, or may they also play a significant role in its resolution? Fossil fuels are the primary source of short-term profitability; however, if efforts to reduce GHG emissions are not needed, long-term social acceptance and profitability may suffer [2].

Now is the time for the oil and gas industries to convey what cleaner energy transitions mean to them – and what they can do to accelerate the process.

According to statistics, the developed countries consisting of 28% of the world's population utilize 77% of the world's energy production. By the year 2050, these nations will have successfully implemented better energy conservation policies, thereby averting an increase in energy consumption [3]. People in underdeveloped countries, on the other hand, generally want to build their own electricity-generating facilities. As per the estimates provided, fossil fuels will meet around 75% of total energy demand and 67% of total electricity supply in upcoming years [4].

As a result, backup renewable energy sources will be increasingly vital soon. This situation will act as a gateway for the introduction of special job possibilities and the growth of new industries. Because of an alarming increase in number of industries, the environment is becoming contaminated. The utilization of renewable energy, security, cost, energy policy, renewable energy applications, and smart grid technology are all parts of sustainable development.

The essential concepts of most energy programs are conservation and the use of domestic energy sources. Considering that carbon dioxide (CO_2) is the main source of GHGs, reducing carbon emissions is a significant problem. Many strategies, including increasing the use of renewable energy sources and encouraging technological innovation, could be used to reduce carbon emissions in this area. The government could also contribute by providing support through mechanisms like feed-in tariffs, renewable portfolio standards, and tax policies to encourage the production of renewable energy sources while also implementing energy-saving measures.

However, in the future, there will be a link between energy use and the environment. As it has been well recognized that many industries, especially considering the oil and gas industries, are causing adverse effects to the environment, like oil spills, improper drilling, gas leaks, and contaminated water leakages. Meeting up environmental regulations could be bit expensive. The petroleum sector is aware of this reality and has taken the lead in lowering the environmental effect of operations. Because of industry action, GHG emissions from big private oil corporations declined by 13% between 2010 and 2015, whereas methane emissions from natural gas wells reduced by 40% between 1999 and 2012. Moreover, several oil firms have committed targets for further reductions in emissions. Renewable energy technologies, when their costs fall, may become key instruments for addressing increased energy demands and higher pollution rules while decreasing fuel consumption and operating expenses. Many countries have already begun to construct power-producing facilities that use renewable energy sources.

Renewable energy sources have become an emergent need wherein energy supply processes related to different sources of renewable energy (e.g., wind and solar power), and energy effectiveness technologies, which refer to technologies used to improve energy-use efficiency (e.g., virtual power plants [VPP], combined heat and power [CHP], and smart meters), are the two main categories of clean technologies. It is crucial to remember that revamping the energy sector and switching from conventional to renewable energy is a continuous process that considers market development and technological advancement [5].

All industrial plants should be designed and built with their environmental effects in mind for improving the economy, promoting ecology, and conserving energy. Substantial financial resources will be needed for energy initiatives pertaining to environmental protection [4]. One method for addressing increased energy demand and production energy, while meeting emissions is to incorporate renewable generation technology into oil and gas operations. Several components of the oil and gas sector lend themselves to the integration of renewable energy technology.

Production facilities are generally located in rural regions and need huge amounts of electricity, which may be supplied by renewable energy sources (wind, solar). Both EOR (enhanced oil recovery) and oil refining require a significant amount of heat, which renewable thermal technology can deliver (solar thermal, geothermal). In some circumstances, using waste heat or gas to power cogeneration plants can be cost-effective [6]. Both energy prices and emissions may be lowered by combining renewable energy technologies and reducing the quantity of fossil fuels needed to produce, transport, and refine petroleum, while oil and gas resources can be maintained for their highest value uses [6].

11.2 OIL AND GAS INDUSTRY STATUS

Until recently, the United States' oil production had been declining. Oil imports accounted for more than half of total domestic oil use. Natural gas investment was shifting toward costly facilities for importing natural gas. The United States has surpassed Saudi Arabia as the world's top producer of oil and natural gas; it is prepared to export liquefied natural gas (LNG) and is exporting more refined goods.

These significant shifts are mainly the result of technological developments in hydraulic fracturing and horizontal drilling, which have allowed industry to extract oil and gas from "tight" formations, also known as "tight" formations and low-permeability formations like shale, which are frequently referred to as "unconventional resources" [7].

These progressions were made feasible in part by DOE's (Department of Energy) technology ventures in the early 1980s, as well as by industry's continuous research and implementation of such technologies. There has also been a push to limit the environmental impact of oil and gas extraction, notably after public worries over hydraulic fracturing on land and the BP Deepwater Horizon event offshore.

The primary fossil fuels in the United States are coal, oil, propane, and natural gas, which account for more than 80% of total energy utilization. They are inefficient and expensive compared to fossil fuels (or, in the case of nuclear power, altogether unable to expand). As a result, consumers who opt for cleaner renewable energy sources for their houses or vehicles are eligible for substantial government subsidies.

Safety rules and procedures have progressed because of government mandates for increased safety and environmental stewardship. There have been efforts to reduce environmental and safety issues within the industry.

11.3 ONGOING TREND AND POTENTIAL OPPORTUNITIES

The integration of renewable technologies into the oil and gas industry is becoming increasingly profitable through three simultaneous trends. The following tendencies are emerging:

1. Due to the exhaustion of good-quality grade oil sources, petroleum activities are becoming more energy rigorous.
2. The oil and gas sectors have a long history of proving a leading role in the environmental field.
3. Technologies for renewable energy production are becoming more and more profitable.

Due to the depletion of the best-quality oil wells, the petroleum activities are becoming more and more energetic.

11.3.1 REDUCTION OF HIGH-QUALITY OIL RESERVES

There were a lot of shallow oil and gas deposits available in the ancient days of the petroleum industry, and reservoir pressure was high. It is no wonder that the easy-to-extract oil and gas resources are exhausted. Deeper reservoirs, lower pressures, and lesser quality characterize the remaining reserves. In the Gulf of Mexico, for example, Figure 11.1 shows an increase in the average water and well depths on a year-over-year basis. As processes get more complicated, so does their energy intensity. EOR techniques, including the injection of gas, heat, or liquid to enhance field recovery rate, are also indicative of the shift toward production from lower-quality deposits. Even if there are still adequate petroleum resources to fulfill demand for

Water and drilling average depth

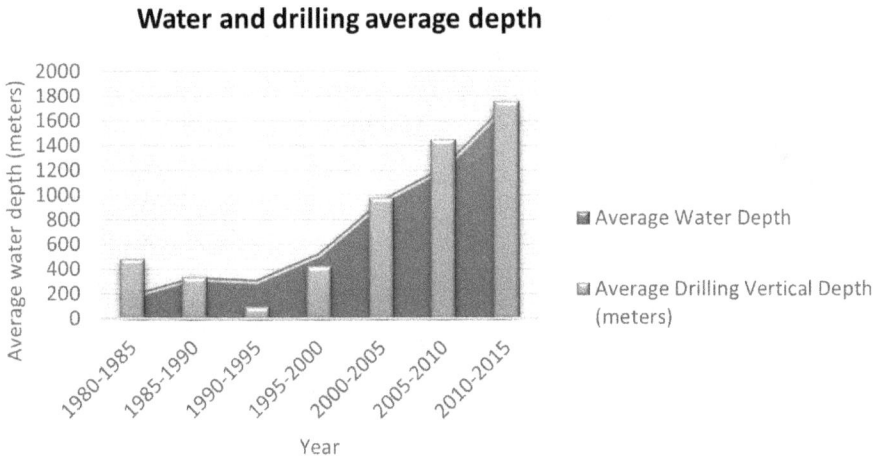

FIGURE 11.1   The average depth of water and drilling in the Gulf of Mexico over time. (Data from BOEM 2017.)

the foreseeable future, the move to marginal reserves raises the energy intensity of production NER (net energy ratio), it is the ratio of energy produced to energy needed to generate it which has decreased from 46% to 88% over the previous four decades, according to a 2017 analysis of five big oil fields [8]. Similar trends are seen in traditional natural gas production, with the ratio of energy output to input approximately halving from 1993 through 2009 [7]. The reduction in NER was caused by a combination of falling field output and rising energy expenditures on advanced recovery technologies [8].

Lower-quality deposits necessitate a greater use of energy in the extraction, transportation, and refining of petroleum. This increases demand for energy at each level of the petroleum supply chain, resulting in greater opportunities for renewable technology integration to minimize energy prices. Renewable technology, in addition to cost savings, may cut pollution, which is becoming an increasingly relevant aspect.

Petroleum production, transportation, and refining are all more energy intensive because of lower-quality reserves. As a result, energy demand rises at every level of the petroleum supply chain, opening up new potential for the integration of renewable technology to reduce energy costs. The reduction of emissions is becoming a more important part of the utilization of renewable technology.

11.3.2   ENVIRONMENTAL IMPACT CAUSED BY THE INDUSTRIES

Oil and gas industry operations have continued to improve in order to comply with or surpass environmental standards in recent years. Adverse occurrences are infrequent, but they can have serious repercussions when they do occur. Continuous or episodic oil and gas operations generate pollutants, which can contribute to environmental deterioration. Oil spills, polluted water spills, and methane leaks can occur because of poorly drilled or finished wells. Compressors and diesel generators, for example,

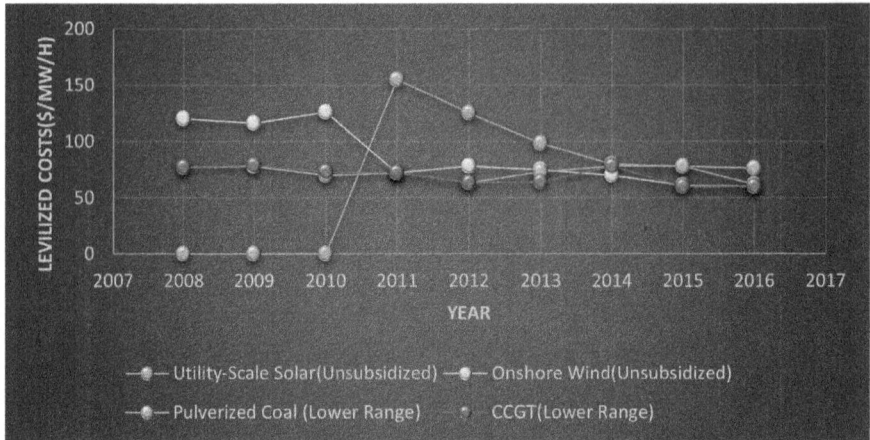

FIGURE 11.2 Various utility-scale technologies' levelized costs. (Data from Lazard Levelized Costs of Energy reports from 2008 to 2017.)

contribute to noise and air pollution. In addition, drilling operations can clog up roadways and pollute ground and surface water supplies. Furthermore, petroleum refining processes emit gases like CO_2, CO, CH_4, SO_2, H_2S, organic compounds, and nitrogen oxides. Noncompliance can result in fines as well as the loss of a company's social license to operate.

11.3.3 DECLINING RENEWABLE ENERGY COST

The economics of renewable energy technology has changed dramatically over the last decade due to price drops. The typical levelized price of solar photovoltaic (PV) power was more than seven times higher in 2009 than in 2017. The rapid decrease in the cost of electricity generated by wind and solar is depicted in Figure 11.2. Wind power and solar power were formerly prohibitively expensive, but in some cases, they are now the most cost-effective source of energy. Wind and solar electricity generation will become even more competitive in the coming years if expected future cost reductions are realized. Furthermore, cost reductions in battery storage and demand-side control technologies, which can compensate for generation uncertainty, have occurred.

11.3.4 RENEWABLE ENERGY OPTIONS FOR THE OIL INDUSTRIES

Until 2030, the chemical and petrochemical industries are expected to need 24 EJ of energy for process heat generation. High-temperature applications will account for roughly half of the region's total procedure heat consumption (12 EJ) [9]. Medium- and low-temperature applications will account for roughly 30% (7 EJ) and 20%

Technical potential of renewable energy technology for the
petrochemical sector in the AmbD scenario(EJ/ yr)

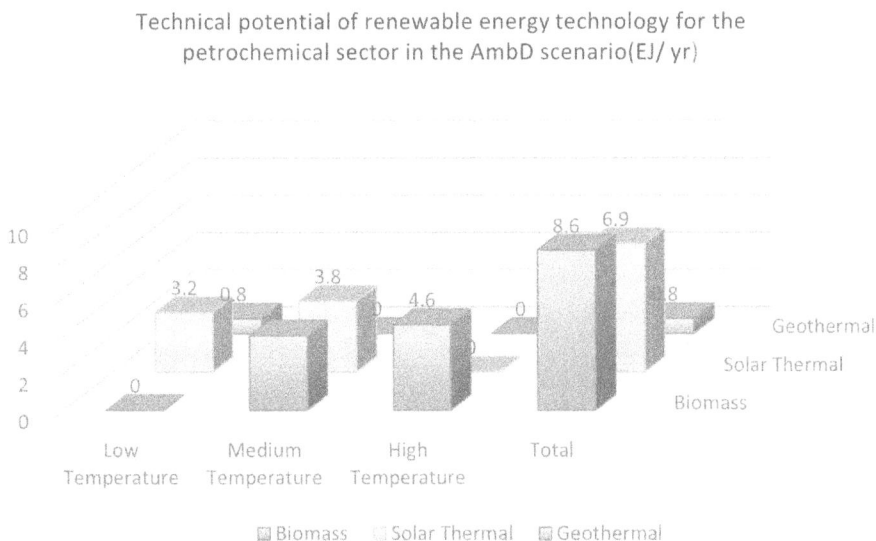

FIGURE 11.3   Technical potential of renewable energy technologies for the chemical and petrochemical industries in the AmbD scenario, expressed in energy joules per year (EJ/Yr).

(6 EJ) of the sector's process heat consumption, respectively. In the chemical and petrochemical industries, steam accounts for the majority of process heat, with direct heat accounting for the remainder. Stand-alone steam boilers, CHP plants (which can meet up to 40% of total demand in some countries [10]), and furnaces supply the whole need.

Scientifically, all low-temperature process heat can be generated by renewable energy systems like solar thermal, geothermal, heat pumps, or biomass. For medium-temperature heating applications, only biomass and solar thermal are considered.

No possibilities for renewable energy technologies are assessed for high-temperature applications associated with the manufacturing of high-value chemicals, such as ammonia and methanol, unless these compounds are produced from bio-based feedstocks. Most units in the industry have a considerable average size, and many are in huge industrial parks that benefit from material and energy integration. As a result, solar thermal systems that require a vast site area or heat pumps with low heat capacity may have limited potential, even if they are technically qualified. Due to the same high degree of process assimilation, retrofitting aging plants with non-fossil fuel-based alternatives will be technically challenging; new capacity potential is indeed prioritized. The biomass potential for medium- and high-temperature heat (excluding the production of the basic chemicals) is 8.6 and 4.3 EJ, respectively, according to the "ambitious development scenario" (AmbD) and "accelerated development scenario" (AccD) scenarios, which are covered in even more detail in Figures 11.3 and 11.4. Additional solar thermal potential of 1.7 EJ for low-temperature applications could be obtained in the current capacity [10].

Technical potential of renewable energy technology for the
petrochemical sector in the AccD scenario(EJ/ yr)

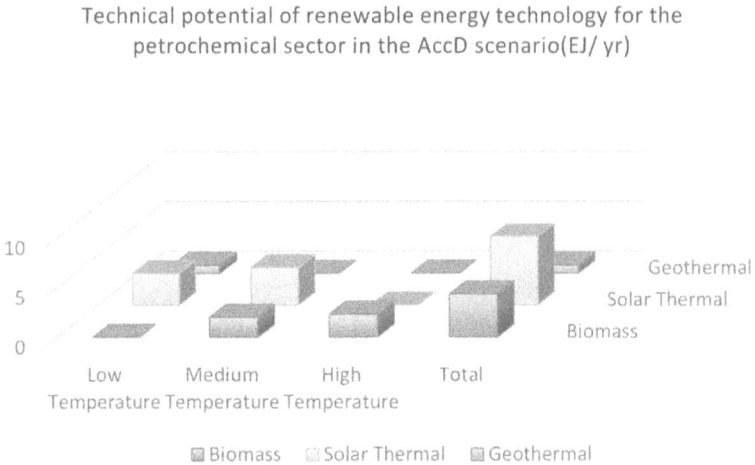

FIGURE 11.4 Summary of the AccD scenario's technical potential for renewable energy technology for the chemical and petrochemical industries (in EJ/Yr).

11.4   RENEWABLE ENERGY OPTIONS FOR THE GAS INDUSTRIES

Instead of using natural gas for electricity generation or as a feedstock for petrochemical products, gas can make the most substantial contribution in the form of "green" or "blue" gas that has been mostly or entirely decarbonized – for better understanding, refer to Figures 11.5 and 11.6. Significant breakthroughs in carbon capture and storage (CCS) technology and economics, as well as the installation or adaptation of infrastructure that supports new fuels, most notably hydrogen, are required for such development [11].

Over other fuels, natural gas seems to have a number of significant advantages. Due to its adaptability, it can be utilized for a variety of purposes, such as heating, cooking, disposing of waste, transportation, and as a feedstock for chemicals, fertilizers, and pharmaceuticals. Large-scale distribution networks already exist in the majority of the UNECE region, allowing for the movement of gas between and among member states. Natural gas can also be utilized as a backup for renewable energy sources that are intermittent. One such instance can be when the wind is not blowing or the sun is not shining, it is possible that natural gas' greatest contribution to the energy transition will be its ability to provide a comparatively low-carbon backup at peak energy demand times. As the costs of renewable energy, particularly solar and wind power, decrease rapidly, the role of gas as a primary fuel for power generation will come under growing pressure in many regions of the UNECE region [11].

11.5   BIOENERGY FOR FUELS

By producing biofuels and other bioproducts, bioenergy can fulfill the demand for liquid fuel while generating less emissions. This requires the establishment, manufacturing, and gathering of sustainable feedstocks in addition to effective conversion

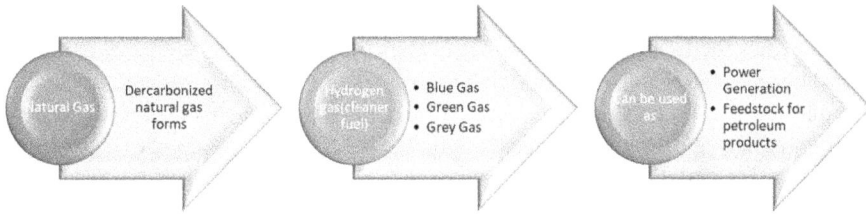

FIGURE 11.5 Flowchart depicting how cleaner hydrogen gas can be extracted from natural gas and can be taken in use.

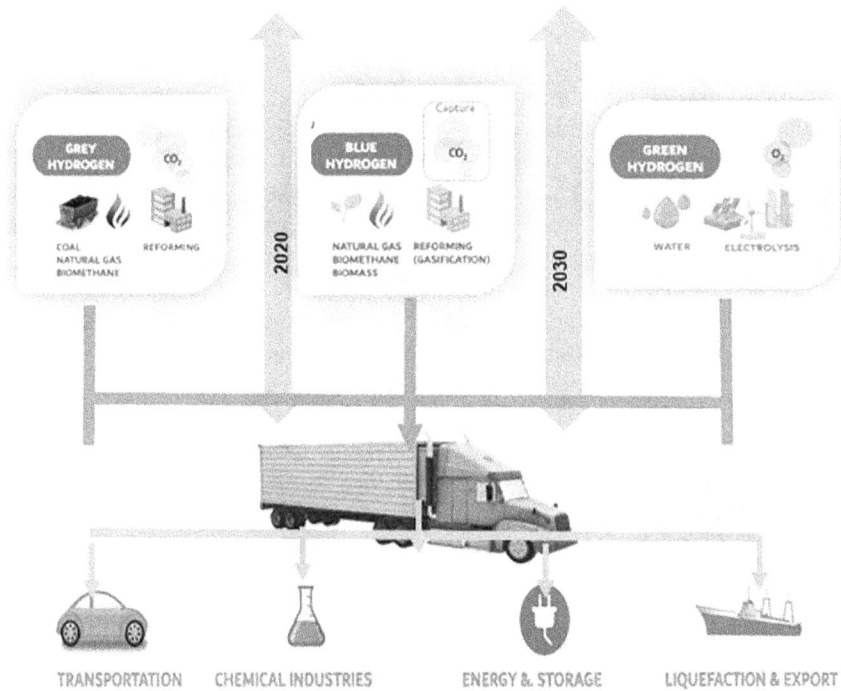

FIGURE 11.6 Visualization of the three hydrogen (H_2) production routes and their corresponding colors.

procedures and a viable final fuel product with the necessary physical and chemical qualities. Appropriate energy content and attributes for use, suitable transport characteristics, the capacity to withstand temperature extremes, and storage appropriateness are all required production and collection of feedstock supplies, the conversion of that feedstock into the desired fuel using a variety of methods, and the broad bioenergy route is divided into many components for distribution in the energy infrastructure for usage. Additionally, biogenic wastes (such as manures, biosolids [treated sewage], food waste, and municipal solid waste) can be converted into liquid products and fuels. There are a few basic steps that make up a bioenergy pathway: cultivation and harvesting raw materials, conversion into fuel using various methods, and distribution to energy infrastructure. Biogenic wastes (such as manures, biosolids [treated sewage], food wastes, and municipal solid waste) can also be transformed into liquid fuels and goods. There are a few basic steps that make up a bioenergy pathway: cultivation and harvesting raw materials, conversion into fuel using various methods, and distribution to energy infrastructure. The following wastes can all be converted into liquid fuels and manufactured goods: fertilizers, biosolids (treated sewage), food wastes, and municipal solid waste [12].

A variety of bioenergy alternatives for industrial process heat generation are already on the market. Fuels generated from biomass, like fossil fuels, are available in solid, liquid, and gaseous forms. Ethanol from sugar fermentation and biodiesel from vegetable oil transesterification are examples of liquid biofuels. Second-generation biofuels are now the topic of intense research and development. A total of 57 kilotonnes (kt) of production capacity is currently in operation and 10 additional units with capacities ranging from 50 to 300 kt annually are expected to start up in the following two years. The most effective coal replacement is projected to be solid bioenergy products. Currently, there are alternative products on the market. For example, charcoal has been used as a coal substitute for a very long time.

Significant volumes of biomass are already cofired with coal in standard coal power plants. For instance, the Dutch Amer 9 CHP power plant now co-fires 35% biomass, primarily wood pellets, with 65% coal to produce 600 MW of electricity and 350 MW of heat. It is projected that the technical advancement of solid biomass fuels would concentrate on raising the energy density of recycled biomass until it can be utilized in furnaces, industrial operations, and coal-burning power plants without modification. Biogas from anaerobic fermentation and producer gas or synthetic gas (syngas) from biomass gasification are the two primary contemporary types of gaseous biofuels [9]. Most bio-based chemicals and polymers are produced with much less nonrenewable energy than their petrochemical counterparts, which is obvious. The savings will be substantially higher if R&D is able to successfully make woody biomass (lignocellulosics) accessible as raw material for biotechnological processes. Due to higher yields and greater volumes of waste that may be converted into heat, energy, or products, using tropical crops (such as sugar cane) and tropical wood types can result in even greater savings [13].

11.6   ALTERNATIVES FOR OIL AND GAS ENERGY

Wind power, nuclear power, solar power, and ethanol are a few of the leading alternatives to oil and gas. As sectors move toward sustainability and more green business practices, there is a significant public push to encourage their use.

The United States has approximately 96 nuclear power reactors, which are providing about 20% of domestic electrical output. With the largest nuclear power in the world, France produces over 80% of its electricity. Nuclear power is the most practical replacement for fossil fuels for future energy use [14]. In comparison with fossil fuels like coal, gas, oil, and ethanol, nuclear power has a very modest effect on the environment. On the other hand, nuclear power is substantially less expensive than other renewable energy sources like solar, wind, or hydropower. While nuclear expansion has been halted for decades in the United States (and many other nations), it has been done partially out of concern for public safety and partly for political reasons. When it comes down to it, clean nuclear energy may potentially power entire cities [15].

Renewable electricity sources, besides nuclear power, include solar and wind energy. Supporters of them argue that it provides a clear break from fossil fuels and is reliant on electricity produced from natural sources, which they assert to be the case. Researchers at the Institute for Energy Research claim that these assertions are false (IER). Most contemporary solar and wind power systems need a consistent source of backup power. These oil substitutes frequently use coal-fired energy to be operational on overcast days or when there is not much wind. On the other hand, installing wind farms and solar panels requires a significant upfront cost.

Current worldwide energy consumption is predicted to be 8%–10% solar/wind, according to the International Energy Agency (IEA). If we are going to encourage the use of these alternatives, we need specialized policy frameworks, such as tax-funded government subsidies and grants. To be sure, the IEA says that "renewable power capacity is expected to grow by 50 percent from now until the year 2024," largely due to solar PVs [10].

11.6.1 SUBSTITUTION OF RENEWABLE TECHNOLOGY IN PRODUCTION STAGE

The oil and gas industry has long used renewable energy in its operations. One of the first commercial uses of solar PV panels was in warning lights for offshore oil facilities beginning in the early 1970s [16]. Power generation can be a considerable expenditure because manufacturing operations demand a lot of electricity. For example, an offshore platform's usual power requirements are between 20 and 35 megawatts (MW) [17].

In upstream operations, renewable energy can be used to cut down on fuel costs and maintenance costs. Additionally, renewable energy can reduce noise, reduce pollution, and enhance safety. Various renewable technologies can be applied at various stages of the life of an oil field, displayed in Figure 11.7.

11.6.2 RENEWABLE ALTERNATIVE IN PRIMARY STAGE

In the near term, drilling uses a lot of energy; one drilling rig can use up to 1 MW of electricity to operate [18]. Drilling and main recovery operations are usually powered by diesel or natural gas generators. Standard oil well construction and drilling typically require between 18,000 and 24,000 gallons of diesel fuel [11]. The energy grid can be utilized to power operations that are close enough to power lines. Electricity can assist minimize noise, pollution, and traffic congestion, which is especially important when operations are near cities.

An electrical link also enables the utilization of renewable energy into numerous places of operation via microgrid. Microgrid technologies enable customers to incorporate distributed generation into a flexible, dependable, and environmentally responsible operation. Equipment used on oil fields and well pads might be converted

Stages in oil fields

Primary Stage-

Oil and gas extraction

- Electric submersible
- Pumps and Rod pumps

Secondary Stage-

Increasing the reservoir pressure

- Use of injection fluids or gases

Tertiary Stage-

Stimulation of oil production using

- Hydraulic fracturing
- Steam injection
- In-situ combustion
- Chemical injection

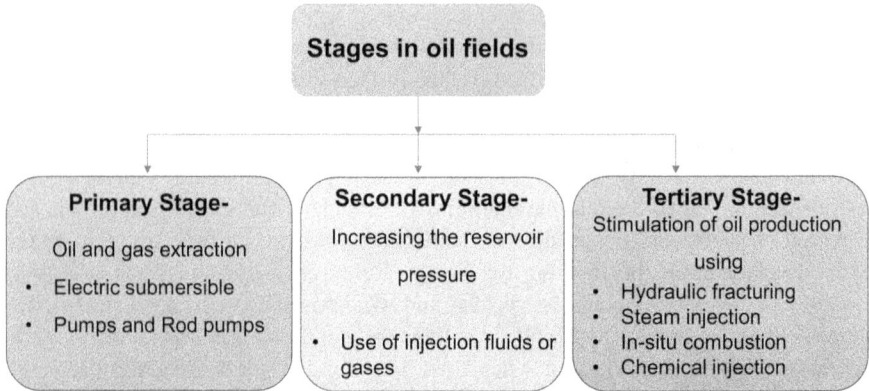

FIGURE 11.7 The different stages of oil extraction where renewable alternatives can be substituted.

to electricity and connected to a microgrid using a controller that maximizes the use of a variety of clean power sources. Energy storage, fuel cells, hydrogen, field gas, solar PV/wind power, and even grid power are examples of sources of electricity. This technique is efficient, reduces leaks and pollution, and improves resilience during power outages. A capacitor and solar-powered rod beam pumps can be utilized to store regenerative energy during the rod down-stroke. It has been proven in tests that there is a lot of potential for energy savings [19].

11.6.3 Renewable Substitution in Secondary Phase of Oil Recovery

Water injection pumps for oil recovery can be powered by renewable energy sources because the output is not impacted by injection fluctuation [15]. The use of offshore wind power to power water injection pumps is one of the most important prospective applications of offshore wind power.

Wind energy can be used to power water injection pumps on offshore oil and gas platforms. In 2012, it was determined that offshore wind was an affordable and environmentally responsible option for powering some of these sites [17]. The economic viability of replacing alternative power from diesel and gas generators with power from wind platforms will increase as wind costs decrease. Utilizing wind energy lessens carbon emissions that can help us achieving our climate goals.

11.6.4 Renewable Substitution for Tertiary Phase of Oil Recovery

There are several prospects for renewable integration in tertiary recovery, or EOR. This process can use renewable technology to create and supply energy. The use of concentrated solar energy for EOR can be cost-effective if the land and solar irradiance conditions are right. From the latent heat in wells, geothermal cogeneration can produce power as well.

Thermal EOR entails injecting steam into the oil reservoir to lower viscosity and allow the oil to flow more freely. Steam is produced using natural gas as an energy source. The reservoir is then filled with steam. Instead of using natural gas to create steam, concentrated solar electricity is used in solar thermal EOR. Since it does not require a turbine to convert steam to electricity, solar thermal EOR offers lower project complexity and costs than CSP for power generation [14].

If solar thermal EOR was compared to natural gas EOR, it was concluded that solar thermal EOR is viable in some circumstances, even if CSP capital costs are substantially greater [20]. Costs of capital and field variables such as solar radiation and estimated field lifetime affect solar thermal EOR's profitability compared to alternatives. Particularly in the Middle East, oil deposits and solar potential tend to coincide. From 19 to 44 gigawatts is estimated as the economic viability of solar thermal EOR projects for petroleum (GW) [14].

Important synergies exist between the use of steam injection and geothermal energy. For typical fields, steam injection increases the temperature of the produced water [20]. The need for energy and heat generated by geothermal power rises for steam injection [20]. An increase in the use of steam injection EOR, as well as a rise in the price and control of carbon emissions, implies that geothermal cogeneration has a bright future.

11.6.5 SUBSTITUTION OF RENEWABLE TECHNOLOGY IN MID-STREAM STAGE

Transportation at the mid-stream stage uses less energy than upstream or downstream activities and has the smallest renewable integration potential. Furthermore, at this time, the application of renewable technology in oil transportation by ship, rail, and truck is limited.

Renewable technology can still have a significant impact on operating costs and emissions, though. Solar energy powers sensors and provides cathodic corrosion prevention in some pipeline operations, for example [21]. However, as the cost of renewable energy decreases and the technologies improve, new opportunities for integration emerge. Electrification of pipeline compressors, compressor heat recovery, and power generation from expansion turbines are some of the possibilities.

11.6.6 COMPRESSOR ELECTRIFICATION

Electric motors are used in many oil pumping stations; thus when the grid greens [21], these systems automatically use more renewable electricity. Nevertheless, natural gas pipes are powered primarily by gas motors or turbines [22].

Electric motors could be used to replace gas engines and turbines, allowing for greater renewable integration while reducing noise, fuel consumption, and pollution. Electric motors have lower operating costs than gas engines or turbines and can manage a larger range of throughput more efficiently. Natural gas demand will fluctuate more when natural gas plants scale up to handle variable generation, which implies pipelines will have to deal with more swings in throughput because of increased unpredictability. Reliability is a major challenge when it comes to employing electric motors in compressor stations. But despite being highly reliable, the electric grid does not always match oil and gas sector operation criteria, especially in rural places where compressor stations are commonly located [22]. As grid dependability is improved and microgrid technologies become more affordable, as well as incentives for electric motors with reduced emissions, the use of electric motors may increase.

11.6.7 COMPRESSOR HEAT RECOVERY

A large amount of heat is generated by the gas turbines that power natural gas compressor plants. It is possible to repurpose this heat for other purposes, such as generating electricity or using it in industrial processes. At natural gas compressor stations in the United States, 12 power generation systems totaling 64 MW of electricity capacity are currently installed [23]. Approximately 40 MW of potential new projects has paybacks predicted in less than five years. Compression station heat recovery could benefit from changes in market conditions, though. About 500 MW to 1.1 GW of power can be generated from compressor station heat recovery, according to estimates [24]. Systems with high utilization rates are more effective because of economies of scale.

The variable output of compressor cogeneration is one of its drawbacks. Not all compressors run; thus, there is not always enough heat or energy to be generated from them [22]. A process that uses extracted heat must be able to operate in a variable-heat environment. It is the same with off-grid power. The consumer must be able to handle intermittent generation.

11.6.8 TURBOEXPANDERS

To transport natural gas across long distances, compressors elevate the gas pressure from 500 to 1,400 pounds per square inch (psi). Power plants and some industrial customers use high-pressure gas. However, most customers require pressures well below 10 psi. By reducing the pressure in the pipeline, the potential energy created by increasing pressure can be wasted in the distribution systems' hubs. You can harness this potential energy by using a turboexpander.

Many different industries employ turboexpanders. Unfortunately, under current market conditions, they are typically regarded uneconomical for use in natural gas pipelines [22,24]. Gas preheating before expansion and pipeline operations producing fluctuation in production are the primary challenges [24]. Compared to coal and gas power plants, turboexpanders have the advantage of being an established technology that produces much fewer pollutants per megawatt-hour generated. There may be commercial prospects for turboexpanders in regions that charge or control emissions.

11.7 SUBSTITUTION OF RENEWABLE TECHNOLOGY IN DOWNSTREAM STAGE

Oil refining – a complex process – requires large capital investments and a lot of energy. To break carbon bonds and remove impurities, heating oil is used in the refinement process. Direct heating and steam generation account for more than 90%

of the energy used in refining. Due to the massive output volumes and high temperatures which are necessary, petroleum refining is the largest consumer of fuel and the major emitter of GHGs in the United States. Refinery energy requirements are rising as a result of the increased usage of heavy and sour crude oil. Refinery emissions deteriorate when oil quality deteriorates, making it more difficult to meet environmental laws [25].

Oil refineries can use renewable energy sources to generate heat and electricity, as well as hydrogen, which is used in the refining process. Additionally, oil refineries can generate electricity by utilizing the excess heat generated during the refining process.

11.7.1 In Generation of Heat and Power

As the refinery process progresses, different temperatures and pressures are required at each stage of the process as mentioned in Table 11.1.

However, contemporary CSP systems can deliver steam generation and heating up to 750°F (400°C), and in some circumstances up to 1,020°F (550°C), and different processes require different temperature displayed in Figure 11.8 [26]. Even though most refineries do not meet the requirement of having both open land and sufficient sunlight, the potential for CSP integration is still quite significant due to the high-energy intensity of oil refining.

Using solar thermal energy in oil refining has an estimated market potential ranging from 21 to 95 gigawatts (GW) [14]. Oil refineries use a lot of electricity because of the high heat requirements. They reduce operating expenses and carbon emissions by increasing the amount of renewable energy used to power gear like compressors and pumps as well as circulators and controllers.

11.7.2 Production of Hydrogen

Hydrogen is used by refineries in a variety of operations, mostly in hydrotreatment process to lower down the sulfur content of the diesel fuel. Refineries have seen an increase in demand for hydrogen due to an increase in demand for diesel fuel and stricter sulfur content requirements [27]. Because the refining process produces more hydrogen than it consumes, demand for hydrogen outpaces the refining industry's ability to create it. Although hydrogen is a co-product from the catalytic reforming of gasoline, the demand for hydrogen far outstrips supply.

TABLE 11.1
Refinery Processes and Their Corresponding Temperature Requirements

S. No.	Process	Optimal Temperature (°C)
1.	Removal of salt and suspended particles	90–150
2.	Processing of crude oil	390
3.	Processing of heavy fuels	390 and 450
4.	Hydrotreatment	260 and 460

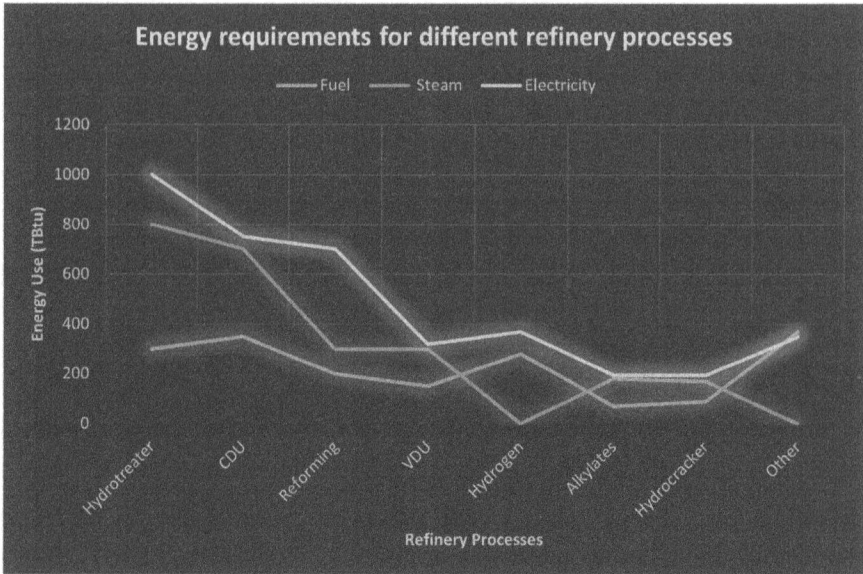

FIGURE 11.8 The different energy requirements for different refinery processes generating fuel, steam, and electricity.

Natural gas is generally used to make hydrogen by steam reforming it. As a heating source for this process, concentrated solar energy can supplement or replace conventional fuel. In addition to meeting oil refinery needs, electrolysis fueled by renewable energy sources can produce hydrogen. As of now, the cost of producing hydrogen from renewable sources is higher than that of conventional methods [28]. However, if the cost of renewable generation continues to decline and rules on GHG emissions are tightened, hydrogen production could become an efficient way to leverage renewable technology to lower oil and gas operations' costs and emission.

11.7.3 FOR POWER COGENERATION

High temperatures and high pressures are used in oil refining. As a result, there are multiple options for heat and power recovery. Cogeneration is already ubiquitous in the refining business, accounting for roughly 13% of all industrial cogenerated electricity.

However, there is still a long way to go before the full potential of cogeneration can be realized. The energy intensity of a refinery can be reduced by 7%–10% by utilizing power recovery devices. The Federal Communications Commission (FCC) uses power recovery the most [27]. Another way to recover power is by the hydrocracking procedure. The pressure difference between the reactor and fractionation phases is what generates the electricity [27]. The Zeeland Refinery in the Netherlands integrates a turboexpander on the hydrocracker to produce 73,000 MWh per year, with a payback period of less than three years. Refining may be done with less energy and carbon intensity because of power recovery. Cogeneration is already bringing in profits for many refineries. Cogeneration may become increasingly advantageous if industry strives to cut air pollution even further.

11.8 PROSPECTS OF GREENER FUEL

Future oil and gas production will shift toward lower-quality and unconventional reserves, which will result in higher energy consumption and operational emissions. Additionally, energy-intensive oil and gas operations could have a negative influence on the environment. Reducing the use of fossil fuels in the production of oil and gas through the integration of renewable energy technology can reduce operational costs and emissions while also preserving petroleum production for higher-value uses [12]. Despite significant advances in technology, safety, and environmental standards, a number of chronic and emergent concerns exist, pointing to a set of research opportunities.

1. Drilling and completion technologies and procedures that are environmentally friendly.
2. Unconventional environmental challenges in the oil and gas industry.
3. Preventing oil spills in the Arctic and on the high seas.
4. Characterization of gas hydrates.

Some opportunities must be addressed soon, owing to the pressing needs of policymakers, regulators, and other public stakeholders. Other alternatives are less obvious, but they have the potential to significantly improve environmental performance or boost resource availability. The National Academy of Sciences, federal advisory bodies like the National Petroleum Council (NPC) and the Secretary of Energy Advisory Board, and environmental organizations like the Environmental Defence Fund and Natural Resources Défense Council are just a few of the organizations in business, government, and academia that have highlighted potential RDD&D (Research Design Development and Demonstration) efforts as a response to these new challenges. To improve technological growth now, the following basic international trends can be identified:

- Technology collaboration.
- Information and communication technology.
- Digitalization.
- The emphasis is on high-tech industries.
- Recognizing the significance of multinational enterprises.

The management of renewable and alternative resources is crucial, even though it is preferred to employ an appropriate energy source in the energy mix. These considerations include technological advancement, cost reduction, energy storage technology, and rising consumer demand. Offshore wind's expanding importance, on the other hand, will attract new investors and may cause more onshore wind suppliers and developers to shift into the industry. The procedures for innovation in goods and services are met by enabling technologies that are diversified and supportive of technology leaders' research activities.

The following are the primary criteria for selecting enabler technologies:

- Addressing global issues such as the need for low-carbon energy or increased resource efficiency.
- To aid in the creation of new products.
- To help the economy expand and to create jobs.

In order to achieve global renewable energy trends, a blend of enabling trends and demand trends is needed to reduce prices and increase integration. The following are examples of current enabling technologies:

1. Photonics.
2. Nanotechnology.
3. Industrial biotechnology.
4. Micro- and nano-electronics.

11.9 CONCLUSION

Most of the energy still comes from fossil fuels, and its use is increasing globally. Although renewable energy facilities are not directly responsible, environmental harm is inevitable in this situation.

Future energy sources are anticipated to primarily emanate from new and renewable sources. Renewable energy sources will be more crucial because fossil fuels will eventually run out. They are efficient in many areas, such as ongoing cost reductions, increased employment possibilities, the creation of new enterprises, and meeting energy and environmental criteria. Simply put, the question of sustainability in terms of social, economic, and environmental aspects revolves upon energy. As a result, the switch to sustainable energy sources and systems is linked to several requirements for the environment, the economy, and development. Local renewable resources, installation costs, and policy framework will be the main determinants. Energy consumption and production have local environmental effects, but the main effects of pollution mobility in the atmosphere can have regional, continental, and even transcontinental effects.

Energy policy objectives that take into account energy mix, efficiency, market, and environmental standards should also be established in order to provide several rehabilitations on unlicensed electricity generation and renewable energy resources, even though the world's demand for electricity and sustainable development are both rising quickly. The following are some of the policies' primary component:

- To ensure that free market prices are higher than feed-in tariff rates.
- To provide further incentives for the selling of locally produced parts of renewable energy power plants.
- Priority should be given to renewable energy when connecting to the grid.

This could help in following ways:

i. Conserving groundwater, increasing water supply, and maintaining air quality in order to reduce the negative effects unconventional oil and gas production will have on safety and the environment.
ii. Reducing methane leaks caused by pipelines and compressors.
iii. Understanding induced seismicity.
iv. Development of understanding required for commercial production of natural gas from natural hydrate deposits.
v. Controlling carbon emissions with CCS for scalability.
vi. Although there are many obstacles to overcome when integrating renewable energy technology into oil and gas operations, there are also significant economic opportunities for renewable technologies in the petroleum sector. More strong reasons are emerging to use renewable energy sources, including declining costs, rising energy intensity, and environmental concerns. Further research of where renewable technologies can be beneficially integrated, as well as efforts to further integrate renewable technology into existing and future operations, would benefit oil and gas operators.

REFERENCES

[1] Energy Information Administration. (2015). *Annual Energy Outlook*; Table A2. Note: For industry and buildings, most of the energy not directly supplied by fuels is from electricity, for which upstream electricity-related generation and other losses are included in the total for energy use by the sector and in the calculation for the share of energy that direct fuel use provides.
[2] IEA. (2020). *The Oil and Gas Industry in Energy Transitions: Insights from IEA Analysis*, OECD Publishing, Paris, https://doi.org/10.1787/aef89fbd-en.
[3] United Nations. (2019). *World Population Prospects 2019: Highlights*, Department of Economic and Social Affairs, Population Division, New York.
[4] Salvarli, M. S., & Salvarli, H. (2020). For sustainable development: future trends in renewable energy and enabling technologies. In Mansour Al Qubeissi, Ahmad El-kharouf and Hakan Serhad Soyhan (Eds.) *Renewable Energy-Resources, Challenges and Applications*, IntechOpen.
[5] Heshmati, A., & Altmann, J. (2014). A Review of Renewable Energy Supply and Energy Efficiency Technologies. Discussion Paper Series, IZA DP No. 8145, Bonn.
[6] Ericson, S. J., Engel-Cox, J., & Arent, D. J. (2019). *Approaches for Integrating Renewable Energy Technologies in Oil and Gas Operations* (No. NREL/TP-6A50–72842), National Renewable Energy Lab. (NREL), Golden, CO.
[7] Freise, J. (2011). The EROI of conventional Canadian natural gas production. *Sustainability*, 3(11), 2080–2104.
[8] Tripathi, V. S., & Brandt, A. R. (2017). Estimating decades-long trends in petroleum field energy return on investment (EROI) with an engineering-based model. *PloS One*, 12(2), e0171083.
[9] Taibi, E., Bazilian, M., & Gielen, D. J. (2010). Renewable Energy in Industrial Applications. An assessment of the 2050 potential. Netherlands.
[10] IEA. (2020), *The Oil and Gas Industry in Energy Transitions: Insights from IEA Analysis*, OECD Publishing, Paris, https://doi.org/10.1787/aef89fbd-en.

[11] Clark, C. E., Han, J., Burnham, A., Dunn, J. B., & Wang, M. (2012). *Life-Cycle Analysis of Shale Gas and Natural Gas* (No. ANL/ESD/11-11), Argonne National Lab.(ANL), Argonne, IL.

[12] Cao, K., Siddhamshetty, P., Ahn, Y., El-Halwagi, M. M., & Kwon, J. S. I. (2020). Evaluating the spatiotemporal variability of water recovery ratios of shale gas wells and their effects on shale gas development. *Journal of Cleaner Production, 276*, 123171.

[13] Patel, M., Crank, M., Dornberg, V., Hermann, B., Roes, L., Huesing, B., & Recchia, E. (2006). *Medium and Long-Term Opportunities and Risk of the Biotechnological Production of Bulk Chemicals from Renewable Resources-The Potential of White Biotechnology*, Utrecht University, Department of Science, Technology and Society (STS)/Copernicus Institute, Netherlands.

[14] Wang, J., O'Donnell, J., & Brandt, A. R. (2017). Potential solar energy use in the global petroleum sector. *Energy, 118*, 884–892.

[15] Feller, F. (2017). WIN WIN-wind-powered water injection. In *Offshore Mediterranean Conference and Exhibition*. OnePetrol, Ravenna.

[16] Halabi, M. A., Al-Qattan, A., & Al-Otaibi, A. (2015). Application of solar energy in the oil industry—Current status and future prospects. *Renewable and Sustainable Energy Reviews, 43*, 296–314.

[17] Korpås, M., Warland, L., He, W., & Tande, J. O. G. (2012). A case-study on offshore wind power supply to oil and gas rigs. *Energy Procedia, 24*, 18–26.

[18] Quinlan, E., Kuilenburg, R., Williams, T., & Thonhauser, G. (2011). The impact of rig design and drilling methods on the environmental impact of drilling operations. In *AADE National Technical Conference and Exhibition, Houston, TX* (pp. 12–14).

[19] Endurthy, A. R., Kialashaki, A., & Gupta, Y. (2016). *Solar Jack Emerging Technologies Technical Assessment*. 10.13140/RG.2.2.16142.36166.

[20] Sandler, J., Fowler, G., Cheng, K., & Kovscek, A. R. (2014). Solar-generated steam for oil recovery: Reservoir simulation, economic analysis, and life cycle assessment. *Energy Conversion and Management, 77*, 721–732.

[21] Pharris, T. C., & Kolpa, R. L. (2008). *Overview of the Design, Construction, and Operation of Interstate Liquid Petroleum Pipelines* (No. ANL/EVS/TM/08-1), Argonne National Lab.(ANL), Argonne, IL.

[22] Greenblatt, J. (2014). Opportunities for efficiency improvements in the us natural gas transmission, storage and distribution system. *Lawrence Berkeley National Laboratory*. LBNL Report #: LBNL-6990E. Retrieved from https://escholarship.org/uc/item/4481k9kw

[23] Elson, A., Tidball, R., & Hampson, A. (2015). *Waste Heat to Power Market Assessment*, Building Technologies Research and Integration Center (BTRIC), Oak Ridge National Laboratory (ORNL), Oak Ridge, TN.

[24] Hedman, B. A. (2008). *Waste Energy Recovery Opportunities for Interstate Natural Gas Pipelines*, Interstate Natural Gas Association of United states of America.

[25] Karras, G. (2010). Combustion emissions from refining lower quality oil: What is the global warming potential. *Environmental Science & Technology, 44*(24), 9584–9589.

[26] Kurup, P., & Turchi, C. (2015). *Initial Investigation into the Potential of CSP Industrial Process Heat for the Southwest United States* (No. NREL/TP-6A20–64709), National Renewable Energy Lab. (NREL), Golden, CO.

[27] Hicks, S., & Gross, P. (2016). *Hydrogen for Refineries is Increasingly Provided by Industrial Suppliers*, US Energy Information Administration, Washington, DC.

[28] Likkasit, C., Maroufmashat, A., Elkamel, A., & Ku, H. M. (2016). Integration of renewable energy into oil & gas industries: Solar-aided hydrogen production. In *International Conference on Industrial Engineering and Operations Management*. Detroit, Michigan.

12 Functional Clay Minerals Application in Oil and Gas Industries

*Tahereh Jafary, Anteneh Mesfin Yeneneh,
Luqman Abidoye, Thirumalai Kumar,
Abdulhameed Khalifullah, Jimoh Adewole,
Khadija Al Balushi, and Maryam Al Buraiki*
National University of Science and Technology

CONTENTS

ABBREVIATIONS

BTX   Benzene, toluene, and xylene
FTIR   Fourier-transform infrared spectroscopy
I-S   Illite-smectite
MMT   Montmorillonites
NP   Nanoparticle
OMS   Organo-montmorillonites
VOC   Volatile organic compound

DOI: 10.1201/9781003242550-12

12.1   INTRODUCTION

Clay is endowed with a multitude of minerals and properties that enable its role in the formation, accumulation, and migration of hydrocarbons [1,2]. Myriads of minerals found in it include palygorskite, sepiolite, halloysite, imogolite, kaolinite, montmorillonite (MMT), mica, vermiculite, chlorite, rectorite, and illite/smectite together with elements such as Si, O, Al, and Mg [1]. Its unique structure and diverse morphologies make clay amenable for various applications in different industrial processes. Particularly, their nanoscale pores have been immense in the generation, storage, and production of shale gas. These rich varieties of minerals strongly influence the physical and chemical properties of conventional sandstone, carbonate, and unconventional shale. In the oil industries, clay minerals can signify evolution of proliferous sedimentary basins undergoing a series of changes in composition and crystal structure. Its signature presence in the reservoir will point to the diagenetic transformation that had occurred in prehistoric times, culminating in hydrocarbon accumulation. Clay minerals can be used to infer tectonic/structural regime, basin evolution history, and the timing of various geologic events because its content can indicate the possibility of potential rock that can generate hydrocarbon at various geological depths [3].

Organo-clay composites play a crucial role in the generation of hydrocarbon [1]. The total organic carbon in sediments freely bind with ubiquitous clay, forming heterogeneous occurrence at the intersurface areas of shale. The composites have been linked to hydrocarbon generation; particularly, sedimentary rocks often contain both clay minerals and organic materials, and ultrafine clay minerals are responsive to alterations in the rocks caused by the production and loss of hydrocarbons. Because the temperatures for the transition from random to ordered I/S correspond with those for the commencement of peak oil generation, changes in the ordering of illite/smectite (I/S) are particularly valuable in researching the hydrocarbon generation. Thus, the predominant and stable minerals of clay offered unique fingerprints for monitoring hydrocarbon source, generation mechanism, and migration dynamics. This characteristic is particularly useful in the investigations of oil spill sources and tracking of hydrocarbon distributions and supplies [4].

12.2   FUNCTIONS OF CLAY

Clays are becoming more prominent as a source of environmentally acceptable or sustainable products. Clay minerals such as bentonite and kaolinite have high specific surface area, excellent adsorption capacity, surface charge, richness, chemical and physical stabilities, and cation exchange ability, as well as being abundant in nature and inexpensive, making them ideal for several industrial applications. The clay is absorbed by the drilling compound due to the complex's secondary valence forces. Because these adsorption phenomena are not dependent on the classical cation exchange reactions, it is able to build an environment around the clay surfaces to lessen the contact of water with the drill chip [5]. The potassium ion in the mud system is held with a larger bonding energy than other exchangeable cations due to a cationic exchange with the sodium or calcium ion in the expanding clay. The drilling mud's potassium ion is mostly responsible for the dramatic decrease in clay swelling. In the mud system, the potassium ion exchanges places with the sodium or calcium ion in the expanding clay, resulting in a cationic exchange. In comparison with other exchangeable cations, the bonding energy of the potassium ion appears to be higher. Potash inputs of 25–40 pounds per Billion Barrels are typical in osmotic dehydration-regulated systems. In the presence of such high potassium concentrations, all attempts to regulate and reduce water loss were fruitless, and the method was discarded. Despite its efficacy in dehydrating clays, organo-aluminum salts do not stimulate the flocculation of a clay dispersion in the same way that simple cations like calcium and magnesium ions do. There has been extensive study on the topic of altering mixed-layer clay minerals. This shift will occur in two stages. In order to make natural clay minerals more compatible at the interface with polymer matrices, the first step is to alter their hydrophilicity to lipophilicity. In the next stage, the gap between the layers will be increased to facilitate access to the polymer chains [6].

12.3 ADSORPTION CHARACTERISTICS OF PETROLEUM FRACTIONS ON CLAYS

Adsorption is another functional characteristic of clay that can affect the migration dynamics of hydrocarbons. This is because the minerals of clay and the ions present in organic clay composites may preferentially attract different fractions of hydrocarbon for adsorption, which will consequently alter the initial flow dynamics. The mobility of different fractions in crude petroleum can be altered by the presence of different minerals in clay. The double-bond group among the petroleum aromatics and the carboxyl group in polar resins and asphaltenes are the major contributors to adhesion, and consequently reduction of mobility, of hydrocarbons in clay [7].

12.4 RESERVOIR QUALITY IN CLAY PRESENCE

Clay affects the properties of reservoir in a number of ways; the presence of clay can cause a reduction in the presence of reservoir petrophysical parameters, such as porosity, permeability, mineral balance, water saturation, and hydrocarbon saturation. The electrical activity of clays is significantly influenced by their composition, internal structure, extremely high surface-to-volume ratios (most clays are between 100 and 1,000 times as porous as water), and charge imbalance along the surface of clay minerals. Each of these clay mineral varieties has its own effect on the interpretative petrophysical parameters determined from well-logging data. For instance, when studying how hydrocarbons form, it is useful to account for the degree of ordering in illite-smectite (I-S) clay minerals [8].

12.5 CLAY MINERALS IN GAS SORPTION AND CARBON SEQUESTRATION

Carbon capture is one of the most environmentally friendly methods for reducing carbon dioxide emissions, the main cause of global warming and the devastation of ecosystems caused by humans. A new approach is needed to lower industrial emissions in the face of this mounting danger. Amorphous solid-based carbon capture is a feasible option since it uses less energy than competing methods and costs less to develop and maintain. Consequently, carbon dioxide adsorption by solids is a method that may be employed both in the long term and on a large industrial scale. In addition, the fact that some sorbents may be reused makes their application in carbon capture economically viable and paves the way for the complete and sustainable acquisition of natural resources [9].

Among these sorbents are clay minerals, a versatile and widely available substance. These materials have the ability to adsorb carbon dioxide both between their layers and on their surfaces. In addition to this, the change of their capacities enables an even greater improvement in their capacity to absorb carbon dioxide. The alteration of clay minerals can increase their suitability for selective carbon dioxide adsorption; some proposals for the design of clay-based carbon dioxide sorbents include immobilization of active sorbents onto clay minerals and modification of the clay surface [10]. The valency of the geomaterials has an effect on the sorption efficiency of different gases. For example, the divalent cations of clay particles affected more of CH_4 adsorption, unlike CO_2. The cation exchange could distinctly increase the CO_2/ CH_4 adsorption ratio so that the first layer of CH_4 molecules can be displaced by CO_2 molecule [11]. Textural factors are the main control on CH_4 adsorption capacity in clay minerals [12]. This fact was corroborated by Klewiah et al. [13], reporting preferential adsorption for CO_2 over CH_4 by clay [14]. According to Chen et al. [15], the presence of quartz, feldspar, illite, and mixed layer contributes to gas storage effectiveness, while the abundance of calcite shows opposite correlation. The adsorption of gases, especially CO_2 and CH_4, is dependent on the geochemical characteristics of the clay materials. Therefore, the presence of organic materials and other chemical in the clay influences the adsorption and consequently, the carbon storage in clay-rich reservoir. However, there is negative correlation between the increased temperature and moisture content [16].

12.6 FUNCTIONAL CLAY MINERALS FOR OIL AND GAS WASTEWATER TREATMENT

Generation of petroleum wastewater is increasing in an exponential rate due to a rapid growth of industrial activities and high global energy demand. The generation volume of this kind of wastewater was estimated to be between 0.4 and 1.6 times of the processed oil [17]. The wastewater from petroleum industry mainly contains oil and grease, heavy metals, high level of total solids, volatile organic compounds (VOCs), ammonia, nitrate, and sulfides [18]. Discharging the generated wastewater without removing the pollutants could result in environmental and health issues, e.g., reducing the level of dissolved oxygen in aqueous environment and threatening marine ecosystem, polluting soil and surface/underground water sources [9,19]. Various potential technologies have been studied and established for the removal of organic and inorganic pollutants existing in the wastewater, e.g., membrane separation, photocatalytic degradation, flocculation and coagulation, and biological treatment. Among the existing technologies, adsorption suggests unique features due to the feasibility of using natural and synthetic materials for physical or chemical adsorption process. Low operating and capital cost, simplicity of operation and design, and high efficiency of pollutant removal in wide range of concentration are other benefits offered by this technology. Moreover, adsorption is a suitable process for removal of both organic and inorganic contaminants [20]. Clay minerals are one of the effective adsorbent materials among those that have been assessed for pollutant removal in wastewater streams. Clay minerals are phyllosilicate minerals that can be found in both geological and marine environments. They are generated by the weathering of silicate minerals, which are abundant in naturalistic environments [21]. They offer various benefits, including low cost, availability, high surface area, high adsorption capacity, surface charge, high porosity, and having different types of active sites [22]. Many types of contaminants have been removed with the use of natural clay minerals. A negative charge is neutralized by inorganic cations in raw clay mineral's layered structure. Raw clays acquire a hydrophilic surface because these cations absorb a lot of water during the manufacturing process. Poor removal efficiency is typically caused by the hydrophilic surface of original clays. In order to improve their adsorption ability, the clays could be modified to increase their pore size, surface area, and active sites. Natural clay minerals have been modified through acid, base, salt activation, heat treatment, polymer and surfactant modification, and pillaring by different polyhydroxy cations [21].

12.6.1 Benzene, Toluene, and Xylene

VOCs found in petroleum and industrial solvents include benzene, toluene, and xylene (BTX). Due to leakage from subterranean oil storage tanks and refinery discharge, BTX compounds are among the most common contaminants in petroleum wastewater. The potential of BTX chemicals to cause persistent toxicity and mutagenesis is well understood. Even at low concentrations, BTX chemicals are of the most dangerous compounds, causing hematological illnesses such as leukemia. Acute exposure to high doses of BTX mostly affects the central nervous system. Clay minerals are of the most inexpensive materials for removal of BTX. Intercalation changing of clay minerals (mainly smectites) with organic cations changes their interlayer surface from hydrophilic to hydrophobic. The produced intercalates are known as organo-clays with high affinity toward hydrophobic organic contaminates [23]. To study the competition of BTX components for the active sites of organo-clay, Lima et al. [24] investigated mono- and multicomponent adsorption of BTX onto adsorbent. Organo-clay variants with organophilic characteristics were created by replacing hydrophobic quaternary ammonium cations for exchangeable inorganic cations. The organo-clay was functionalized by the dialkyldimethylammonium surfactant. The results of selectivity coefficient and absorption study revealed that xylene was the most competitive containment for the active sites of organo-clay than toluene and benzene in mono- and multicomponent tests with 38.3% and 36.6% removal, respectively. Carvalho et al. [25] performed ion exchange between smectite clay and quaternary ammonium salt to make the organo-clay. Similarly, toluene and xylene showed high affinity for the organo-clay with higher removal efficiencies between 30% and 90%. In terms of benzene removal, Deng et al. [26] investigated the impact of clay microstructure on adsorption of benzene using three common types of clay minerals: MMT, kaolinite, and halloysite. Heating treatment was used to adjust the interlayer space and porosity of the modified clays. Calcium-based and heat-treated MMT clay showed the highest benzene adsorption, while the kaolinite showed the poorest. Development of interlayer pores as the function of interlayer cation types and heating conditions and the specific surface area were discussed as the contributing factors on adsorption capacity. In general, the literature results confirm the potential and effectivity of clay minerals as BTX adsorbents. Modifying the volume of micropores and adsorption sites of clay minerals could be promising approaches for enhancing BTX adsorption performance of clay minerals [24–26].

12.6.2 PHENOLIC COMPOUNDS

Phenol and its compounds are recognized to be hazardous, causing harm to both people and the environment. Long exposure to phenolic chemicals can have acute and chronic health consequences affecting the liver, kidneys, neurological system, respiratory disease, and muscle contractions/fatigue [27]. Ganigar et al. [28] studied the adsorption of trinitrophenol and trichlorophenol on MMT- and sepiolite-modified clays with polydiallyl dimethylammonium chloride and poly-4-vinylpyridine-co-styrene (PVPcoS) polycations. Both phenolic derivatives showed higher affinity for PVPcoS-modified MMT, while trinitrophenol showed almost complete removal versus 60% of trichlorophenol.

The hydrophilic besides electrostatic interactions were suggested as the main reasons of this higher affinity. In another study, phenol and 4-nitrophenol were successfully adsorbed on the surface of a composite, which was made of clay, coconut shell, and lime and was activated by KOH. High phenol and 4-nitrophenol adsorption capacities of in turn 1,665 and 477 mg/g were linked to hydrogen bonding, hydrophobic interaction, and electron acceptor-donor mechanisms [29]. Acid and thermal activation of Moroccan clay was reported to enhance the phenol removal by more than three times due to increasing the surface area [30]. Mirmohamadsadeghi et al. [31] reported a modification method for bentonite clay for higher adsorption capacity of phenol. The first step of modification was HCL treatment for surface area and basal spacing modification followed by a substitution of inorganic cations with hexadecyltrimethylammonium bromide (HDTMA) to provide smectite surface organophilic. The modification resulted in shortened removal time due to a multilayered adsorption. Unlimited adsorption capacity was reported as the main advantage of the aforementioned modification technique. Chloro-derivatives of phenol are other pollutants associated with the wastewater generated in petroleum industry. Slomkiewicz et al. [32] studied the adsorption of phenol and chlorophenols by HDTMA-modified halloysite nanocomposite. FTIR spectra and inverse gas chromatographic analyses verified that the modification of halloysite nanotubes by HDTMA enhanced the halloysite surface properties. This enabled the adsorption of phenol and chlorophenols from negative to positive charge and more hydrophobic [32].

12.6.3 HEAVY METALS

Worldwide, heavy metal contamination poses serious ecological threats. Common vectors of transmission include polluted water and soil, which in turn infect animals and humans. The buildup of heavy metals to dangerous levels poses major hazards to human health, although trace amounts of metals like manganese, copper, iron, zinc, and molybdenum are essential for normal metabolic activity in all organisms. Arsenic, cadmium, chromium, copper, lead, mercury, nickel, and zinc are the most common heavy metal pollutants (Zn). Lengthy-term health issues for humans exposed to heavy metal-contaminated surroundings include, but are not limited to, breathing problems, cancer, skin disorders, paralysis, tooth loss, eye problems, kidney and lung malfunction, muscle and joint discomfort, and a long list of others [14,20]. Heavy metals can be removed from wastewater with the use of clay minerals, which have been shown to adsorb divalent metal ions at a high efficiency rate in a number of previous studies. In a molecular dynamic simulation study, Anitha et al. [33] investigated the adsorption behavior of divalent metal cations (Cd^{2+}, Cu^{2+}, Pb^{2+}, and Hg^{2+}). They have calculated the normalized radial density profiles of metal ions with and without the presence of functional groups (such as $-COO-$, $-OH$, and $-CONH_2$) on clay surface. They have also found that the adsorption kinetic mostly followed the Langmuir isotherm model. The maximum adsorption capacities of the heavy metal ions were found in the following order: $Pb^{2+} > Cu^{2+} > Cd^{2+} > Hg^{2+}$ Also, it was resulted that clays with functionalization of carboxylic group ($-COO-$) act a better adsorbent in comparison with other functional groups ($-OH$ and $-CONH_2$).

Awad and colleagues' experimental findings demonstrated that carboxyl-carbon sites are over 20 times more energetic for zinc sorption than unoxidized carbon sites. These findings were based on the results of their research. Clay was treated by Salam et al. using 8-hydroxyquinoline, which may remove Cu^{2+}, Pb^{2+}, Cd^{2+}, and Zn^{2+} ions. During the course of that investigation, many adsorption factors, such as the quantity of clay, the working temperature, the pH value, the concentration of metal ions, and the ionic strength, were explored and optimized. Their findings demonstrated that the majority of the metals were extracted from the aqueous solutions in which they were dissolved, and functionalizing the clay surface with 8-hydroxyquinoline led to a considerable increase in the removal efficiency [21].

Bentonite MMT [6] and attapulgite [34] are the most extensively utilized and effective clay minerals for sorption of various types of heavy metals, due to their high specific surface area, swelling/expanding capacity, and cation exchange capacity. More than 99% of concurrent sorption of Cu^{2+}, Co^{2+}, Pb^{2+}, Ni^{2+}, and Zn^{2+} ions was achieved in bentonite clay in acidic wastewater. The removal of metals was governed by chemisorption as the rate-limiting chemical reaction, according to adsorption kinetics. The effect of long-range intraparticle diffusion on particle percolation across bentonite clay interlayers was also addressed [35]. The effectiveness of MMT clay minerals in the adsorption of various heavy metal ions was shown in different research. Sdiri et al. [36] showed more than 95% removal of Pb^{2+}, Hg^{2+}, Cr^{3+}, and Cd^{2+} with adsorption capacity in the range of 6–131 mg/g. Arsenic [5], mercury [6], arsenate, and arsenite [37] are other heavy metals that showed in turn 64%, 74%, 99%, and 68% of adsorption rate on MMT. Acid, organic, thermal, and nanoscale zero-valent iron modifications are the modification methods applied in clays for heavy metal removal. The modification of attapulgite with nanoscale zero-valent iron composite increased the removal efficiency of Cr^{6+} from 63% to almost 91% [38]. Kostenko et al. [16] investigated the effect that organic alteration of bentonite clay has on the removal of copper, cadmium, and lead. It has been observed that the poly-condensation of silanes results in an increase in the interplanar space of bentonite, which in turn results in an increase in the removal effectiveness of heavy metals on the modified clay minerals by approximately four times. It is usual practice to make use of acids in the process of modifying the surface of clay minerals and removing impurities. It has been demonstrated that expanding the pore size and edges of clay minerals through acid activation can increase their ability to adsorb substances. Additionally, it eliminates cations such as magnesium and calcium, in addition to any metal oxides that might be present in the clay. As a consequence, there are more active sites for the adsorption of heavy metals. The alteration of MMT clay with acid was reported to result in an increase in surface area and pore volume that was almost twice as great [39]. As a result, clay minerals have the potential to be utilized on a global scale as a strategy to remediation of heavy metals in industrial and petroleum effluent that is both cost-effective and kind to the environment.

12.6.4 HEAVY METAL REMOVAL BY THE SMECTITE GROUP

Due to their shared characteristic of an expanding lattice, the Clay Minerals Group of the Mineralogical Society of Great Britain proposed the name "smectite" for the class of clay minerals, which includes MMT, beidellite, nontronite, saponite, and bentonite [40]. Clays belonging to the smectite family have a cation exchange capacity, specific surface area, and adsorption capacity that are significantly greater than those of any other clay family [41]. Compositional and structural differences between the various clay minerals in this class are the most important distinctions between them. In the natural world, smectite clays such as MMTs are frequently seen. It was in the French region of Montmorillon that the smectite clay that is now known as montmorillonite was first identified [42]. Smectite has been given the name "montmorillonite" due to the abundance of current literature and the priorities of study. This name has been widely embraced for the mineral [43]. The fact that water can be absorbed in the interlayer zones between the silicate sheets makes it impossible for hydrophilic MMT microparticles to be used on a wide scale as an adsorbent. The surface area of MMTs has been treated or manipulated chemically to increase in order to generate highly porous composites. Extensive research into treatment, physical properties, and adsorption, as well as examinations of the modified forms of MMT, has been carried out. In recent years, organo-modified montmorillonites (OMMT) have become increasingly popular for usage in polymer/clay nanocomposites. Remove Cu (II) from a solution using organically modified montmorillonite clay (OMHP-MMT), and learn how the solution's pH, stirring time, common ion effects, eluent type, concentration, and volume played a role in the process. Over a pH range of 3.0–8.0 and a stirring time of ten minutes, we observed efficient removal and selectivity toward Cu (II). In a 6.0 pH environment, elimination efficiency was 99.20.9%. Elimination was unaffected when Cu (II) was supplied as sulfate or nitrate

TABLE 12.1
Clay Minerals' Maximum Adsorption Performance for Various Metal Ions

Adsorbate	Adsorbent	Maximum Adsorption Capacity, qm (mg/g)	Reference
Cadmium (II)	Smectite	971.00	[21]
Lead (II)	Illitic clay	238.98	[21]
Chromium (III/VI)	Montmorillonite composite/polyaniline	308.60	[41]
Mercury (II)	Montmorillonite	385.50	[45]
Cobalt (II)	Chemically treated bentonite	138.10	[23]
Copper (II)	Immobilized bentonite	54.07	[23]
Zinc (II)	Kaolinite	250.00	[40]
Nickel (II)	Kaolinite	140.84	[40]
Manganese (II)	Kaolinite	149.25	[40]

instead of chloride, indicating that OMHP was the predominant function. The results show that OMHP-MMT can be employed effectively for Cu (II) recovery in a wide range of samples. Recent research has focused on modifying organo-montmorillon-ites (OMts) using cationic and zwitterionic surfactants to enhance OMHP-MMT and remove copper [43].

Zwitterionic surfactant (Z16)-modified MMT has a similar adsorption capacity for Cu (II) as unaltered MMT. The research findings could yield useful new knowledge for creating superior heavy metal adsorbents. Organo-clays have gained limited interest as adsorbents of heavy metals because organic cations compete with metals for adsorption sites on the surfaces of clay minerals. Researchers have discovered that MMTs that have had naturally occurring organic cations added to them have the ability to adsorb heavy metals such as lead and mercury [44]. The highest adsorption capabilities of various clay minerals are detailed in Table 12.1. According to the findings of a study that compared the maximum adsorption capabilities of a variety of clay minerals toward a number of different adsorbates, it can be confirmed that various clay minerals have high adsorption capacity of heavy metals, making them competitive for application in the oil and gas sector.

12.7 CHALLENGES AND FUTURE PERSPECTIVES

The modification of clay minerals could greatly improve their adsorption efficiency; nevertheless, the addition of chemicals and the accompanying increase in cost are also factors that must be taken into consideration. The selection of the best organic materials for application on a real-world scale is the primary problem that is related to the organic alteration of the clay minerals. The accumulation of ammonium salt as a result of excessive use of quaternary ammonium surfactant compounds may provide a danger of harm to the natural environment. The use of nanoscale zero-valent iron has some drawbacks, one of which is that it is extremely reactive and possesses a high reducing capacity. This is due to the fact that it can rapidly oxidize in its surrounding environment, resulting in the formation of ferrous oxide layers. It is possible that the synthesis of magnetite and goethite in groundwater will be triggered by the use of nanoscale zero-valent iron. In addition, because of their little size, nanoparticles (NPs) are able to penetrate the environment as well as live cells with relative ease; nevertheless, the long-term effects of utilizing nanomaterials have not yet been investigated. Even though thermal modification does not call for the utilization of any chemicals, it may be challenging to determine the appropriate temperature for calcination of the many different clay minerals. The structural integrity of clay materials can be compromised by exposure to temperatures over their optimal range. Although acid treatment can increase the surface area and adsorption capacity of clay minerals, this is not always the case. Because of the potential for cation loss inside the clay structure, the adsorption capacity of clays may be diminished. The adsorption ability of clay is easily diminished by strong acids at high concentrations [14]. Transferring research from the laboratory to practical use is also difficult and calls for extensive testing. There is growing interest in cleaning water with clay minerals; however, most of the studies have been conducted only in the laboratory. Synthetic wastewater has been used extensively in experiments performed in a laboratory setting (initial concentration, pH, temperature, and coexisting ions). Wastewater treatment plants, on the other hand, are intricate machines wherein several organic and biological species, as well as interference species, might show up in high concentrations. As a result, the viability of using modified clay minerals in practical applications must be investigated through an additional pilot-scale study with real or simulated contaminated wastewater [46].

Several clay materials have shown an excellent potential for removing metal contamination even without any alterations; however, their removal capacities can be improved. Because of this, the adsorption capacities also depend on the extent to which the adsorbent has been subjected to chemical treatment, activation, and modification. Some of the technical challenges of water filtration may be alleviated by the use of nanotechnology. Because of their nanoscale dimensions, nanoclays, which are composed of a variety of layered mineral silicates, are finding an expanding range of applications in the field of soil science. Hybrid organic-inorganic nanomaterials such nanoclay composites (also known as organically modified montmorillonites or organo-clays, polymer-clay nanocomposites, etc.) have shown great promise. Due to its ability to selectively adsorb molecules and provide practical solutions to water problems, nanoclay has recently garnered a great deal of interest from researchers throughout the world. This cutting-edge equipment has made important contributions to the field of clay mineralogy and has spawned new fields of study at a time when many new avenues of inquiry are being pursued. As a result of rising environmental awareness, one area of interest in clay mineral research is the potential for the development of unique clay materials through the application of nanotechnology. More research is needed to improve the accuracy, practicality, and reliability of analytical procedures for determining quantitative mineral analysis. Future applications for water remediation could include the use of nanoclays, metal-oxide delaminated and pillared clays, chemically treated clay mineral surfaces, and composite clay structures [41].

12.8 CLAY MINERALS AS ADDITIVES FOR OIL/GAS DRILLING

In the drilling industry, clay swelling is a frequent occurrence it has a significant impact on the quality of the cementing. Due to increasing clay layers, insufficient cementing is a common issue in bore holes. As a result, the perforation's lifetime is reduced, exposing the company to further costs. It also leads to certain other problems, such as the loss of drilling fluids; one of the most serious issues in the drilling sector is shale instability. Water adsorption, osmotic swelling, and cation exchange are the main causes of this condition. Clay stabilizers can help to reduce the swelling of clay; cations like salt and potassium chloride are utilized because potassium ions are efficient at reducing edema. Rheological characterization of drilling fluids is required due to rapidly changing ambient conditions such as temperature, pressure, and conduit geometrical forms. Rheological models are useful tools for analyzing dispersions' rheological properties. As the temperature rises, the viscosity of the liquid phase decreases and the ionic activity of the drilling fluid increases. The significance of pH in mud is rarely recognized until it has been stabilized over a wide range [45].

Drilling fluid properties like apparent viscosity, plastic viscosity, and yield point are critical for designing efficient and optimized drilling operations that clean rock fragments from beneath the bit and transport them to the surface while also cooling and lubricating the rotating drill string and bit. Clay lithology investigations and knowledge of the area's geological history have assisted in defining a certain kind of drilling difficulties to also be expected, as well as the drilling fluids required to overcome them. There are multiple opportunities to modify drilling fluids with NP additives, and the study of the properties of resulting fluids is ongoing. The prevailing challenges and future directions in drilling fluids research include safety, green processes, and high-temperature and high-pressure-resistant clay minerals.

REFERENCES

[1] X. Zhu, J. Cai, G. Wang, and M. Song, "Role of organo-clay composites in hydrocarbon generation of shale," *Int. J. Coal Geol.*, vol. 192, pp. 83–90, 2018, doi: 10.1016/j.coal.2018.04.002.

[2] L. M. Wu, C. H. Zhou, J. Keeling, D. S. Tong, and W. H. Yu, "Towards an understanding of the role of clay minerals in crude oil formation, migration and accumulation," *Earth-Sci. Rev.*, vol. 115, no. 4, pp. 373–386, 2012, doi: 10.1016/j.earscirev.2012.10.001.

[3] Y. Gu, Q. Wan, W. Yu, X. Li, and Z. Yu, "The effects of clay minerals and organic matter on nanoscale pores in Lower Paleozoic shale gas reservoirs, Guizhou, China," *Acta Geochim.*, vol. 37, no. 6, pp. 791–804, 2018.

[4] G. Hu, S. Mohammadiun, A. A. Gharahbagh, J. Li, K. Hewage, and R. Sadiq, "Selection of oil spill response method in Arctic offshore waters: A fuzzy decision tree based framework," *Mar. Pollut. Bull.*, vol. 161, p. 111705, 2020, doi: 10.1016/j.marpolbul.2020.111705.

[5] D. Mohapatra, D. Mishra, G. R. Chaudhury, and R. P. Das, "Arsenic adsorption mechanism on clay minerals and its dependence on temperature," *Korean J. Chem. Eng.*, vol. 24, no. 3, pp. 426–430, 2007, doi: 10.1007/S11814-007-0073-Z.

[6] C. Green-Ruiz, "Effect of salinity and temperature on the adsorption of Hg(II) from aqueous solutions by a Ca-montmorillonite," *Environ. Technol.*, vol. 30, no. 1, pp. 63–68, 2009, doi: 10.1080/09593330802503859.

[7] Y. Liu et al., "Quantitative measurement of interaction strength between kaolinite and different oil fractions via atomic force microscopy: Implications for clay-controlled oil mobility," *Mar. Pet. Geol.*, vol. 133, p. 105296, 2021, doi: 10.1016/j.marpetgeo.2021.105296.

[8] D. V. Barshep and R. H. Worden, "Reservoir quality and sedimentology in shallow marine sandstones: Interplay between sand accumulation and carbonate and clay minerals," *Mar. Pet. Geol.*, vol. 135, p. 105398, 2022.

[9] D. Aljuboury, P. Palaniandy, H. B. Abdul Aziz, and S. Feroz, "Treatment of petroleum wastewater by conventional and new technologies—A review," *Glob. Nest J*, vol. 19, pp. 439–452, 2017.

[10] J. Krūmiņš, M. Kļaviņš, R. Ozola-davidāne, and L. Ansone-bērtiņa, "The prospects of clay minerals from the baltic states for industrial-scale carbon capture: A review," *Minerals*, vol. 12, no. 3, p. 349, 2022, doi: 10.3390/min12030349.

[11] X. Hu, H. Deng, C. Lu, Y. Tian, and Z. Jin, "Characterization of CO_2/CH_4 competitive adsorption in various clay minerals in relation to shale gas recovery from molecular simulation," *Energy Fuels*, vol. 33, no. 9, pp. 8202–8214, 2019, doi: 10.1021/ACS.ENERGYFUELS.9B01610/SUPPL_FILE/EF9B01610_SI_001.PDF.

[12] P. P. Ziemiański, A. Derkowski, J. Szczurowski, and M. Kozieł, "The structural versus textural control on the methane sorption capacity of clay minerals," *Int. J. Coal Geol.*, vol. 224, p. 103483, 2020, doi: 10.1016/j.coal.2020.103483.

[13] I. Klewiah, D. S. Berawala, H. C. Alexander Walker, P. Andersen, and P. H. Nadeau, "Review of experimental sorption studies of CO_2 and CH_4 in shales," *J. Nat. Gas Sci. Eng.*, vol. 73, p. 103045, 2020, doi: 10.1016/J.JNGSE.2019.103045.

[14] B. O. Otunola and O. O. Ololade, "A review on the application of clay minerals as heavy metal adsorbents for remediation purposes," *Environ. Technol. Innov.*, vol. 18, p. 100692, 2020, doi: 10.1016/j.eti.2020.100692.

[15] G. Chen et al., "GCMC simulations on the adsorption mechanisms of CH_4 and CO_2 in K-illite and their implications for shale gas exploration and development," *Fuel*, vol.≈224, pp. 521–528, 2018, doi: 10.1016/J.FUEL.2018.03.061.

[16] L. S. Kostenko, I. I. Tomashchuk, T. V. Kovalchuk, and O. A. Zaporozhets, "Bentonites with grafted aminogroups: Synthesis, protolytic properties and assessing Cu(II), Cd(II) and Pb(II) adsorption capacity," *Appl. Clay Sci.*, vol. 172, pp. 49–56, 2019, doi: 10.1016/j.clay.2019.02.009.

[17] A. Coelho, A. V. Castro, M. Dezotti, and G. L. Sant'Anna, "Treatment of petroleum refinery sourwater by advanced oxidation processes," *J. Hazard. Mater.*, vol. 137, no. 1, pp. 178–184, 2006, doi: 10.1016/J.JHAZMAT.2006.01.051.

[18] S. Varjani, R. Joshi, V. K. Srivastava, H. H. Ngo, and W. Guo, "Treatment of wastewater from petroleum industry: Current practices and perspectives," *Environ. Sci. Pollut. Res.*, vol. 27, no. 22, pp. 27172–27180, 2020, doi: 10.1007/S11356-019-04725-X/TABLES/2.

[19] B. Peng, Z. Yao, X. Wang, M. Crombeen, D. G. Sweeney, and K. C. Tam, "Cellulose-based materials in wastewater treatment of petroleum industry," *Green Energy Environ.*, vol. 5, no. 1, pp. 37–49, 2020, doi: 10.1016/J.GEE.2019.09.003.

[20] M. K. Uddin, "A review on the adsorption of heavy metals by clay minerals, with special focus on the past decade," *Chem. Eng. J.*, vol. 308, pp. 438–462, 2017, doi: 10.1016/J.CEJ.2016.09.029.

[21] A. M. Awad et al., "Adsorption of organic pollutants by natural and modified clays: A comprehensive review," *Sep. Purif. Technol.*, vol. 228, p. 115719, 2019, doi: 10.1016/J.SEPPUR.2019.115719.

[22] S. Barakan and V. Aghazadeh, "The advantages of clay mineral modification methods for enhancing adsorption efficiency in wastewater treatment: A review," *Environ. Sci. Pollut. Res.*, vol. 28, no. 3, pp. 2572–2599, 2020, doi: 10.1007/S11356-020-10985-9.

[23] R. Zhu, Q. Zhou, J. Zhu, Y. Xi, and H. He, "Organo-clays as sorbents of hydrophobic organic contaminants: Sorptive characteristics and approaches to enhancing sorption capacity," *Clays Clay Miner.*, vol. 63, no. 3, pp. 199–221, 2015, doi: 10.1346/CCMN.2015.0630304.

[24] L. F. Lima, J. R. De Andrade, M. G. C. Da Silva, and M. G. A. Vieira, "Fixed bed adsorption of benzene, toluene, and xylene (BTX) contaminants from monocomponent and multicomponent solutions using a commercial organoclay," *Ind. Eng. Chem. Res.*, vol. 56, no. 21, pp. 6326–6336, 2017, doi: 10.1021/ACS.IECR.7B00173.

[25] M. N. Carvalho, M. da Motta, M. Benachour, D. C. S. Sales, and C. A. M. Abreu, "Evaluation of BTEX and phenol removal from aqueous solution by multi-solute adsorption onto smectite organoclay," *J. Hazard. Mater.*, vol. 239–240, pp. 95–101, 2012, doi: 10.1016/J.JHAZMAT.2012.07.057.

[26] L. Deng et al., "Effects of microstructure of clay minerals, montmorillonite, kaolinite and halloysite, on their benzene adsorption behaviors," *Appl. Clay Sci.*, vol. 143, pp. 184–191, 2017, doi: 10.1016/J.CLAY.2017.03.035.

[27] M. Tamang and K. K. Paul, "Adsorptive treatment of phenol from aqueous solution using chitosan/calcined eggshell adsorbent: Optimization of preparation process using Taguchi statistical analysis," *J. Indian Chem. Soc.*, vol. 99, no. 1, p. 100251, 2022, doi: 10.1016/J.JICS.2021.100251.

[28] R. Ganigar, G. Rytwo, Y. Gonen, A. Radian, and Y. G. Mishael, "Polymer–clay nanocomposites for the removal of trichlorophenol and trinitrophenol from water," *Appl. Clay Sci.*, vol. 49, no. 3, pp. 311–316, 2010, doi: 10.1016/J.CLAY.2010.06.015.

[29] M. A. Adebayo and F. I. Areo, "Removal of phenol and 4-nitrophenol from wastewater using a composite prepared from clay and Cocos nucifera shell: Kinetic, equilibrium and thermodynamic studies," *Resour. Environ. Sustain.*, vol. 3, p. 100020, 2021, doi: 10.1016/J.RESENV.2021.100020.

[30] Y. Dehmani et al., "Kinetic, thermodynamic and mechanism study of the adsorption of phenol on Moroccan clay," *J. Mol. Liq.*, vol. 312, p. 113383, 2020, doi: 10.1016/J.MOLLIQ.2020.113383.

[31] S. Mirmohamadsadeghi, T. Kaghazchi, M. Soleimani, and N. Asasian, "An efficient method for clay modification and its application for phenol removal from wastewater," *Appl. Clay Sci.*, vol. 59–60, pp. 8–12, 2012, doi: 10.1016/J.CLAY.2012.02.016.

[32] P. Słomkiewicz, B. Szczepanik, and M. Czaplicka, "Adsorption of phenol and chlorophenols by HDTMA modified halloysite nanotubes," *Materials*, vol. 13, no. 15, p. 3309, 2020, doi: 10.3390/MA13153309.

[33] K. Anitha, S. Namsani, and J. K. Singh, "Removal of heavy metal ions using a functionalized single-walled carbon nanotube: A molecular dynamics study," *J. Phys. Chem. A*, vol. 119, no. 30, pp. 8349–8358, 2015.

[34] V. Zotiadis and A. Argyraki, "Development of innovative environmental applications of attapulgite clay," *Int J. Environ. Res. Public Health*, vol. 47, no. 2, pp. 992–1001, 2013, doi: 10.12681/bgsg.11139.

[35] M. Vhahangwele and G. W. Mugera, "The potential of ball-milled South African bentonite clay for attenuation of heavy metals from acidic wastewaters: Simultaneous sorption of Co2+, Cu2+, Ni2+, Pb2+, and Zn2+ ions," *J. Environ. Chem. Eng.*, vol. 3, no. 4, pp. 2416–2425, 2015, doi: 10.1016/J.JECE.2015.08.016.

[36] A. Sdiri, M. Khairy, S. Bouaziz, and S. El-Safty, "A natural clayey adsorbent for selective removal of lead from aqueous solutions," *Appl. Clay Sci.*, vol. 126, pp. 89–97, 2016, doi: 10.1016/J.CLAY.2016.03.003.

[37] N. Ghorbanzadeh, W. Jung, A. Halajnia, A. Lakzian, A. N. Kabra, and B.-H. Jeon, "Removal of arsenate and arsenite from aqueous solution by adsorption on clay minerals," *Taylor Fr.*, vol. 18, no. 6, pp. 302–311, 2015, doi: 10.1080/12269328.2015.1062436.

[38] W. Zhang, L. Qian, D. Ouyang, Y. Chen, L. Han, and M. Chen, "Effective removal of Cr(VI) by attapulgite-supported nanoscale zero-valent iron from aqueous solution: Enhanced adsorption and crystallization," *Chemosphere*, vol. 221, pp. 683–692, 2019, doi: 10.1016/J.CHEMOSPHERE.2019.01.070.

[39] K. G. Akpomie and F. A. Dawodu, "Acid-modified montmorillonite for sorption of heavy metals from automobile effluent," *Beni-Suef Univ. J. Basic Appl. Sci.*, vol. 5, no. 1, pp. 1–12, 2016, doi: 10.1016/J.BJBAS.2016.01.003.

[40] Ç. Arpa, R. Say, N. Şatiroğlu, S. Bektaş, Y. Yürüm, and Ö. Genç, "Heavy metal removal from aquatic systems by northern Anatolian smectites," *Turkish J. Chem.*, vol. 24, no. 2, pp. 209–215, 2000.

[41] I. K. Kinoti, E. M. Karanja, E. W. Nthiga, C. M. M'thiruaine, and J. M. Marangu, "Review of clay-based nanocomposites as adsorbents for the removal of heavy metals," *J. Chem.*, vol. 2022, pp. 1–25, 2022, doi: 10.1155/2022/7504626.

[42] J. R. C. Banta and M. N. Lunag, "Adsorption of heavy metals from small-scale gold processing in Baguio Mining District, Philippines," *Int. J. Environ. Res.*, vol. 16, no. 5, pp. 1–14, 2022, doi: 10.1007/s41742-022-00450-5.

[43] I. Kovalchuk, B. Kornilovych, V. Tobilko, A. Bondarieva, and Y. Kholodko, "Adsorption removal of heavy metal ions from multi-component aqueous system by clay-supported nanoscale zero-valent iron," *J. Dispers. Sci. Technol.*, pp. 1–12, 2022, doi: 10.1080/01932691.2022.2127754.

[44] R. Novikau and G. Lujaniene, "Adsorption behaviour of pollutants: Heavy metals, radionuclides, organic pollutants, on clays and their minerals (raw, modified and treated): A review," *J. Environ. Manag.*, vol. 309, p. 114685, 2022, doi: 10.1016/j.jenvman.2022.114685.

[45] G. Barast, A. R. Razakamanantsoa, I. Djeran-Maigre, T. Nicholson, and D. Williams, "Swelling properties of natural and modified bentonites by rheological description," *Appl. Clay Sci.*, vol. 142, pp. 60–68, 2017, doi: 10.1016/j.clay.2016.01.008.

[46] R. Mukhopadhyay et al., "Clay–polymer nanocomposites: Progress and challenges for use in sustainable water treatment," *J. Hazard. Mater.*, vol. 383, p. 121125, 2020, doi: 10.1016/J.JHAZMAT.2019.121125.

Index

For Product Safety Concerns and Information please contact our EU
representative GPSR@taylorandfrancis.com
Taylor & Francis Verlag GmbH, Kaufingerstraße 24, 80331 München, Germany

www.ingramcontent.com/pod-product-compliance
Lightning Source LLC
Chambersburg PA
CBHW060403220326
41598CB00023B/3009

9 7 8 1 0 3 2 1 5 1 0 1 4